RNA–Protein Interactions

Frontiers in Molecular Biology

SERIES EDITORS

B. D. Hames
*Department of Biochemistry
and Molecular Biology
University of Leeds, Leeds LS2 9JT, UK*

D. M. Glover
*Cancer Research Laboratories,
Department of Anatomy and Physiology,
University of Dundee, Dundee DD1 4HN, UK*

TITLES IN THE SERIES

1. **Human Retroviruses**
   Bryan R. Cullen

2. **Steroid Hormone Action**
   Malcolm G. Parker

3. **Mechanisms of Protein Folding**
   Roger H. Pain

4. **Molecular Glycobiology**
   Minoru Fukuda and Ole Hindsgaul

5. **Protein Kinases**
   Jim Woodgett

6. **RNA-Protein Interactions**
   Kyoshi Nagai and Iain W. Mattaj

7. **DNA-Protein: Structural Interactions**
   David M. J. Lilley

8. **Mobile Genetic Elements**
   David J. Sherratt

9. **Chromatin Structure and Gene Expression**
   Sarah C. R. Elgin

10. **Cell Cycle Control**
    Chris Hutchison and David M. Glover

11. **Molecular Immunology (Second Edn)**
    B. David Hames and David M. Glover

12. **Eukaryotic Gene Transcription**
    Stephen Goodbourn

13. **Molecular Biology of Parasitic Protozoa**
    Deborah F. Smith and Marilyn Parsons

14. **Molecular Genetics of Photosynthesis**
    *Bertil Andersson, A. Hugh Salter,
    and James Barber*

15. **Eukaryotic DNA Replication**
    J. Julian Blow

16. **Protein Targeting**
    Stella M. Hurtley

17. **Eukaryotic mRNA Processing**
    Adrian Krainer

18. **Genomic Imprinting**
    Wolf Reik and Azim Surani

RNA–Protein Interactions

EDITED BY

Kiyoshi Nagai
Medical Research Council
Laboratory of Molecular Biology, Cambridge

and

Iain W. Mattaj
European Molecular Biology Laboratory
Heidelberg, Germany

at
OXFORD UNIVERSITY PRESS
Oxford New York Tokyo

This book has been printed digitally in order to ensure its continuing availability

OXFORD
UNIVERSITY PRESS

Great Clarendon Street, Oxford OX2 6DP

Oxford University Press is a department of the University of Oxford.
It furthers the University's objective of excellence in research, scholarship,
and education by publishing worldwide in

Oxford New York

Auckland Bangkok Buenos Aires Cape Town Chennai
Dar es Salaam Delhi Hong Kong Istanbul Karachi Kolkata
Kuala Lumpur Madrid Melbourne Mexico City Mumbai Nairobi
São Paulo Shanghai Singapore Taipei Tokyo Toronto

with an associated company in Berlin

Oxford is a registered trade mark of Oxford University Press
in the UK and in certain other countries

Published in the United States
by Oxford University Press Inc., New York

© Oxford University Press, 1994

The moral rights of the author have been asserted
Database right Oxford University Press (maker)

First published 1994

Reprinted (with corrections) 1996, 2002

All rights reserved. No part of this publication may be reproduced,
stored in a retrieval system, or transmitted, in any form or by any means,
without the prior permission in writing of Oxford University Press,
or as expressly permitted by law, or under terms agreed with the appropriate
reprographics rights organization. Enquiries concerning reproduction
outside the scope of the above should be sent to the Rights Department,
Oxford University Press, at the address above

You must not circulate this book in any other binding or cover
and you must impose this same condition on any acquirer

A catalogue record for this book is available from the British Library

Library of Congress Cataloging in Publication Data
RNA-protein interactions / edited by Kiyoshi Nagai and Iain W. Mattaj,
— 1st ed.
(Frontiers in molecular biology)
Includes bibliographical references and index.
1. RNA-protein interactions. I. Nagai, Kiyoshi. II. Mattaj,
Iain W. III. Series.
QP624.75.P74R57   1994   574.87'3283—dc20   94-21987

ISBN 0-19-963504-8 (Pbk)

ISBN 0-19-963505-6 (Hbk)

Preface

Besides the well-known roles of RNA as a mediator of genetic information from DNA to protein (mRNA), as an adapter molecule (tRNA), and as an integral component of the ribosome (rRNA) it has become clear that RNA is involved in a great diversity of other functions in the cell. Furthermore, the discovery of catalytic RNA completely changed our thinking about the origin of life. RNA is the only macromolecule that can flexibly fulfil the requirements as a carrier of both genetic information and catalytic activity. It would thus have the capacity to produce itself at a very early stage of evolution and be acted upon by selective pressure, potentially leading to the emergence of current biological diversity. In living cells, RNA is rarely present on its own and is normally complexed with proteins to form ribonucleoprotein particles (RNPs). The ribosome, RNAse P, telomerase, etc., all consist of RNA and protein components. In this first monograph on RNA–protein interactions we try to cover some important aspects of this broad topic.

RNA is chemically very similar to DNA but small chemical differences give rise to large differences in both structural and chemical properties. RNA is normally single stranded but folds back through base pairing of complementary stretches to form secondary structures such as hairpins, bulges, and internal loops. The first chapter by Varani and Pardi deals with small secondary structural elements of RNA and their investigations using nuclear magnetic resonance (NMR). It may come as a surprise to many that the only methods thus far used to derive secondary and tertiary structural information of large RNA molecules rely on phylogenetic comparison of related RNA sequences and computer modelling. In the second chapter Michel and Westhof describe this approach with particular reference to their work on self-splicing introns. The chapter by Arnez and Moras provides an excellent example of the scope for interplay between X-ray crystallography and biochemical studies. Precise recognition between aminoacyl-tRNA synthetases and their cognate tRNA is essential to maintain the fidelity of protein synthesis. Based on the crystal structure of two aminoacyl-tRNA synthetases complexed with their cognate tRNAs and a wealth of biochemical data, the authors discuss the enzymology and evolution of tRNA-synthetases. Moving towards more purely biochemical approaches we find the chapters on the interactions of ribosomal proteins with rRNA by Draper and the roles and interactions of the RNA and protein components of RNAse P (Chapter 5), an RNP enzyme which cleaves tRNA from its precursor transcript. The RNA component of *Escherichia coli* RNAse P is catalytically active on its own but for the enzyme to achieve optimal kinetic properties the protein component is essential. This gives an interesting insight into the evolution of ribonucleoprotein particles. The study of hnRNPs (Chapter 6) and snRNPs (Chapter 7) has provided numerous excellent examples of the roles of RNA–

protein and protein–protein interactions in forming RNPs. SnRNPs are involved in the splicing of mRNA precursors. These precursors are heterogeneous in size and were thus called heterogeneous nuclear RNAs (hnRNAs). They are complexed with numerous proteins in the form of hnRNP. The sequences of hnRNP proteins have proved to be a treasure trove of different conserved structural elements involved in binding RNA that, subsequent to their discovery in the hnRNPs, have been found in many other proteins. One such element not (yet) found in hnRNP proteins is the classical zinc finger. This motif is found in proteins that bind either to DNA, RNA, or both and Pieler's chapter describes the few cases of interaction between zinc-finger proteins and RNA thus far characterized. The interaction between two HIV proteins, tat and rev, with RNA is described by Frankel and by Karn and colleagues. These proteins are essential in the life cycle of HIV, the virus that causes AIDS and have been studied extensively as targets for anti-HIV drugs. Frankel describes an extreme reductionist approach to the analysis of the RNA interaction with HIV tat and rev in these cases NMR and biochemical studies of binding of target RNAs to short peptides and even, in the case of tat, to the amino acid arginine, have provided insight into the basis of specificity of these interactions. Karn and his co-workers, in contrast, used chemically synthesized RNA containing modified bases to characterise the detailed stereochemistry of interactions between tat and rev and their target RNA sites. This approach will undoubtedly become widely used in conjunction with footprinting and chemical modification experiments as it provides detailed information that is complementary to these techniques. The final chapter is a preview of what we and many others feel is going to be a major aspect of the future of all manner of studies involving RNA. The ability to select RNA molecules capable of binding to essentially any ligand of choice or, if you are smart enough to design a selection scheme, of carrying out any desired chemical reaction, has opened up new horizons for RNA research. These methods, described by Bartel and Szostak, will, among many other things, provide enormous stimulation to the study of RNA–protein interactions.

In a rapidly growing field such as RNA–protein interactions a major worry in editing a book is that it will be obsolete by the time it leaves press. In preparing this, the first monograph dedicated to the subject of RNA–protein interactions, we tried to avoid this fate in three ways. First, we chose contributors whose work is at the leading edge of the field. Second, we asked them to provide background and advice on the methods they employed in their research. Not only should this make the book more useful for newcomers to this field, but these aspects of the subject are by their nature also less prone to rapid obsolescence. Third we harried both contributors and production staff in an effort to keep as closely as possible to the publishing schedule, and we take this opportunity to thank all the contributors for their (generally positive) response which has kept the delay from writing to publishing respectable.

Despite the fears voiced above, much of the book is not (yet) outdated. However, several publications of direct relevance to themes presented in various chapters have appeared that we will briefly refer to here. These novel observations come from X-ray crystallographic studies. The crystal structure of a third aminoacyl tRNA synthetase complex has been solved (Biou *et al.* (1994). *Science*, **263**, 1404–10) and provides an excellent example of how the shape of an RNA in three dimensions can be specifically recognized in a novel way.

Another recent paper (Lindahl *et al.*, (1994). *EMBO J.*, **13**, 1249–54) speaks to the evolutionary origin of a widely distributed β–α–β–β–α–β structural motif found in many RNA-binding proteins (discussed in Chapters 6 and 7). This RNP motif has now been found in ribosomal protein S6 (Lindahl *et al.*, *op. cit*). Since ribosomes must have arisen early in evolution it is not unlikely that the many members of this family are derived from an ancestral ribosomal protein that is also the progenitor of the present-day S6.

A second conserved structural motif found in various proteins, among them RNA- or more generally single-stranded nucleic acid-binding proteins, is a five-stranded β-barrel. These include the bacterial cold-shock proteins (Schindelin *et al.* (1993). *Nature*, **364**, 164–9; Schnuchel *et al.*, (1993). *Nature*, **364**, 169–71) and the N-terminal domain of aspartyl tRNA-synthetase, whose interaction with the anti-codon loop of aspartyl-tRNA has been studied at atomic resolution (see the cover illustration). Remarkably, one of the β-strands that forms this structure in a subset of the proteins of this family is strongly conserved in amino acid sequence with a β-strand formed by the most conserved eight amino acids of the RNP motif referred to above. Given the lack of structural conservation between the two protein families, this likely represents an example of convergent evolution, perhaps due to selection for a β-strand capable of forming a tight interaction with single-stranded nucleic acid. Recently the crystal structure of the complex of U1A protein with its cognate RNA hairpin has been solved (Oubridge *et al.* (1994) Nature, in press). This clearly shows how the β-strands of the RNP motif interacts with single-stranded RNA within the hairpin loop in a sequence-specific manner.

We hope this book will help to bring molecular and structural biologists together to work on interesting and important problems in the field of RNA–protein interactions. This area is rapidly growing and may not yet be covered in depth in undergraduate and postgraduate courses. We therefore hope that this book will also serve as an introduction to this important subject for young scientists.

Cambridge   K.N.
Heidelberg   I.W.M.
September 1994

Contents

List of contributors	xvi

1 Structure of RNA   1
GABRIELE VARANI and ARTHUR PARDI

1. Introduction	1
2. Techniques	3
2.1 RNA synthesis	3
2.2 NMR solution-structure determination	4
3. RNA secondary structure motifs	9
3.1 Helices	9
3.2 Bulges and internal loops	11
3.3 Hairpins	14
4. Tertiary structure	17
4.1 Base triples	17
4.2 Pseudoknots	18
4.3 G-quartets	18
4.4 Helical junctions	19
5. Concluding remarks	19
Acknowledgements	20
References	20

2 Prediction and experimental investigation of RNA secondary and tertiary foldings   25
ERIC WESTHOF and FRANÇOIS MICHEL

1. Introduction	25
2. Hierarchy in structure	26
2.1 The definitions of secondary structure	26
2.2 Tertiary motifs	29
3. Hierarchy of folding	31

	3.1 Initiation of folding: hairpins	31
	3.2 Subdomain formation: local motifs with co-axiality	32
	3.3 Final 3-D anchors: motifs without co-axiality	35
4.	**Identification of interacting partners**	37
	4.1 Secondary structure prediction from energy minimization	37
	4.2 Comparative sequence analysis	37
	4.3 Experimental identification of interacting partners	42
5.	**Modelling tertiary structure**	44
	5.1 The assembly of 3-D architecture from secondary structure motifs	44
	5.2 Accuracy and limits of molecular modelling	45
6.	**Cooperativity of folding**	45
	Acknowledgement	47
	References	47

3 Aminoacyl-tRNA synthetase–tRNA recognition 52

JOHN G. ARNEZ and DINO MORAS

1.	**Introduction**	52
2.	**Transfer RNAs**	53
3.	**Aminoacyl-tRNA synthetases**	58
4.	**tRNA recognition by class I aminoacyl-tRNA synthetases**	61
	4.1 Glutaminyl-tRNA synthetase	61
	4.2 Methionyl-tRNA synthetase	65
	4.3 Tyrosyl-tRNA synthetase	66
5.	**tRNA recognition by class II aminoacyl-tRNA synthetases**	67
	5.1 Aspartyl-tRNA synthetase	67
	5.2 Asparaginyl- and lysyl-tRNA synthetases	74
	5.3 Seryl-tRNA synthetase	74
	5.4 Phenylalanyl-tRNA synthetase	74
6.	**Conformational changes in tRNA**	74
7.	**Conclusion**	75
	References	76

4 RNA–protein interactions in ribosomes 82

DAVID E. DRAPER

1.	**Introduction**	82

	2. Nomenclature	82
	3. Thermodynamic measurements	83
	4. Methods for studying RNA–protein complexes	85
	4.1 Definition of the RNA-binding site	85
	4.2 Localization of RNA contacts	85
	4.3 Interpretation of footprint-type experiments	86
	5. Ribosomal protein–RNA complexes	88
	5.1 S8 and other proteins recognizing an irregular helix	88
	5.2 L11 and other proteins requiring rRNA tertiary structures	90
	5.3 S4 recognition of a large rRNA domain	93
	6. Protein structure	95
	Acknowledgement	96
	References	96

5 RNA–protein interactions in RNase P   103
VENKAT GOPALAN, SIMON J. TALBOT, and SIDNEY ALTMAN

1. Introduction	103
2. Effects of C5 protein on catalysis by M1 RNA	104
2.1 Efficiency	104
2.2 Versatility	106
2.3 Does C5 protein influence the choice of cleavage site in substrates?	107
3. Gel-retardation analysis of RNA–protein interactions in the RNase P holoenzyme	108
3.1 Ionic strength dependence and ion requirement for formation of RNase P holoenzyme	109
3.2 Kinetic and thermodynamic parameters affecting the C5 protein–M1 RNA interaction	111
4. Binding site for C5 protein on M1 RNA	112
4.1 Footprint analyses	113
4.2 Chemical mutagenesis	113
4.3 Gel-retardation analyses	115
5. Demarcation of the role of different regions in M1 RNA	115
6. Model for the formation of the RNase P holoenzyme	116
7. RNase P: evolutionary considerations	119
7.1 Role of protein subunit in RNase P function	119

    7.2 Conservation of surface epitopes in various RNase P holoenzymes   121
    7.3 Relationship of C5 protein to other RNA-binding proteins   122
Acknowledgements   122
References   122

6 Structure and function of hnRNP proteins   127

MEGERDITCH KILEDJIAN, CHRISTOPHER G. BURD, MATTHIAS GÖRLACH, DOUGLAS S. PORTMAN, and GIDEON DREYFUSS

1. Introduction   127
2. Structural motifs in hnRNP proteins   130
    2.1 The RNP consensus sequence (RNP-CS) RNA-binding domain (RBD)   130
    2.2 The RGG box   134
    2.3 The KH domain   137
    2.4 Auxiliary domains in hnRNP proteins   137
3. RNA-binding specificities of hnRNP proteins   138
4. Functions of hnRNP proteins   140
References   141

7 RNA–protein interactions in the splicing snRNPs   150

KIYOSHI NAGAI and IAIN W. MATTAJ

1. Introduction   150
2. RNA components of spliceosomal snRNPs   152
3. Protein components of snRNPs   152
4. Identification of protein-binding sites on snRNAs   158
    4.1 Common protein-binding: the Sm-binding site   158
    4.2 Specific protein-binding sites within U1 snRNP   160
    4.3 Specific protein-binding sites within U2 snRNP   162
    4.4 U4/U6 snRNP   164
    4.5 Specific protein-binding sites within U5 snRNP   164
5. Electron microscopic studies of U snRNPs   164
6. Structural basis for two specific RNA–protein interactions   166
    6.1 Binding of U1A and U2B" proteins   167
7. A spin-off with structural feedback   172
References   173

8 Interaction of 5S RNA with TFIIIA — 178
TOMAS PIELER

1. Introduction — 178
2. Structural features of 5S ribosomal RNA — 179
3. RNA structural requirements for the binding of TFIIIA — 182
4. Functional domains of TFIIIA — 185
5. Other RNA-binding zinc finger proteins — 188
References — 188

9 Control of human immunodeficiency virus gene expression by the RNA-binding proteins tat and rev — 192
JONATHAN KARN, MICHAEL J. GAIT, MARK J. CHURCHER, DEREK A. MANN, IVAN MIKAÉLIAN, and CLARE PRITCHARD

1. Introduction — 192
2. Stimulation of transcription by tat — 193
   2.1 TAR is functional as an RNA transcript — 193
   2.2 tat is an RNA-binding protein — 194
   2.3 *Trans*-activation mechanism — 198
3. Control of late mRNA expression by rev — 202
   3.1 The rev-response and INS elements — 202
   3.2 rev binds directly to the RRE — 203
   3.3 Activity of rev *in vivo* — 208
4. Biological implications — 211
   4.1 Principles of RNA recognition — 211
   4.2 Thresholds and kinetics controlling HIV gene expression — 211
   4.3 Future challenges — 213
Acknowledgements — 213
References — 213

10 Using peptides to study RNA–protein recognition — 221
ALAN D. FRANKEL

1. Introduction — 221
2. The arginine-rich domain in RNA-binding proteins — 221

3. The tat–TAR interaction — 222
    3.1 Identification of TAR — 222
    3.2 Identification of a tat RNA-binding peptide — 223
    3.3 Determinants of specificity in TAR RNA — 224
    3.4 Determinants of specificity in the tat peptide — 225
    3.5 NMR structure of a TAR–arginine complex — 230
4. The rev–RRE interaction — 234
    4.1 Identification of RRE — 234
    4.2 Identification of a rev RNA-binding peptide — 234
    4.3 Determinants of specificity in RRE RNA — 235
    4.4 Determinants of specificity in the rev peptide — 236
    4.5 Model of a rev peptide–RRE complex — 238
5. Other peptide–RNA model systems: what might be learned? — 240
References — 242

11 Study of RNA–protein recognition by *in vitro* selection — 248
DAVID P. BARTEL and JACK W. SZOSTAK

1. Overview — 248
    1.1 Fundamentals of *in vitro* selection — 248
    1.2 Development of *in vitro* selection methods — 249
    1.3 Comparison of *in vitro* and *in vivo* genetic selections — 250
2. Choosing the length and degree of randomization — 252
    2.1 Completely random pools — 252
    2.2 Partially random pools — 255
3. Creating randomized pools — 257
    3.1 Constructing very large, complex pools — 257
    3.2 Large-scale PCR — 258
    3.3 Base distribution — 259
4. Enrichment and amplification of functional sequences — 259
    4.1 Enrichment techniques — 259
    4.2 Selection stringency — 261
    4.3 Amplification — 261
    4.4 Selection and amplification artefacts — 263

5. Analysis of selected sequences 264
6. Future prospects 265
References 265

Index 269

Contributors

SIDNEY ALTMAN
Department of Biology, Yale University, New Haven, CT 06520, USA.

JOHN G. ARNEZ
Institute de Biologie Moléculaire et Cellulaire du CNRS, 15 rue René Descartes, 67084 Strasbourg-cedex, France.

DAVID P. BARTEL
Department of Molecular Biology, Massachusetts General Hospital, Boston, MA 02114, USA.

CHRISTOPHER G. BURD
Howard Hughes Medical Institute and Department of Biochemistry and Biophysics, University of Pennsylvania School of Medicine, Philadelphia, PA 19104-6148, USA.

MARK J. CHURCHER
MRC, Laboratory of Molecular Biology, Hills Road, Cambridge CB2 2QH, UK.

DAVID E. DRAPER
Department of Chemistry, Johns Hopkins University, Baltimore, MD 21218, USA.

GIDEON DREYFUSS
Howard Hughes Medical Institute and Department of Biochemistry and Biophysics, University of Pennsylvania School of Medicine, Philadelphia, PA 19104-6148, USA.

ALAN D. FRANKEL
Department of Biochemistry and Biophysics and Gladstone Institute of Virology and Immunology, University of California at San Francisco, PO Box 419100, San Francisco, CA 94141, USA.

MICHAEL J. GAIT
MRC, Laboratory of Molecular Biology, Hills Road, Cambridge CB2 2QH, UK.

VENKAT GOPALAN
Department of Biology, Yale University, New Haven, CT 06520, USA.

MATTHIAS GÖRLACH
Howard Hughes Medical Institute and Department of Biochemistry and Biophysics, University of Pennsylvania School of Medicine, Philadelphia, PA 19104-6148, USA.

JONATHAN KARN
MRC, Laboratory of Molecular Biology, Hills Road, Cambridge CB2 2QH, UK.

MEGERDITCH KILEDJIAN
Howard Hughes Medical Institute and Department of Biochemistry and Biophysics, University of Pennsylvania School of Medicine, Philadelphia, PA 19104-6148, USA.

DEREK A. MANN
MRC, Laboratory of Molecular Biology, Hills Road, Cambridge CB2 2QH, UK.

IAIN W. MATTAJ
EMBL, Meyerhofstrasse 1, 69117 Heidelberg, Germany.

FRANÇOIS MICHEL
Centre de Génétique Moléculaire du CNRS, Laboratoire associé à l'université Pierre et Marie Curie, 91198 Gif-sur-Yvette, France.

IVAN MIKAÉLIAN
MRC, Laboratory of Molecular Biology, Hills Road, Cambridge CB2 2QH, UK.

DINO MORAS
Institute de Biologie Moléculaire et Cellulaire du CNRS, 15 rue René Descartes, 67084 Strasbourg-cedex, France.

KIYOSHI NAGAI
MRC, Laboratory of Molecular Biology, Hills Road, Cambridge CB2 2QH, UK.

ARTHUR PARDI
Department of Chemistry and Biochemistry, University of Colorado at Boulder, Boulder, CO 80309-0215, USA.

TOMAS PIELER
Georg-August-Universität Göttingen, Institut für Biochemie und Molekulare Zellbiologie, Humboldtallee 23, D-37073 Göttingen, Germany.

DOUGLAS S. PORTMAN
Howard Hughes Medical Institute and Department of Biochemistry and Biophysics, University of Pennsylvania School of Medicine, Philadelphia, PA 19104-6148, USA.

CLARE PRITCHARD
MRC, Laboratory of Molecular Biology, Hills Road, Cambridge CB2 2QH, UK.

JACK W. SZOSTAK
Department of Molecular Biology, Massachusetts General Hospital, Boston, MA 02114, USA.

SIMON J. TALBOT
Department of Biology, Yale University, New Haven, CT 06520, USA.

GABRIELE VARANI
MRC, Laboratory of Molecular Biology, Hills Road, Cambridge CB2 2QH, UK.

ERIC WESTHOF
Equipe de Modélisation et de Simulation des Acides Nucléiques, UPR Structure des Macromolécules Biologiques et Mécanismes de Reconnaissance, Institut de Biologie Moléculaire et Cellulaire du CNRS, 15 rue René Descartes, 67084 Strasbourg-cedex, France.

1 | Structure of RNA

GABRIELE VARANI and ARTHUR PARDI

1. Introduction

Our understanding of RNA's central role in many cellular functions has increased dramatically since the discovery of RNA enzymatic activity (1). The field of structural biology has also made tremendous advances in the last decade and X-ray crystallography and nuclear magnetic resonance (NMR) spectroscopy have provided three-dimensional (3-D) structures of many proteins. However, there has been much less progress in structure determinations of RNA, with a major limitation having been the production of sufficient quantities of RNAs of defined sequence. While there have been few new X-ray structure determinations of RNAs because of the difficulties in obtaining suitably diffracting crystals, advances in the methods for large-scale synthesis (2) and in spectroscopic techniques have made it possible to apply multidimensional NMR to the solution-structure determinations of RNA. Tremendous progress in NMR techniques permits solution-structure determination of nucleic acids at a level of resolution comparable to crystallographic studies. This chapter will review recent methodological developments and structural studies of RNA secondary and tertiary structure domains and Chapter 2 will discuss phylogenetic, chemical, and enzymatic probing methods that have been used to determine the structure of larger RNAs.

DNA and RNA have very similar covalent structures, the only differences being the change from a 2'-deoxyribose sugar to a ribose sugar and from a methyl group in thymine to a hydrogen in uracil. However, there are very substantial differences in the biological functions of these two nucleic acids. Early fibre diffraction studies of nucleic acid double helices indicated that DNAs could take on a wider range of structures than RNAs, leading to the belief that RNAs were less conformationally flexible than DNAs (3). In contrast, the much wider range of RNA biological activities requires a correspondingly wider range of structures. An interesting question is whether the different structures and functions of RNA and DNA are a direct result of the presence of the 2'-hydroxyl on the sugar or whether these different structures arose because DNA and RNA evolved to perform different functions. Recent results seem to indicate that the conformational variability of RNA and DNA is more similar than generally believed (4, 5). However, more structural data are needed to predict conformational or functional changes induced by replacing deoxyribonucleotides with ribonucleotides (or vice versa) and to fully understand the similarities and differences in the conformational variability of nucleic acids.

The much more varied structures of RNAs permit diverse mechanisms for protein recognition. In the case of DNA, sequence-specific recognition often occurs by insertion of an α-helix in the wide DNA major groove. On the other hand, RNA-binding proteins seldom target fully double-stranded tracts for recognition (6). Although a new protein structural motif has been shown to recognize double helical RNA (7), there is at present no evidence for any sequence specificity. As shown in *Figure 1*, RNA-binding proteins generally target single-stranded regions within secondary structure domains (hairpin loops, internal loops, bulges, etc.), where the functional groups on the bases may be easily available for sequence-

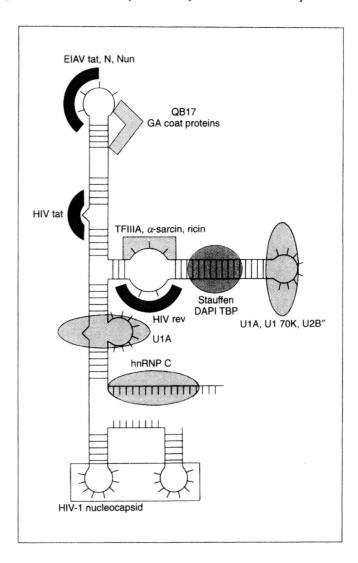

Fig. 1 RNA-binding proteins often recognize simple RNA secondary structural elements. A few representative RNA–protein contacts are shown here to highlight RNA recognition at hairpin loops (EIAV tat, N, Nun, U1A, . . .), bulge loops (HIV tat, U1A), internal loops (HIV rev, TFIIIA, . . .), or single-stranded regions (hnRNP C).

specific recognition. As discussed in this chapter, developments in NMR spectroscopy in the past few years have provided a much more detailed knowledge of the structure of those same structural elements. These studies will prove invaluable in understanding the structural basis of RNA recognition by RNA-binding proteins.

2. Techniques
2.1 RNA synthesis

Structural studies of RNA have been limited in the past by difficulties in the preparation of large amounts (up to tens of milligrammes) of pure RNAs of defined sequence, but RNA can now be efficiently prepared either enzymatically (2) or chemically (8). Large-scale chemical synthesis of RNA is only practical for relatively short RNAs (<30 nucleotides (nt)), due to the relatively low coupling efficiency at each step of the chain elongation and the high cost of the protected mononucleotides. Efficient deprotection of the 2'-hydroxyl also makes the purification of chemically synthesized RNA laborious (8). The enzymatic synthesis of RNA involves *in vitro* transcription with T3, T7, or SP6 RNA polymerases using either chemically synthesized single- or double-stranded templates or double-stranded templates from linearized plasmid DNA (2). Milligrammes of very high purity and biologically active RNAs can be produced by *in vitro* transcription, but the efficiency of the transcription reaction is length and sequence dependent (2). Synthetic single-stranded templates generally work well for RNAs of 10–30 nt, whereas double-stranded templates are often required for high yields of RNAs of >30 nt. One difficulty is the tendency of the RNA polymerases to add one or more nucleotides to the 3' end of the transcribed RNA (2). For RNAs of <50 nt, polyacrylamide gel electrophoresis is used to purify the correct length RNA, but for larger RNAs it is not always possible to separate these different species. RNAs with homogeneous 3' ends can be synthesized by site-specific self-cleavage of the RNA (in *cis*) with a hammerhead ribozyme (9). Using this set of procedures, it is now possible to produce milligramme quantities of RNAs of almost any length or sequence.

Perhaps the most important recent technical advance for NMR studies of RNA has been the development of methodology for the routine production of ^{13}C- and ^{15}N-labelled molecules (10, 11). Cells (for instance *Escherichia coli*) are grown on minimal media containing ^{13}C-labelled glucose and ^{15}N-labelled ammonium as the sole carbon and nitrogen sources. The RNA is extracted, hydrolysed to mononucleotide monophosphates using RNase P1 and the mononucleotides are enzymatically phosphorylated to mononucleotide triphosphates, which are then used in the enzymatic synthesis of RNA. Thus, 99% ^{13}C-, ^{15}N-, or ^{13}C/^{15}N-labelled RNA oligonucleotides can now be routinely synthesized using biosynthetically prepared labelled NTPs. It is also possible to prepare RNAs selectively labelled by residue type, where only a subset of the four nucleotides are labelled (11, 12). In addition, since many biologically active RNAs, for example the hammerhead ribozyme (13), can be made in several fragments and still retain full activity, it is possible to make functional RNAs where only certain regions of the molecule are isotopically labelled. Another approach being developed for selective labelling of RNAs involves the chemical synthesis of mononucleotides labelled at only specific positions of the bases or sugars (J. Santa Lucia and I. Tinoco, Jr, unpublished results). Although in principle it is possible to prepare appropriately protected ^{13}C-labelled phosphoramidites from labelled monomers for chemical synthesis of RNAs, such procedures have not yet been implemented due to the high cost of ^{13}C precursors and the low level of incorporation of the phosphoramidites in the solid-phase synthesis of RNA. The ability to selectively or uniformly label RNAs or certain pieces of functional RNAs, provides a great deal of flexibility in the design of the systems that can be studied by NMR.

2.2 NMR solution-structure determination

NMR solution-structure determination of a biopolymer involves the following steps

1. Resonance assignment of the NMR spectra.

2. Extraction of structural restraints in the form of proton–proton distances from nuclear Overhauser effect (NOE) data and dihedral angle restraints from spin–spin coupling constant data, respectively.

3. Structure calculations employing distance geometry or molecular dynamics algorithms that generate 3-D structures consistent with the covalent structure and the NMR restraints (14).

This procedure was originally developed for the determination of the structure of peptides and proteins, but it has also been successfully used to determine the solution structure of RNAs. The more recent application of multidimensional, multinuclear NMR spectroscopy to isotopically labelled proteins has allowed high-resolution solution-structure determinations of proteins of up to 20 kDa, at a resolution that rivals that obtained by X-ray crystallography (for a recent review see Clore and Gronenborn (15)).

2.2.1 Spectral assignments

The first step in NMR solution-structure determination is obtaining as complete a set of proton resonance assignments as possible (14). The main difficulty in assignment (and structure determination) of RNA is the overlap of the sugar proton resonances, as shown in the proton NMR spectrum of *Figure 2* (top). Whereas the base protons (between ~7 and 8 p.p.m.) and the sugar H1' and pyrimidine H5 protons (between 5 and 6 p.p.m.) are relatively dispersed, five sugar protons per residue resonate between 4 and 5 p.p.m. (the 2' through to 5'/5" protons) (14,16). Observation of non-diagonal peaks ('cross-peaks') in two-dimensional (2-D) spectra

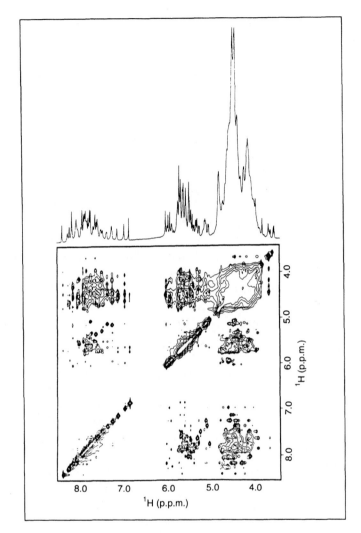

Fig. 2 1-D and 2-D proton NMR spectra of a 30 nt RNA that has lead-dependent catalytic activity (48). The lower part of the figure is a contour plot of a 2-D NOE spectrum of the RNA. The 1-D proton spectrum is plotted above and illustrates the significant amount of spectral overlap inherent in proton spectra of RNAs.

(*Figure 2*, bottom), indicate that the two protons involved are close in space (in nuclear Overhauser effect spectroscopy (NOESY) experiments such as this) or separated by only a few chemical bonds in correlated experiments such as correlation spectroscopy (COSY) or TOCSY. This type of structural information is crucial in obtaining spectral assignments. Furthermore, spreading the information into additional dimensions greatly increases the resolution of the NMR experiment. Using 2-D NMR, it has been possible to assign most base and sugar proton resonances for oligonucleotides up to ~30 nt (17–22), but complete assignments of all ^{1}H, protonated ^{13}C, and ^{31}P resonances have only been obtained for very low molecular weight RNAs (<15 nt) (18, 19). This size limitation has now been overcome by the use of ^{13}C- and/or ^{15}N-labelled oligonucleotides. Application of

3-D and 4-D NMR experiments eliminates the residual overlap by spreading out the off-diagonal peaks into additional dimensions (23, 24).

Resonance assignments of RNAs start with the identification of all proton resonances belonging to each sugar using scalar coupling experiments that provide information on all protons separated by only a few chemical bonds (16). However, the small (<2 Hz) H1′–H2′ scalar coupling in A-form helices limits the utility of homonuclear experiments because it is impossible to transfer magnetization efficiently between the well-resolved H1′ resonances and other sugar proton resonances. Potentially ambiguous NOEs (providing information on which protons are close in space) are generally used alone or in conjunction with scalar couplings to make sugar proton assignments (16). This problem is overcome in ^{13}C-labelled RNA, since it is possible to transfer magnetization very efficiently by using the large one-bond ^{1}H–^{13}C (145–160 Hz) and ^{13}C–^{13}C (38–45 Hz) couplings. Methods have been reported for unambiguous identification of the sugar proton and carbon resonances in ^{13}C-labelled RNAs (25) and for stereospecific assignments of the H5′ and H5″ protons (26).

The major advantage of heteronuclear NMR experiments is the large chemical shift dispersion of the carbon and nitrogen resonances in RNA. Although it is impossible to distinguish the sugar proton resonances from their chemical shifts (*Figure 2*), each type of sugar carbon generally resonates in a different spectral region, providing a great increase in spectral resolution (*Figure 3*) (23, 27). Similarly, it is possible to distinguish between G and U imino protons by analysis of ^{15}N chemical shifts in 2-D (^{15}N, ^{1}H) NMR spectra of ^{15}N-labelled RNAs (23). Cross-peaks that overlap even in 2-D experiments can often be resolved by using ^{13}C or ^{15}N to edit the proton resonances into additional dimensions in heteronuclear 3-D and 4-D experiments (23, 24, 28). Application of higher dimensionality heteronuclear NMR experiments on isotopically labelled RNAs enormously facilitates resonance assignments and, therefore, structure determination.

The sequential resonance assignment procedure for nucleic acids has proven very successful for DNA and RNA duplexes of <30 nt (16). The application of 3-D heteronuclear NMR experiments to labelled RNAs results in a spectacular increase in the ease and the reliability of the sequential resonance assignments and will allow reliable resonance assignment of much larger RNAs (28). However, since the NOE-based sequential resonance assignment procedure for nucleic acids was developed primarily for assignment of helical regions in nucleic acids, assignments are more tenuous in non-helical regions. Up to perhaps 30 nt, ^{31}P–^{1}H through-bond correlation provides an unambiguous procedure for sequential assignments (29, 30). However, the efficiency of these through-bond correlations is often quite low, will decrease as the size of the molecule increases, and is also conformation dependent. For larger RNAs, the resonance assignment for residues in non-helical regions at present needs to be approached *de novo*, but it is clear that once all of the helical regions are assigned, identification of the remaining nucleotides is greatly simplified. The present results indicate that the application of heteronuclear multidimensional NMR experiments to ^{13}C/^{15}N-labelled RNAs should lead

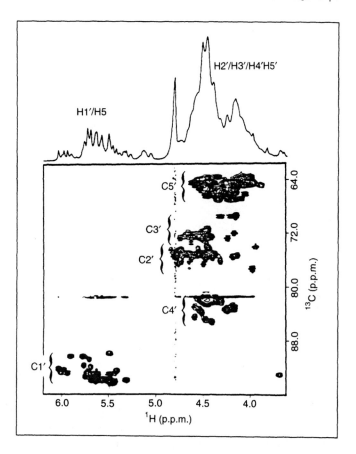

Fig. 3 A contour plot of the sugar region of a 2-D heteronuclear (^{13}C, ^{1}H) correlation (HMQC) experiment on a 99% ^{13}C-labelled sample of the lead-dependent ribozyme discussed in *Figure 2* (48). The 1-D proton spectrum is plotted above showing the overlap of the H2' through to H5' resonances. As discussed in the text, each type of sugar carbon generally resonates in a different spectral region and, therefore, the carbon chemical shifts in this spectrum can be used to help assign the sugar resonances in RNAs.

to complete assignment of the ^{1}H, ^{13}C, and ^{15}N spectra of at least 50-mer oligonucleotides.

2.2.2 Structure determination

Once spectral assignments are obtained, cross-peaks in NOESY experiments provide direct information on which protons (in a well-defined position of the RNA molecule) are close to each other in space. The quality of structures determined by NMR depends primarily upon the number of interproton distance restraints obtained from NOESY spectra. The higher the number of assigned resonances, the higher the number of unambiguously identified NOESY cross-peaks and the higher the quality of the structure. Scalar coupling constants between nuclei separated by three chemical bonds provide additional information on the torsion angle between those bonds using generalized Karplus equations that have been calibrated on model compounds to correlate with the observed scalar couplings with torsion angles for nucleic acids (31). Since the chemical composition (configuration) of RNA is known, the 3-D structure of RNA is determined if seven torsion angles (*Figure 4*) defining the backbone conformation and the relative position of the base

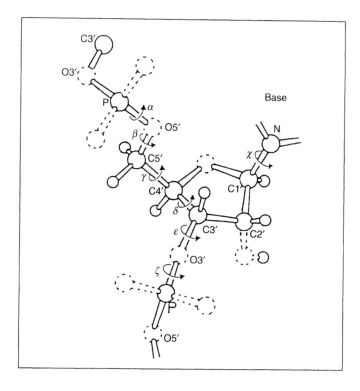

Fig. 4 The structure of an RNA is defined by seven degrees of freedom that specify the backbone conformation (α to ζ) and the relative position of bases and sugars (the glycosidic angle χ).

and the sugar are determined. With the exception of oxygen, all other nuclei (^{1}H, ^{13}C, ^{15}N, and ^{31}P) can provide structural information, either at natural abundance or in isotopically enriched samples.

Because of the much lower density of protons in nucleic acids as compared with proteins, the information available from interproton distances is not as useful. Furthermore, most of the NOEs observed in helical regions are either intranucleotide or between nucleotides on neighbouring base pairs (16, 32). Thus, proton–proton distance restraints derived from NOE data are combined with information on torsion angles derived from homonuclear (^{1}H, ^{1}H) and heteronuclear (^{1}H, ^{31}P) coupling constants to determine the 3-D structure of a molecule. Although there is some controversy as to how precisely and accurately the structure of a DNA or RNA double helix can be determined by NMR (33), the glycosidic angle, the sugar pucker, and the stacking geometry of consecutive nucleotides are the most precisely defined aspects of the structure and are in general quite well defined. Other properties such as helix curvature or bending, the groove size and width, and the conformation of the phosphodiester backbone are generally much less precisely defined. The high resolution of the structure of several RNA hairpins containing so-called tetraloops (see below) was made possible in part by the availability of a large number of coupling constants, resulting from the complete assignment of the ^{1}H and ^{31}P resonances. Those studies demonstrated that low molecular weight ($<\sim$5 kDa) RNA solution structures can be determined to a level of precision comparable to that of small proteins. For larger RNAs, many aspects of the structures, particularly global parameters such as many helical properties, are much more difficult to define (20–22, 30). Isotopic labelling will greatly increase the number of observable NOEs, undoubtedly leading to a significant improvement in the precision of the structure.

To summarize, the availability of ^{13}C/^{15}N-labelled RNA should enable spectral assignment and ultimately structure determination of RNAs up to at least 15 kDa or ~50 nt. ^{13}C- and ^{15}N-labelling will also be a powerful tool in the structural studies of RNA complexes with proteins and other ligands. Application of other techniques, such as selective labelling by nucleotide type or analysis of RNAs that have been assembled from subfragments, each of which can be individually labelled in a separate experiment, will probably be required to determine the structure of RNAs of up to 30–35 kDa or ~100 nt.

3. RNA secondary structure motifs
3.1 Helices

DNA double helices assume B-form-type structures in solution, whereas RNA helices are found in the A-form (3). The main differences between the two most common helical forms are the sugar conformation and the displacement of the bases from the helix axis, resulting in very different groove depths and widths. Although deoxyoligonucleotides often crystallize in conformations close to the A-form (34), all the available evidence suggests that in solution DNA adopts a conformation close to the B-type family of structures (32). There is no example of B-form RNA, but both RNA and DNA can adopt the unusual left-handed Z-form (3, 35), although it is not clear whether this structure is biologically relevant.

One difference between the A- and B-form structures is the sugar conformation. In RNA helices, the sugar conformation is always found close to the N-type family (3'-*endo*). In B-form helices, the sugar pucker is close to the S-type (2'-*endo*). In the 3'-*endo* sugar conformation, the C3' carbon is displaced above the plane defined by the other four atoms by ~0.4Å, but in 2'-*endo* sugars it is the C2' carbon that is displaced, resulting in very different nucleotide conformations (3). Both N- and S-type sugar conformations are nearly equally populated in mononucleotides and short unstructured oligonucleotides (31) and energy calculations also show that the differences in energy between the two conformers are very small.

The most significant difference between the A- and B-type helices, the displacement of the base pairs from the helix axis, defines very different groove sizes and widths (*Figure 5*). In classical B-form helices, the centre of the base pair is close to the axis of the helix, whereas it is displaced by ~4 Å in the A-form. As a consequence, the minor groove is wide (~11 Å) and shallow (~3 Å) in the A-form and the major groove very narrow (~3 Å) and deep (~13.5 Å). By contrast, the major groove is wide (~12 Å) but less deep (~9 Å) in the B-form (3). This classical picture has been confirmed in the recent investigation of RNA double-helical tracts by crystallography (36, 37) and NMR (20, 29).

10 | STRUCTURE OF RNA

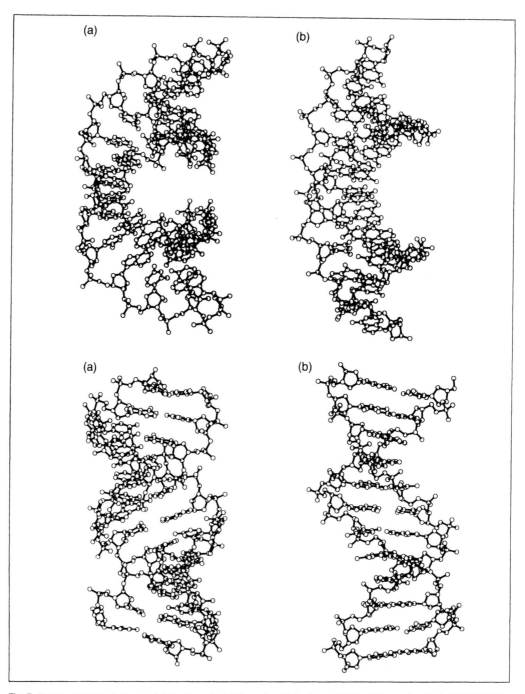

Fig. 5 The comparison between (a) the classical A-form double helix and (b) the B-form double helix highlights the very different groove size and widths of the B-form DNA and A-form RNA. Lower set, a view perpendicular to the helix axis; upper set, the helix axis is tilted to reveal the differences in the groove size between the (a) A-form and (b) B-form double helices.

Since sequence-dependent variations within each helical family or thermal motion may blur the differences between the families, a rigid distinction between the A- and B-form helices may be somewhat artificial. However, while local aspects of the structure, such as base tilt or propeller twist, do not differ significantly between the two helical families, the differences in groove size and shape allow a clear structural and functional distinction. Many DNA-binding proteins recognize specific DNA sequences by forming specific hydrogen bonds between the bases and the side chains of an α-helix inserted into the accessible DNA major groove (38). The much narrower and deeper A-form RNA major groove is not accessible to an α-helix. DNA-binding proteins do not generally recognize RNA, one well-known exception being transcription factor IIIA (TFIIIA). The differences in groove size and width between DNA and RNA double helices may prevent the recognition of RNA double helices by proteins in a manner analogous to the way proteins recognize DNA.

3.2 Bulges and internal loops

Bulges occur when one or more nucleotides on one strand are not base paired, while all nucleotides on the opposite strand are base paired (*Figure 1*). The unpaired nucleotides may be stacked within the helix or bulged outside of the helix (39). These nucleotides represent inviting targets for site-specific recognition of RNA by proteins and may also form tertiary contacts between distant regions of the RNA secondary structure. For example, several phage coat proteins, including R17, GA, and Qβ, recognize an RNA hairpin loop separated by a few base pairs from a single-bulged adenine (40) and an A-rich bulge is required for proper folding of an independent domain of the *Tetrahymena* group I ribozyme (41). A bulged adenosine also plays a crucial role in group II self-splicing introns and nuclear pre-mRNA splicing.

For single-nucleotide bulges in DNA, purines tend to stack within the helix, but pyrimidines tend to be outside (39). However, this rule has many exceptions, since the sequence of the base pairs neighbouring the bulge as well as temperature can influence the structure. In the only RNA case examined, a single-bulged U between two Gs was extrahelical.

The bulged loop in the human immunodeficiency virus (HIV) TAR RNA, which is the binding site for the *trans*-activator protein tat, has been the subject of a great deal of investigation. Although a high resolution structure is not yet available, the NMR data indicate that the first base in the bulge, a U absolutely required for tat binding, stacks on to the last base pair of the stem but adopts the 2'-*endo* sugar conformation (42). A-form-type stacking is continued across the loop on the opposite strand, but the conformation of the two remaining bulged nucleotides is poorly defined by NMR and they are likely flexible in the absence of tat. It has been proposed that sequence-specific recognition of RNA by tat or other RNA-binding proteins is facilitated by bulges and internal loops that may locally open the narrow RNA major groove and allow recognition of the functional groups on the bases

(43). The NMR data indicate a minor groove geometry very close to the classical A-form on either side of the TAR bulge for the free RNA, but a more detailed structural investigation is required to prove or disprove this appealing suggestion.

Bulges significantly bend DNA or RNA double helices and this property has been used to measure the helical repeat of RNA in solution (44). The extent of bending depends both on the number of bulged nucleotides and on the identity of the unpaired bases. The kink induced by DNA bulges may be due to the distortions needed to accommodate the intrahelical stacking of the bases (45). Bulge-induced bending may be one mechanism for recognition of RNA by proteins. For example, the TAR bulge loop bends the RNA (46). It is possible that, besides providing sequence-specific contacts with tat, the bulge may help correctly position tat relative to other factors that may recognize tat and the apical hairpin loop in TAR (see Chapters 9 and 10).

Internal loops form when two double-stranded helices are separated on each strand by several non-Watson–Crick-paired nucleotides (*Figure 1*). The absence of Watson–Crick base pairs does not necessarily imply the absence of base pairing. One crystal structure (37) and several NMR structures (21, 23, 30) have demonstrated that non-Watson–Crick base pairs can readily form within internal loops. Several ribosomal proteins, the eukaryotic transcription factor TFIIIA (chapter 8), and the HIV regulatory protein rev (Chapters 9 and 10) recognize RNA internal loops. Internal loops are also found in the catalytic site of ribozymes (47, 48) and represent potential sites for long-range RNA–RNA tertiary contacts that stabilize RNA 3-D structure (41).

The first structure of an internal loop was the crystal structure of a UUCG-containing sequence, GGAC(UUCG)GUCC, which crystallizes as a helix containing an internal loop of four consecutive mismatches (although it forms a very stable hairpin in solution) (37). The UUCG sequence in the centre of the molecule forms four non-Watson–Crick pairs and two consecutive G–U and U–C base pairs (*Figure 6*). Both the U–C and G–U base pairs are stabilized by water molecules bridging the two sides of the stem. In the case of the C–U base pair, a water molecule

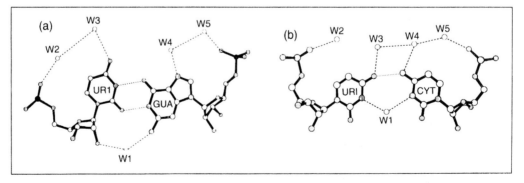

Fig. 6 Structure of the non-Watson–Crick base pairs found in the UUCG-containing duplex (from Holbrook *et al.* (37)). Tightly bound water molecules play an important role in stabilizing the two wobble G–U pairs (a) and the two consecutive U–C pairs (b).

coordinates the N3 positions of the uracil and cytosine, thereby providing a water-mediated hydrogen bond in addition to the single direct base–base contact. For the G–U base pair, a bound water molecule bridges the guanine carbonyl and the 2'-hydroxyl group of the opposing sugar. The observation that the major groove was considerably widened by the non-Watson–Crick pairs was consistent with the proposed mechanism for the facilitated recognition of base pairs by RNA-binding proteins.

A second well-characterized internal loop is loop E from 5S RNA, part of the recognition site for TFIIIA and the subject of many investigations by low-resolution techniques, such as chemical and enzymatic mapping. The loop E sequence is strikingly conserved in a variety of internal loops, including a tomato virusoid and the binding site for the cytotoxins α-sarcin and ricin on 23S ribosomal RNA (21). Upon irradiation with UV light, a cross-strand cross-link was found in several loop E-like internal loop sequences including those from *Xenopus* 5S RNA, a tomato viroid, the hepatitis delta genome, and a small oligonucleotide model. The cross-linking results and sequence similarities indicate that this internal loop represents a common RNA structural motif. In fact, the NMR structures of the α-sarcin loop in the large subunit of ribosomal RNA is very similar to the loop E structure, with small differences probably due to the presence of a further mismatch at the end of the loop (21, 30).

Although loop E and the α-sarcin loop could potentially form several Watson–Crick pairs, the NMR structures of these loops show two well-defined non-Watson–Crick base pairs, an *anti–anti* G–A pair and a reverse-Hoogsteen A–U base pair next to a bulged guanosine (21, 30). Two other non-Watson–Crick base pairs were also proposed for the less well-conserved region of the loop, but the quality of the NMR data was insufficient to define the A–A and U–U base pair unambiguously. The *anti–anti* G–A base pair has now been observed in model DNA duplexes (49) and in several RNA structures (19). The reverse-Hoogsteen A–U base pair and the bulged guanosine next to it, perhaps the most surprising features of the structure, produced a highly distorted backbone (*Figure 7*). On the other hand, the conformation of the strand opposite to the bulged guanosine was very close to the ideal A-form, despite the absence of any Watson–Crick pairing. The UV-induced cross-link was nicely explained by the near superposition of the heteroaromatic rings of the cross-linked G (in the G–A pair) and U (in the reverse-Hoogsteen A–U pair). This unusual interstrand stacking is a consequence of the consecutive G–A and A–U mismatches. While the general feature of a closed loop was correctly predicted using only chemical or enzymatic probing, the detailed geometry of the base pairs was not.

The conformation of the loop E-like internal loops provides several potential signals for recognition of these RNAs by TFIIIA or α-sarcin. These include the bulged guanosine and the distorted backbone required to accommodate this bulge and the numerous non-Watson–Crick base pairs that expose functional groups to the solvent that are not normally accessible in standard double helices. Structural data on the mechanism by which diverse proteins, such as TFIIIA or α-sarcin, recognize this RNA structural element would be extremely valuable.

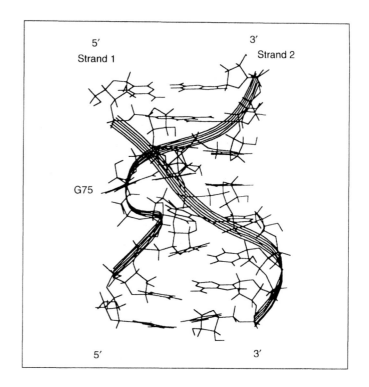

Fig. 7 The ribbon diagram of the internal loop E from 5S ribosomal RNA emphasizes the irregularity in the backbone that is needed to accommodate the bulged guanosine, in stark contrast with the nearly ideal A-form conformation of the opposite strand (from Wimberly et al. (21)).

3.3 Hairpins

Hairpins (or stem–loops) are structures where the phosphodiester backbone folds back on itself to form a double-helical tract (stem) leaving unpaired nucleotides (loop) (*Figure 1*). Inspection of phylogenetically determined RNA secondary structures suggests that hairpins represent the dominant secondary structure element in RNA (50). Besides this structural role, stem–loops may also represent sites for initiation of RNA folding and are involved in many biological functions including transcription termination and viral packaging. The loops in hairpins also commonly define the binding site for RNA-binding proteins, including phage coat proteins (40), the U1A and U2B" proteins from snRNPs (Chapter 7), several ribosomal proteins and transcription terminators/antiterminators. Excluding double helices, hairpins represent the most extensively studied of all RNA-folding motifs.

The Watson–Crick base-paired stem generally adopts an A-form structure, whereas the (nominally) unpaired nucleotides in the loop are potentially exposed for recognition by proteins and other RNA structural elements. As described below, NMR studies have shown that some small hairpin loops contain a surprisingly high degree of structure (18, 19), but larger loops are more poorly structured (51, 52). It is tempting to suggest that the absence of a rigid loop structure favours recognition of unpaired nucleotides by proteins, while the stem may help to position correctly the protein by non-sequence-specific contacts with the backbone.

The prototypical 'large' hairpin loop is represented by the tRNA anticodon loop. Three of the four hairpin loops in the tRNA clover-leaf are large, containing five to ten unpaired nucleotides. The crystal structures of tRNA have revealed that even in the absence of base pairing, the loop nucleotides are involved in extensive stacking interactions (3). In the anticodon loop, stacking is continued on the 3'-side of the stem until a sharp turn just after the anticodon (a so-called π-turn (3)) reverses the direction of the phosphodiester to bridge the two sides of the stem. This turn requires changes in the sugar and phosphate conformations and is stabilized by a hydrogen bond between a base in the loop and the turning phosphate. However, this structure is modified in the tRNA-synthetase complex where two bases that stack in free RNA are unstacked in the complex, presumably to facilitate recognition by the synthetase (53) (see also Chapter 3).

Structural studies of hairpin loops containing a large number of unpaired nucleotides are difficult, due to the presence of multiple conformations or significant dynamics that lead to poorly defined structural features. For instance, the loop at the top of the HIV TAR recognition element was not well ordered in the NMR studies on the free RNA (51). In some cases, however, non-Watson–Crick base pairs can form within the nominally unpaired loop regions, significantly reducing the size of the loop (54).

The very stable and compact conformation of small hairpin loops provided an ideal opportunity for NMR studies of RNA. When hundreds of ribosomal RNA sequences were compared, a striking preference was found for hairpins containing four nucleotide loops ('tetraloops') (50). Rather surprisingly, three sequence families, the UNCG, GNRA, and CUUG tetraloops (where N is any of the four nucleotides and R is a purine), accounted for the majority of all tetraloops. This size and sequence preference is not limited to rRNA, but extends to *E. coli* transcription terminators and to hairpin loops in self-splicing introns and small nuclear RNAs. Thus, these three tetraloop families appear to be important RNA structural motifs that represent ubiquitous building blocks of RNA structure. Even though there is no sequence similarity between the three families, closely related organisms will sometimes mutate a UNCG-type tetraloop to a GNRA tetraloop (50), suggesting structural similarities (55). For instance, the highly conserved GNRA sequence in the eukaryotic signal recognition particle RNA can be mutated as a unit to UNCG with conservation of function, whereas other non-GNRA sequences were not viable (56).

The tetraloop hairpins have unusually high thermodynamic stability (57). The UNCG family is the most stable of the tetraloops and this loop sequence is equivalent thermodynamically to extending the stem by two base pairs. The NMR structures of five members of these three families have revealed extensive similarities between the three families of tetraloops (18, 19, 29) (see *Figure 8*). In the UNCG and GNRA families, a non-Watson–Crick base pair is formed between the first and last nucleotides in the tetraloop and similarly a C–G base pair is formed in CUUG family (F. M. Jucker and A. Pardi, unpublished results), thereby leaving only two nucleotides unpaired in all these tetraloops. In an A-form helix, there is an 11 Å

16 | STRUCTURE OF RNA

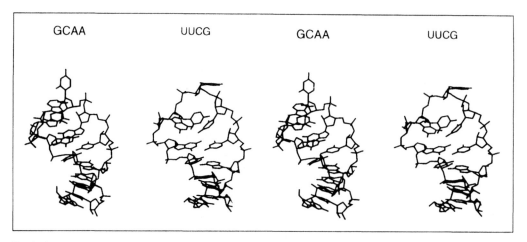

Fig. 8 Stereoviews of the NMR solution structures for the GCAA (19) and the UUCG (18) tetraloops. Only the heavy atoms are shown for these structures.

gap between the two sides of the stem and this relatively large distance led to the suggestion that three to four unpaired nucleotides would be the minimum size of a hairpin loop. However, there are now examples of stable hairpins with either two (18, 19, 29) or three (54, 58) unpaired nucleotides. In order to bridge the 11 Å gap, the two unpaired nucleotides assume the more extended 2'-*endo* sugar conformation (3). This conformational change appears to be critical for the extremely stable UNCG family, whereas the GNRA family seems to be able to tolerate a more dynamic equilibrium between the 2'-*endo* and 3'-*endo* conformations. The lower thermodynamic stability of the GNRA hairpins could be a result of the dynamics that are observed in the NMR studies of these loops (19), whereas the NMR data on UUCG loops do not show this dynamic behaviour. GNRA loops are more frequently used for tertiary contacts in RNA- or protein-binding sites than UNCG loops. The less stable structure may allow proteins or other RNA structural elements to interact specifically with the GNRA motif, whereas the less accessible structure of the UNCG-type tetraloop may not permit efficient recognition of specific hydrogen-bond donors and acceptors.

Analysis of tRNA structure and comparison with NMR studies on a few small DNA hairpins led to the proposal that stacking on the 3'-side of the stem, such as that observed in the anticodon loop of tRNAs, is the dominant feature of hairpin loop folding in nucleic acids (59). As seen in *Figure 8*, the structures of the GNRA and UUCG tetraloops and the more qualitative studies on other RNA hairpins, have revealed the much greater complexity of RNA hairpin folding. For many loops, non-Watson–Crick base pairs can readily form between loop nucleotides (18, 19, 52, 54), but base–base hydrogen bonding may not be sufficient to stabilize these non-Watson–Crick base pairs. In the UNCG tetraloops, the reverse wobble G–U base pair closing the stem is not present when the 2-amino group on the guanosine is removed to make an I–U base pair (60), or an isomorphous A–C base pair (B. Wimberly and I. Tinoco, Jr, unpublished results). Further interactions may be critical for stabilizing the loop-closing base pair, probably involving the phosphodiester backbone. Base–phosphate interactions, such as those observed in the tRNA anticodon loop and the UUCG tetraloop and proposed for the GNRA loop, may be a general way to stabilize sharp turns. Some such interactions may involve bridging water molecules such as those observed in the crystal structures of the UUCG duplex. A more general folding rule may be the switch of sugar conformation to 2'-endo whenever the backbone needs to be extended. The phosphate–phosphate distance in an RNA increases by >1 Å in switching from a 3'- to a 2'-endo sugar conformation. A very interesting question yet to be answered is the structural role of the 2'-hydroxyl groups in stabilizing the UNCG and other RNA structures. Both DNA and RNA oligomers containing GNRA tetraloops are unusually stable, but only RNA, not DNA, UNCG loops have been shown to be unusually stable (57).

4. Tertiary structure
4.1 Base triples

Base triples occur when single-stranded nucleotides hydrogen bond with Watson–Crick base pairs. RNA homopolymers readily form triple-stranded structures in solution and in fibres and triple-stranded nucleic acids have been characterized structurally and thermodynamically since the early 1960s (3). In homopolymers, the base triple occurs by insertion of the third strand into the major groove and several isomorphous base triples have been identified. One example of isomorphism in RNA base triples is the study of the binding of arginine to the TAR RNA element, where a proposed AUU triple was replaced with an isomorphic GCC^+ triple under conditions of acidic pH (61).

In tRNAs, three base interactions between distal hairpin loops are responsible for folding the clover-leaf into its 3-D shape. A combination of phylogeny, genetic, and comparative sequence analysis led to the proposal of several base triples in the catalytic core of group I self-splicing introns (62) and these data have been used to generate a 3-D model of the intron core (63). NMR studies have recently been performed on small oligonucleotide models of the junction of the P4 and P6 helices in the core of group I introns which were designed to contain one of the proposed base triples (22). In agreement with the model of the ribozyme core, the two helices stacked on to one another and base triples were found in the minor groove for sequences which are very often represented in group I introns, but not for other mutant sequences. However, the geometry of the base triple observed in the NMR studies was not the same as that proposed by the phylogenetic studies. The minor groove 'triple' contacts in the NMR study were from the single-stranded A to a 2'-hydroxyl in the Watson–Crick base pair and a less well-defined U to ribose contact was also proposed. Since the NMR study was on a small oligonucleotide, it is not clear whether this RNA represents an appropriate model for this region of the intact group I intron. The reliable identification of these and other base triple contacts in large RNAs will provide very important restraints for modelling RNA structure.

4.2 Pseudoknots

Pseudoknots form when nucleotides in a single-stranded loop form Watson–Crick base pairs with a complementary sequence outside this loop. This structural motif was originally proposed to explain how the 3' end of turnip yellow mosaic virus RNA could adopt a 3-D structure similar to tRNA (64). Since then, pseudoknots have been associated with a variety of structural roles in ribosomal RNA and self-splicing introns, are thought to play important functional roles in the control of protein synthesis (65, 66), and are recognized by several RNA-binding proteins (67).

The most common type of pseudoknot and the only example on which physical data are available is where unpaired nucleotides from a hairpin loop form a second helix by hydrogen bonding the single-stranded nucleotides outside the hairpin. Thermodynamic and NMR studies on a model pseudoknot have revealed several important features of pseudoknot folding (68, 69). This type of pseudoknot consists of two helical stems, sometimes separated by just a few nucleotides and two single-stranded (loop) regions. The two helical stems stack on to one another, while the loop nucleotides adopt the extended 2'-*endo* pucker conformation to connect the two helical stems (68). The length of the single-stranded regions is critical for pseudoknot formation, since the two non-equivalent loops are required to join the two sides of the helical stems by bridging either the major or the minor grooves (69). The length to be spanned depends on the number of base pairs in the helical stems. In this model pseudoknot, only a few (two to three) nucleotides are required to cross the minor groove, whereas a much larger loop is necessary to span the major groove. Although it is difficult to draw general rules from this one example, these results emphasize the steric constraints that need to be accommodated to fold pseudoknotted structures. The number of single-stranded nucleotides crossing the minor groove is an important determinant of pseudoknot structure and function.

Pseudoknots are only marginally more stable than the constituent hairpins and their formation requires the presence of Mg^{2+} ions, presumably to screen the negative charges on the phosphates at the junction between the two co-axially stacked stems. This result emphasizes the importance of divalent ions in stabilizing RNA tertiary structure, as already observed for tRNA (3). It has also been observed that divalent ions can specifically bind to a G–U base pair in the acceptor stem of tRNA (70) and may therefore also play a role in RNA secondary structure stabilization.

4.3 G-quartets

So-called G-quartets are unusual nucleic acid structures originally observed in fibre diffraction studies (3), that have been proposed to describe the structure of telomeric DNA (71). The G-quartet model proposed from gel-shift and chemical probing data consists of a planar arrangement where each guanosine is hydrogen bonded by Hoogsteen pairing to another guanosine in the quartet (72, 73). The symmetric array of guanosines in each plane is stabilized by stacking of successive G-quartets and by K^+ ions, which fit between neighbouring planes. NMR (74, 75) and crystallographic (76) structures of DNA sequences that form G-quartets showed that the original models had correct overall structures, but did not correctly predict the sequential pattern of glycosidic angles in a particular strand of the G-quartets. The structure of RNA quartets comprising both G_4- and U_4-quartets has also been determined by NMR (77). There is some evidence that the G-quartet structure may be involved in the packaging of HIV RNA (78).

4.4 Helical junctions

The hammerhead ribozyme and the catalytic core of group I introns are formed by three or more helices coming together at a common junction region and helical junctions also play important structural roles, for instance in 5S rRNA. It is important to understand how helices stack at junctions and to identify any interactions that occur between the 'single-stranded' nucleotides. Co-axial stacking, such as that observed for tRNA and in the model pseudoknot described above, may be an important factor in stabilizing helical junctions. One of the best candidates for identifying these types of interaction is the hammerhead ribozyme, which is a very well-characterized RNA enzyme (13). In the hammerhead ribozyme, nine of the 11 conserved nucleotides that are required for catalytic activity are in helical junctions. Homonuclear NMR studies on unlabelled hammerheads have confirmed the secondary structure proposed from phylogenetic and biochemical studies (79–81). Because of its large size (~50 nt), the homonuclear NMR spectra were poorly resolved and did not provide information on any tertiary contacts within the catalytic core. Multidimensional heteronuclear NMR studies on a $^{13}C/^{15}N$-labelled hammerhead ribozyme should soon allow resonance assignment and structure determination of the catalytic core of the hammerhead ribozyme (P. Legault and A. Pardi, unpublished results).

5. Concluding remarks

Watson–Crick base pairing has provided a simple structural code for investigating RNA secondary structure and helical pairing has been very successfully predicted by phylogeny and chemical and enzymatic probing (Chapter 2). For tRNA, tertiary interactions between the secondary structural elements fold the molecule into its 3-D structure. Presumably, other RNAs will follow similar rules: the helical regions will generally define the RNA secondary structure and tertiary contacts between the secondary structure regions of the RNA, often involving nucleotides in single-stranded regions, then fold the RNA into its 3-D structure.

Recent advances in synthetic methodologies and NMR spectroscopy have allowed a closer look into RNA structure, revealing a far more complex picture. The results have shown that non-Watson–Crick base pairs and sequence-specific contacts between the bases and the sugar–phosphate backbone are very common in non-perfectly paired regions of RNA, with the inevitable conclusion that the simple structural code of Watson–Crick base pairing fails to describe the structure of non-helical regions of RNA. The catalytic activity of ribozymes is mediated by the crucial tertiary contacts that allow RNA to assume its active 3-D shape (or to act as a protein binding site); these are localized in non-helical regions and probably often involve non-Watson–Crick base pairing. Did the studies on RNA structural elements described here provide enough information to derive a new set of rules which can predict unknown RNA structures? Not yet, but these studies have revealed the complexity of RNA structure even at the level of its simpler structural elements. More importantly, new structural motifs for biochemists to understand RNA structure–function have been identified (the example of the tetraloop hairpins comes to mind). Nevertheless, tRNA remains the only example of the 3-D structure of a folded globular RNA. Despite the significant progress in understanding the structural building blocks of RNA (hairpin loops, internal loops, bulges, pseudoknots, etc.), our knowledge of RNA globular structure is still limited to the one known example of an RNA structure with well-defined tertiary interactions. The synthetic and NMR techniques developed in the past few years and described in the initial section of this review, will significantly accelerate the structure determinations of many more RNAs and RNA–protein complexes in the near future. The availability of techniques for the isotopic labelling of both the protein and the RNA provides an obvious strategy for structure determination of RNA–protein complexes and promises structures of very high resolution.

Acknowledgements

It is a pleasure to acknowledge Professor I. Tinoco, Jr and Professor Peter Moore for sharing with us the results of their laboratories prior to publication. A.P. wishes to acknowledge partial support by NIH grants AI30726, AI33098, and AI01051 and to thank the W. M. Keck Foundation for their generous support of RNA science on the Boulder campus.

References

1. Cech, T. R. (1987) The chemistry of self-splicing RNA and RNA enzymes. *Science*, **236**, 1532.
2. Milligan, J. F., Groebe, D. R., Witherell, G. W., and Uhlenbeck, O. C. (1987) Oligoribonucleotide synthesis using T7 RNA polymerase and synthetic DNA templates. *Nucleic Acids Res.*, **15**, 8783.
3. Saenger, W. (1984) *Principles of Nucleic Acid Structure*. Springer Verlag, New York.
4. Ellington, A. D. and Szostak, J. W. (1990) *In vitro* selection of an RNA molecule that binds specific ligands. *Nature*, **346**, 818.

5. Ellington, A. D. and Szostak, J. W. (1992) Selection *in vitro* of a single stranded DNA molecule that folds into specific ligand-binding structures. *Nature*, **355**, 850.
6. Mattaj, I. W. (1993) RNA recognition: a family matter? *Cell*, **73**, 837.
7. St Johnston, D., Brown, N. H., Gall, J. G., and Jantsch, M. (1992) A conserved double-stranded RNA binding domain. *Proc. Natl Acad. Sci. USA*, **89**, 10 979.
8. Chou, S.-H., Flynn, P., and Reid, B. (1989) Solid-phase and high-resolution NMR studies of two synthetic double helical RNA dodecamers: r(CGCGAAUUCGCG) and r(CGCGUAUACGCG). *Biochemistry*, **28**, 2422.
9. Grosshans, C. A. and Cech, T. R. (1991) A hammerhead ribozyme allows synthesis of a new form of the *Tetrahymena* ribozyme homogeneous in length with a 3' blocked for transesterification. *Nucleic Acids Res.*, **19**, 3875.
10. Batey, R. T., Inada, M., Kujawinski, E., Puglisi, J. D., and Williamson, J. R. (1992) Preparation of isotopically labeled ribonucleotides for multidimensional NMR spectroscopy of RNA. *Nucleic Acids Res.*, **20**, 4515.
11. Nickonowicz, E. P., Sirr, A., Legault, P., Jucker, F. M., Baer, L. M., and Pardi, A. (1992) Preparation of ^{13}C and ^{15}N labeled RNA for heteronuclear multidimensional NMR studies. *Nucleic Acids Res.*, **20**, 4507.
12. Michnicka, M. J., Harper, J. W., and King, G. C. (1993) Selective isotopic enrichment of RNA: application to the HIV-1 TAR element. *Biochemistry*, **32**, 395.
13. Uhlenbeck, O. C. (1987) A small catalytic oligoribonucleotide. *Nature*, **328**, 596.
14. Wüthrich, K. (1986) *NMR of Proteins and Nucleic Acids*. John Wiley & Sons, New York.
15. Clore, G. M. and Gronenborn, A. M. (1991) Structures of larger proteins in solution: three- and four-dimensional heteronuclear NMR. *Science*, **252**, 1390.
16. Varani, G. and Tinoco, I., Jr (1991) RNA structure and NMR spectroscopy. *Q. Rev. Biophys.*, **24**, 479.
17. Varani, G., Wimberly, B., and Tinoco, I., Jr (1989) Conformation and dynamics of an RNA internal loop. *Biochemistry*, **28**, 7760.
18. Cheong, C., Varani, G., and Tinoco, I., Jr (1990) Solution structure of an unusually stable RNA hairpin, 5'GGAC(UUCG)GUCC. *Nature*, **346**, 680.
19. Heus, H. A. and Pardi, A. (1991) Structural features that give rise to the unusual stability for RNA hairpins containing GNRA loops. *Science*, **253**, 191.
20. White, S. A., Nilges, M., Huang, A., Brunger, A. T., and Moore, P. B. (1992) NMR analysis of helix I from the 5S RNA of *Escherichia coli*. *Biochemistry*, **31**, 1610.
21. Wimberly, B., Varani, G., and Tinoco, I., Jr (1993) The conformation of loop E of eukaryotic 5S ribosomal RNA. *Biochemistry*, **32**, 1078.
22. Chastain, M. and Tinoco, I., Jr (1992) A base triple structural domain in RNA. *Biochemistry*, **31**, 12 733.
23. Nikonowicz, E. P. and Pardi, A. (1992) Three-dimensional heteronuclear NMR studies of RNA. *Nature*, **355**, 184.
24. Nikonowicz, E. P. and Pardi, A. (1992) Application of four-dimensional heteronuclear NMR to the structure determination of a uniformly ^{13}C-labeled RNA. *J. Am. Chem. Soc.*, **114**, 1082.
25. Pardi, A. and Nikonowicz, E. P. (1992) A simple procedure for resonance assignment of the sugar protons in ^{13}C-labeled RNA. *J. Am. Chem. Soc.*, **114**, 9202.
26. Hines, J. V., Varani, G., Landry, S. M., and Tinoco, I., Jr (1993) The stereospecific assignment of H5' and H5" in RNA using the sign of two-bond carbon–proton scalar coupling. *J. Am. Chem. Soc.*, **115**, 11 002.
27. Varani, G. and Tinoco, I., Jr (1991) Carbon assignments and heteronuclear coupling

constants for an RNA oligonucleotide from natural abundance ^{13}C–^{1}H correlated experiments. *J. Am. Chem. Soc.*, **113**, 9349.
28. Nikonowicz, E. P. and Pardi, A. (1993) An efficient procedure for assignment of the proton, carbon and nitrogen resonances in ^{13}C/^{15}N labeled nucleic acids. *J. Mol. Biol.*, **232**, 1141.
29. Varani, G., Cheong, C., and Tinoco, I., Jr (1991) Structure of an unusually stable RNA hairpin. *Biochemistry*, **30**, 3280.
30. Szewczak, A. A., Moore, P. B., Chan, Y.-L., and Wool, I. G. (1993) The conformation of the sarcin/ricin loop from 28S ribosomal RNA. *Proc. Natl Acad. Sci. USA*, **90**, 9581.
31. Altona, C. (1982) Conformational analysis of nucleic acids. Determination of backbone geometry of single-helical RNA and DNA in aqueous solutions. *Rev. Trav. Chem. Pays-Bas*, **101**, 413.
32. van de Ven, F. J. M. and Hilbers, C. W. (1988) Nucleic acid and magnetic resonance. *Eur. J. Biochem.*, **178**, 1.
33. Metzler, W. J., Wang, C., Kitchen, D. B., Levy, R. M., and Pardi, A. (1990) Determining local conformational variations in DNA. Nuclear magnetic resonance structures of the DNA duplexes d(CGCCTAATCG) and d(CGTCACGCGC) generated using back-calculation of the nuclear Overhauser effect spectra, a distance geometry algorithm and constrained molecular dynamics. *J. Mol. Biol.*, **214**, 711.
34. Kennard, O. and Hunter, W. N. (1989) Oligonucleotide structure: a decade of results from single crystal X-ray diffraction studies. *Q. Rev. Biophys.*, **22**, 327.
35. Hall, K., Cruz, P., Tinoco, I., Jr, Jovin, T. M., and Van de Sande, J. H. (1984) Z-RNA – a left-handed RNA double helix. *Nature*, **311**, 584.
36. Dock-Bregeon, A. C., Chevrie, B., Podjarni, A., Moras, D., de Bear, J. S., Gough, G. R., Gilham, P. T. and Johnson, J. E. (1988) High resolution structure of the RNA duplex [U(U–A)$_6$A]$_2$. *Nature*, **335**, 375.
37. Holbrook, S. R., Cheong, C., Tinoco, I., Jr, and Kim, S.-H. (1991) Crystal structure of an RNA double helix incorporating a track of non-Watson–Crick base pairs. *Nature*, **353**, 579.
38. Steitz, T. A. (1990) Structural studies of protein–nucleic acid interactions: the sources of sequence-specific binding. *Q. Rev. Biophys.*, **23**, 205.
39. Turner, D. H. (1992) Bulges in nucleic acids. *Curr. Opin. Struct. Biol.*, **2**, 334.
40. Witherell, G. W., Gott, J. M., and Uhlenbeck, O. C. (1991) Specific interactions between RNA phage coat proteins and RNA. *Prog. Nucleic Acid Res. Mol. Biol.*, **40**, 185.
41. Murphy, F. L. and Cech, T. R. (1993) An independent folding domain of RNA tertiary structure within the *Tetrahymena* ribozyme. *Biochemistry*, **32**, 5291.
42. Puglisi, J. D., Tan, R., Canlan, B. J., Frankel, A. D., and Williamson, J. R. (1992) Conformation of the TAR–arginine complex by NMR. *Science*, **257**, 76.
43. Weeks, K. M. and Crothers, D. M. (1992) RNA binding assays for tat-derived peptides: implications for specificity. *Biochemistry*, **31**, 10 281.
44. Bhattacharya, A., Murchie, A. I. H., and Lilley, D. M. J. (1990) RNA bulges and the helical periodicity of double stranded RNA. *Nature*, **343**, 484.
45. Aboul-Ela, F., Murchie, A. I. H., Homans, S. W., and Lilley, D. M. J. (1993) Nuclear magnetic resonance study of a deoxynucleotide duplex containing a three-base bulge. *J. Mol. Biol.*, **229**, 173.
46. Riordan, F. A., Bhattacharya, A., McAteer, S., and Lilley, D. M. J. (1992) Kinking of RNA helices by bulged bases, and the structure of the human immunodeficiency virus type I *trans*-activation response element. *J. Mol. Biol.*, **226**, 305.

47. Hampel, A., Tritz, R., Hicks, M., and Cruz, P. (1990) Hairpin catalytic RNA model: evidence for helices and sequence requirements for substrate RNA. *Nucleic Acids Res.*, **18**, 299.
48. Pan, T. and Uhlenbeck, O. C. (1992) A small metalloribozyme with a two-step mechanism. *Nature*, **358**, 560.
49. Li, Y., Zon, G., and Wilson, W. D. (1991) NMR and molecular modeling evidence for a GA mismatch in a purine-rich DNA duplex. *Proc. Natl Acad. Sci. USA*, **88**, 26.
50. Woese, C. R., Winker, S., and Guttell, R. R. (1990) Architecture of ribosomal RNA: constraints on the sequence of 'tetraloops'. *Proc. Natl Acad. Sci. USA*, **87**, 8467.
51. Colvin, R. A., White, S. W., Garcia-Blanco, M. A., and Hoffman, D. W. (1993) Structural features of an RNA containing the CUGGGA loop of the human immunodeficiency type-1 *trans* activation response element. *Biochemistry*, **32**, 1105.
52. Hoffman, D. W., Colvin, R. A., Garcia-Blanco, M. A., and White, S. W. (1993) Structural features of the *trans*-activation response RNA element of equine infectious anemia virus. *Biochemistry*, **32**, 1096.
53. Rould, M. A., Perona, J. J., Soll, D., and Steitz, T. A. (1989) Structure of *E. coli* glutaminyl tRNA synthetase complexes with tRNA^{Glu} and ATP at 2.8 Å resolution. *Science*, **246**, 1135.
54. Puglisi, J. D., Wyatt, J. R., and Tinoco, I., Jr (1990) Solution conformation of an RNA hairpin loop. *Biochemistry*, **29**, 4215.
55. Uhlenbeck, O. C. (1990) Nucleic acid structures—tetraloops and RNA folding. *Nature*, **346**, 613.
56. Selinger, D., Liao, X., and Wise, J. A. (1993) Functional interchangeability of the structurally similar tetranucleotide loops GAAA and UUCG in fission yeast signal recognition particle RNA. *Proc. Natl Acad. Sci. USA*, **90**, 5409.
57. Antao, V. P. and Tinoco, I., Jr (1992) Thermodynamic parameters for loop formation in RNA and DNA hairpin loops. *Nucleic Acids Res.*, **20**, 819.
58. Davis, P. W., Thurmes, W., and Tinoco, I., Jr (1993) Structure of a small RNA hairpin. *Nucleic Acids Res.*, **21**, 537.
59. Haasnoot, C. A. G., Hilbers, C. W., van der Marel, G. A., van Boom, J. H., Singh, U. C., Pattabiraman, N., and Kollmann, P. A. (1986) On loop folding in nucleic acid type structures. *J. Biomol. Struct. Dyn.*, **3**, 843.
60. Sakata, T., Hiroaki, H., Oda, Y., Tanaka, T., Ikehara, M., and Uesugi, S. (1990) Studies on the structure and stabilizing factor of the CUUCGG hairpin RNA using chemically synthesized oligonucleotides. *Nucleic Acids Res.*, **18**, 3831.
61. Puglisi, J. D., Chen, L., Frankel, A. D., and Williamson, J. R. (1993) Role of RNA structure in arginine recognition of TAR RNA. *Proc. Natl Acad. Sci. USA*, **90**, 3680.
62. Michel, F., Ellington, A. D., Couture, S., and Szostak, J. W. (1990) Phylogenetic and genetic evidence for base triples in the catalytic domain of group I introns. *Nature*, **347**, 578.
63. Michel, F. and Westhof, E. (1990) Modelling the three-dimensional architecture of group I catalytic introns based on comparative sequence analysis. *J. Mol. Biol.*, **216**, 585.
64. Pleij, C. W. A. (1990) Pseudoknots—a new motif in the RNA game. *TIBS*, **15**, 143.
65. Draper, D. E. (1990) Pseudoknots and the control of protein synthesis. *Curr. Opin. Cell Biol.*, **2**, 1022.
66. Farabaugh, P. J. (1993) Alternative readings of the genetic code. *Cell*, **74**, 591.
67. ten Dam, E., Pleij, K., and Draper, D. (1992) Structural and functional aspects of RNA pseudoknots. *Biochemistry*, **31**, 11 665.

68. Puglisi, J. D., Wyatt, J. R., and Tinoco, I. J. (1990) Conformation of an RNA pseudoknot. *J. Mol. Biol.*, **214**, 437.
69. Wyatt, J. R., Puglisi, J. D., and Tinoco, I. J. (1990) RNA pseudoknots: stability and loop size requirements. *J. Mol. Biol.*, **214**, 455.
70. Limmer, S., Hofmann, H.-P., Ott, G., and Sprinzl, M. (1993) The 3'-terminal end (NCCA) of tRNA determines the structure and stability of the aminoacyl acceptor stem. *Proc. Natl Acad. Sci. USA*, **90**, 6199.
71. Sen, D. and Gilbert, W. (1988) Formation of parallel 4-stranded complexes by guanine-rich motifs in DNA and its implications for meiosis. *Nature*, **334**, 364.
72. Sundquist, W. I. and Klug, A. (1989) Telomeric DNA dimerizes by formation of guanine tetrads between hairpin loops. *Nature*, **342**, 825.
73. Williamson, J. R., Raghumaran, M. K., and Cech, T. R. (1989) Monovalent-cation induced structure of telomeric DNA—the G-quartet model. *Cell*, **59**, 871.
74. Smith, F. W. and Feigon, J. (1992) Quadruplex structure of *Oxytricha* telomeric DNA oligonucleotides. *Nature*, **356**, 164.
75. Aboul-Ela, F., Murchie, A. I. H., and Lilley, D. M. J. (1992) NMR study of parallel stranded tetraplex formation by the hexadeoxynucleotide d(TG$_4$T). *Nature*, **360**, 280.
76. Kang, C. H., Zhang, X., Ratliff, R., Mayzis, R., and Rich, A. (1992) Crystal structure of four-stranded *Oxytricha* telomeric DNA. *Nature*, **356**, 126.
77. Cheong, C. and Moore, P. B. (1992) Solution structure of an unusually stable RNA tetraplex containing G- and U-quartet structures. *Biochemistry*, **31**, 8406.
78. Sundquist, W. I. and Heaphy, S. (1993) Evidence for interstrand quadruplex formation in the dimerization of HIV-1 genomic RNA. *Proc. Natl Acad. Sci. USA*, **90**, 3393.
79. Caviani Pease, A. and Wemmer, D. E. (1990) Characterization of the secondary structure and melting of a self-cleaved hammerhead domain by ^{1}H NMR spectroscopy. *Biochemistry*, **29**, 9039.
80. Heus, H. A. and Pardi, A. (1991) Nuclear magnetic resonance studies of the hammerhead ribozyme domain. Secondary structure formation and magnesium dependence. *J. Mol. Biol.*, **217**, 113.
81. Odai, O., Kodama, H., Hiroaki, H., Sakata, T., Tanaka, T., and Uesugi, S. (1990) Synthesis and NMR study of ribooligonucleotides forming a hammerhead-type RNA enzyme system. *Nucleic Acids Res.*, **18**, 5955.

2 | Prediction and experimental investigation of RNA secondary and tertiary foldings

ERIC WESTHOF and FRANÇOIS MICHEL

1. Introduction

Through folding of the polynucleotide sugar–phosphate backbone, the ribonucleic acid bases interact and form hydrogen-bonded double-stranded helices separated by single-stranded regions forming hairpin loops, bulges, and internal or multiple loops (*Figure 1*). The main secondary structural element, the antiparallel double-stranded helix, is maintained via hydrogen bonds involving atoms of the bases, that is, atoms of the side chains and not of the polynucleotide backbone (in contrast to proteins where secondary structure elements such as sheets or helices are held through hydrogen bonds involving atoms of the polypeptide backbone). The establishment of the secondary structure of an RNA molecule is thus feasible either directly from its base sequence on the basis of solution data (chemical and enzymatic probing) coupled with theoretical free energy minimizations or by sequence comparisons of functionally similar molecules across phylogeny.

For the determination of tertiary structure, X-ray crystallography is outstanding, since it yields a wealth of unequalled structural information on the crystallized molecule. However, RNAs are notoriously difficult to crystallize (but see Doudna *et al.* (1)), since they are highly charged molecules which can undergo spontaneous or metal-induced cleavages. Recently, nuclear magnetic resonance (NMR) techniques have also provided useful structures (see Chapter 1). This approach is, however, restricted to relatively small fragments. A serious drawback of these biophysical methods is that RNA molecules can exchange between alternative foldings, a property that is sometimes essential for their biological function. Molecular modelling appears, therefore, to be an attractive approach for obtaining some three-dimensional (3-D) knowledge of large RNAs. Starting with the first model of B-DNA (2), modelling has been used extensively in the nucleic acid field, even before detailed X-ray crystallographic data on fragments were available. The modelling

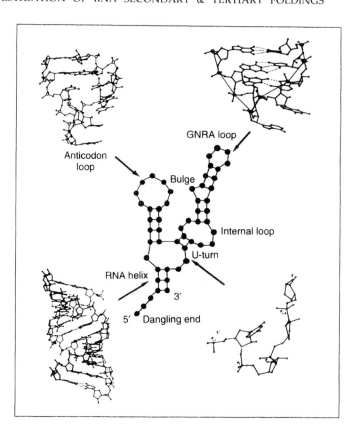

Fig. 1 A hypothetical secondary structure of an RNA molecule illustrating various types of structural elements (hairpins, bulge, internal loop, a 5′ dangling end, and, at the centre of the drawing, a multiple loop). Four 3-D motifs are represented (clockwise from the top): a seven-base anticodon loop, a four-base GNRA loop, the sharp U-turn, and a standard RNA helix. Note that double-stranded RNA helices are normally A-form, unlike double-stranded DNA helices which are generally B-form. In A-form helices, the base pairs are inclined with respect to the helical axis and displaced toward the minor groove side. These changes deepen and narrow the major groove while shallowing and expanding the minor groove. Thus, the A-form grooves are better described as deep and shallow, instead of major and minor (see Chapter 1).

of the codon–anticodon interaction (3) and of a full tRNA (4) constitute remarkable achievements. The structural database for RNA is poor and modelling will therefore have to rely on the variety of biochemical and chemical probes available to explore the accessibility of atoms involved in the secondary and tertiary structures, on the systematic use of the enormous amount of information contained in biological sequences from various organisms, and on site-directed mutagenesis coupled with physicochemical experiments. We will first delineate the conceptual framework on which our present understanding of RNA folding is based.

2. Hierarchy in structure
2.1 The definitions of secondary structure

Probably the most straightforward definition of secondary structure would include any nucleotide such that both itself and at least one of its immediate neighbours in the 5′- or 3′-direction is involved in classical (Watson–Crick and G–U) base pairing with a stretch of nucleotides in an antiparallel orientation. Ideally, one would like an RNA secondary structure to be a planar graph which can be represented as a tree, that is, the lines connecting the paired bases do not intersect. However, there are three ways in which two paired segments can be related to each other in a folded single-stranded RNA molecule: two consecutive hairpins (for example, P2 and P2.1 in *Figure 2a*), two helices separated by an internal loop or bulge (for example, P6a and P6b in *Figure 2a*), and pseudoknots (for example, P3 with P7 (5') and P7(3') in *Figure 2a*, see also *Figure 2b*). The first two motifs can be represented as two-dimensional (2-D) graphs without self-intersections whereas pseudoknots cannot. Thus, pseudoknots are fundamentally 3-D motifs. Indeed, pseudoknots result from Watson–Crick base pairing involving a stretch of bases in a loop between paired strands and a distal single-stranded region (which could belong to a hairpin loop or a bulge). The existence of RNA pseudoknots is made possible by the special geometry of double-stranded RNA helices (see below).

Clearly then, it is not always possible to represent an RNA secondary structure in a plane. This is a real problem because the dynamic programming algorithms commonly used to predict secondary structure *ab initio* (see below) cannot deal with non-tree-like structures (such as the one in *Figure 2b*, as opposed to that in *Figure 2a*): such programs and also most of the ones designed for drawing secondary structures use what we call a strict algorithmic definition of secondary structure. As a consequence, those base-paired segments that cannot be accommodated in a planar representation will tend to be discarded as 'tertiary pairings' or pseudoknots. This may lead to arbitrary decisions, for instance, when pairings are sorted out according to whose 5' strand is first encountered in a 5'- to 3'-direction. *Figure 2* illustrates this point and also the fact that a planar drawing need not be strictly tree-like: compare one of the original secondary structure drawings of group I introns (5) in which the complementary stretches P7(5') and P7(3') are not paired (*Figure 2a*) with the current 'standard' representation (6) which contains explicitly both the P7 and P3 helices (*Figure 2b*).

Alternatively, one may want to rely on the folding program itself to sort out those base-paired segments that are incompatible with a tree-like structure into secondary and tertiary pairings on an energy basis. But is this parsing physically meaningful? In other words, do the secondary and tertiary base pairs really correspond to two different energy levels? And, how well supported by experimental data is the distinction between classical base pairs and all other non-covalent RNA–RNA bonds? Isolated small secondary structure elements (for example, hairpins and pseudoknots) are known to form autonomous entities. The thermodynamic parameters given in *Table 1* show that, for example the T-arms of tRNAs or small hairpins with tetraloops (GNRA or UNCG) melt cooperatively with large entropic and enthalpic changes. However, identification by differential chemical modification of the nucleotides that change state during the initial unfolding step of group I self-splicing introns (7, 8) confirmed earlier studies on tRNA (9–11) in showing that in addition to the residues known to be engaged in tertiary interactions, cooperative denaturation also involves a subset of all classical base pairs. In conclusion, a definition which is operational and physically meaningful would be that the secondary structure represents a subset (to be established in each case) of all

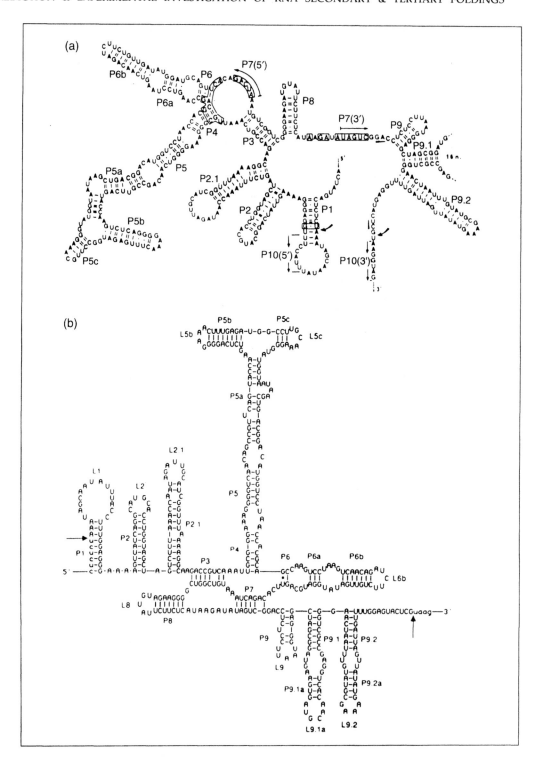

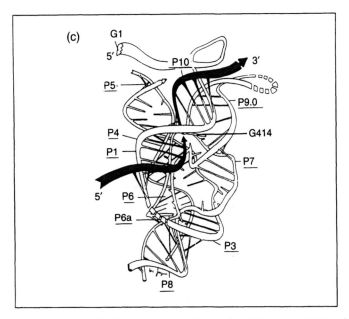

Fig. 2 Models for the secondary and tertiary structures of the self-splicing group I intron from *Tetrahymena thermophila*. The nomenclature of secondary structure elements is P for paired double-stranded helices (when interrupted by a bulge or an internal loop, small letters are added to the number of the helical segment, for example P5a; if constituted of several stem–loop hairpins, a number preceded by a dot is added to the number of the helical segment, for example P9.2); L, for hairpin loops; J, for junction segments connecting two double-stranded helices (for example, J4/5 connects P4 to P5). (a) Model of the secondary structure, drawn according to Michel and Dujon (82). The curved arrows indicate the cleavage sites and the long black arrowheads the P9.0 pairings. The two strands of helix P7 are drawn unpaired. This drawing could be represented by a tree. One could as well choose to separate the strands of helix P3 instead, with helix P7 formed. (b) Model of the secondary structure, drawn according to Burke *et al.* (6). Both P3 and P7 are drawn as paired helices. This drawing cannot be represented by a tree because of the presence of a pseudoknot. (c) 3-D model, according to Michel and Westhof (17). The last residue of the intron, which is invariant and always a guanine, is G414. The step shown precedes the second transesterification reaction and the ligation of the 5' and 3' exons.

classical contiguous base pairs obtained by folding an RNA sequence. It remains to be seen how dependent on experimental conditions are the exact contents, for a given molecule, of such experimentally defined secondary and tertiary structure subsets.

2.2 Tertiary motifs

The crystal structures of tRNA^{Phe} (12–14) and tRNA^{Asp} (15) have shown that numerous isolated interactions between bases, bases and sugars, or bases and phosphate groups occur. However, since the currently available RNA structural database is so limited, it is essential to introduce and identify structural motifs. We picture tertiary motifs as topologically rigid 3-D elements which are recurrent in

Table 1 Measured thermodynamic parameters of RNA folding

	T_m (°C)	ΔH (kcal mol^{-1})	ΔS (e.u.)	ΔE_a (kcal mol^{-1})	ΔG (37°C)
T-arm of yeast tRNA^{Asp} [a]	79.5	−62.0	−44.0	64.0	
T-arm of E. coli tRNA^{fMet} [b]	61.0	−54.0		55.0	
Yeast tRNA^{Asp} [a]	51.4	−101.0	−78.0	48.0	
rGGC–GCAA–GCC [c]	71.0	−30.3	−88.0		−3.0
rGGC–ACAA–GCC [c]	62.7	−29.2	−86.9		−2.2
rGGAC–GAAA–GUCC [d]	65.9	−49.1	−145.0		−4.2
rGGAC–UUCG–GUCC [d]	71.7	−56.5	−163.9		−5.7
rGGAG–UUCG–CUCC [d]	60.1	−48.6	−145.7		−3.4

sunY of T4 [e]					ΔG (52°C)
Wild type	56.2	−227.3	−690.7		−2.8
Mutant 1	50.0	−156.5	−484.4		0.9
Mutant 2	44.3	−141.5	−445.6		3.4
Mutant 3	54.6	−171.2	−522.5		−1.34
Mutant 4	49.6	−161.4	−500.2		1.21
Mutant 5	52.0	−141.5	−435.3		0.03

The parameters show that substructures (such as the T-arms of tRNAs) have an autonomous fold independent of the overall tRNA structure. Similarly, hairpins closed by a tetraloop, such as GNRA or UNCG, have high melting temperatures with large variations in enthalpy and entropy. The parameters given for the autocatalytic group I intron *sunY* of bacteriophage T4 represent some specific tertiary interactions. In P9.0a the wild-type RNA contains a tertiary G–C pair as well as a tertiary pair between residues G131 and C1013 (part of a loop–loop interaction, see section 3.3.2). In the first two mutants, each of those tertiary contacts is broken separately. The wild-type RNA also presents interactions between G3 and A4 of a GUGA loop in L9 and an A–G dinucleotide in P5 (see section 3.3.1 and *Figure 9*). In the third mutant, the L9 loop has been changed to GUAA, with minor loss of contact, in the fourth one to GUUA, with a total loss of contact between U3 and P5 bp 3 (A), and in the last mutant only the G–C pair of P5 has been changed to a C–G pair, with loss of interactions between A4 and P5 bp 4. See Jaeger et al. (26). [a] From Coutts et al. (9). [b] From Crothers et al. (10). [c] From SantaLucia et al. (87). [d] From Antao et al. (88). [e] From Jaeger et al. (8, 26).

RNA structures and can be assumed to constitute autonomously folded units. One can classify RNA tertiary structure as those interactions involving

(1) two helices;

(2) two unpaired regions;

(3) one unpaired region and a double-stranded helix (16).

The interactions between two helices are basically of two types: either two helices with a contiguous strand stack on to each other or two distant helices position themselves so that their minor grooves fit. An unpaired region belongs to either a single-stranded stretch (forming an internal loop or a bulge) or a hairpin loop closing a helix. Interactions between two unpaired regions lead to pseudoknots if a single loop is involved and to loop–loop motifs otherwise. Interactions between an unpaired region and a double-stranded helix can led to various types of motifs.

Pairing of a single-stranded stretch, either in the deep or the narrow groove of a double helix, yields a triple helix. One motif is known in which the unpaired region constitutes a terminal loop: GNRA tetraloops bind the shallow groove of the helix (see below). Those motifs involving single-stranded stretches have great potential to form tertiary structures because they can lead to co-axiality between helices. Depending on their sequence, internal loops or bulges could constitute 3-D motifs, but such motifs have not yet been characterized.

3. Hierarchy of folding

In a structural and hierarchical view of folding, the architecture of highly structured RNA results from the compaction of preformed secondary structure components in 3-D space (16, 17). This working hypothesis will serve here as a framework for organizing our thoughts on RNA secondary structure and for describing 3-D motifs (in fact, a consensus on this hierarchical view of folding is now slowly emerging). According to such a view, secondary structure elements (helices and hairpin loops) form first. Helices then interact locally end-to-end through stacking or by forming pseudoknots or triple helices. Finally, these autonomously folded subdomains associate cooperatively by loop–loop base pairings and by numerous contacts involving loops, bulges, and helices. This theoretical framework stresses the role in the first folding steps of unusually stable hairpin loops such as the tetraloops (GNRA and UNCG). It predicts also that, during the subsequent stages and depending on the environmental conditions, folding of subdomains could lead to kinetic traps and non-native conformers separated from the native conformers by high-energy barriers. And, finally, the last steps should constitute 'all-or-none' processes involving numerous specific and local interactions between preformed motifs (18, 19). We will now discuss briefly some 3-D motifs because of their importance in the understanding of RNA folding, in analysing sequence alignments, and in interpreting probing data. The described motifs will clearly be ideal and numerous deviations from regularity have been found and are to be expected.

3.1 Initiation of folding: hairpins

The structures of hairpins emphasize the role of the non-Watson–Crick base pairs in closing and stabilizing the loop. These base pairs (for example, *trans* Watson–Crick or Hoogsteen pairs; see *Figure 3* for definitions) are shorter and allow a tightening of the loop together with stacking of the other loop residues (*Table 2*). Because non-Watson–Crick base pairs are seldom isosteric (see the example of *trans* base pairs in *Figure 4*), reversal of such a base pair usually leads to a change in the conformation of the loop. Chemical probing experiments are necessary to ascertain the adopted conformation. The tRNA U33 sharp turn (*Figure 1* bottom right) is quite systematically present. Among terminal loops, UNCG and GNRA are unexpectedly frequent in structured RNAs (20) and this has been attributed in part to their thermal stability (for example, Tuerk *et al.* (however, see below) (21)).

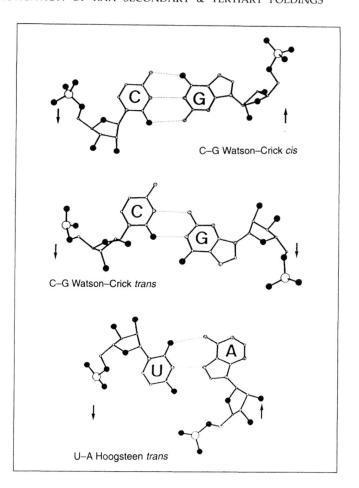

Fig. 3 Some definitions for base pairings. A purine base can hydrogen bond via its Watson–Crick sites (O6, N1, N2 for a guanine) or via its Hoogsteen sites (O6 and N7 for a guanine). Moreover, the glycosyl bonds can be both on the same side (*cis*) of the middle line through the hydrogen bonds or on either side (*trans*). Thus, a C–G pair can be *cis* (the standard or canonical case) or *trans*. The same is true for A–U pairs. Pairings between A and U can also be of the Hoogsteen type with a *cis* (as in standard triple helices) or a *trans* (as in the quasi-invariant U8–A14 of tRNA structure) orientation of the glycosyl bonds. Those parameters, together with the relative orientation about the glycosyl bond (*anti* versus *syn*), dictate the relative local orientations of the pairing strands (antiparallel or parallel). Simple prediction rules can be found in Westhof (83); a more sophisticated treatment is in Lavery *et al.* (84). Only the *cis* Watson–Crick base pairs are isosteric.

3.2 Subdomain formation: local motifs with co-axiality

Those motifs are local and are usually formed from contiguous secondary structure elements. As mentioned above, motifs involving single-stranded stretches of bases may lead to co-axial stacking of two adjacent helices. In those instances, the bases on one strand are contiguous or interrupted by at most one or two unpaired residues. Because of the geometry and right-handedness of RNA helices, triple-helix formation is topologically related to pseudoknotting (*Figures 5* and *6*). The double helix in the triple-helix motif and the 'main' hairpin in the pseudoknot motif might both be considered as 'preformed' structural elements controlling efficient interactions between widely spaced nucleotides (18, 19) (*Figure 7*).

3.2.1 Single-stranded stretch with helix: triple helices

In triple helices, a single-stranded stretch interacts either in the deep (major) groove or in the shallow (minor) groove of a double-stranded RNA helix. When in

Table 2 Small structured loops favouring hairpin formation

Loops with seven residues

1. Anticodon loop[a]
   U-turn after the second residue
   Five stacked bases in the 3' direction with bases 3–4–5 oriented for external pairing
2. Thymine loop[a]
   First base pair in Hoogsteen *trans*
   U-turn after the second residue
   Last two bases rejected toward the exterior
   One intercalation and one Watson–Crick pair possible with another loop

Loops with five residues

1. Ending with a 5'C and a G3'[b]
   U-turn after the 5'C and three stacked residues
   5'C–G3' in Watson–Crick *trans* with G-*syn*
2. Ending with a 5'G and a C3'[c]
   U-turn after the 5'G
   Four unstacked residues
   No 5'G–C3' pair

Loops with four residues

1. 5'GNRA3'-type[d]
   U-turn after 5'G and two stacked residues
   Non-canonical 5'G–A3' pair
   N6/N7 of A*anti* with N3/N2 of G*anti*
2. -5'UUNG3'-type[e]
   U-turn after 5'U without good stacking
   Non-canonical 5'U–G3' pair: Watson–Crick *trans* with G-*syn*

[a] Structure established by X-ray diffraction of single crystals of yeast tRNA[Phe] (12–14) and yeast tRNA[Asp] (15). [b] Structure proposed on the basis of chemical probing and modelling of yeast tRNA[Ser] (89). [c] Structure proposed on the basis of chemical probing and modelling of *Escherichia coli* 5S rRNA (23). [d] Structure proposed on the basis of chemical probing and modelling of *Xenopus laevis* 5S rRNA (22) and established by NMR techniques (90). [e] Structure established by NMR techniques (91).

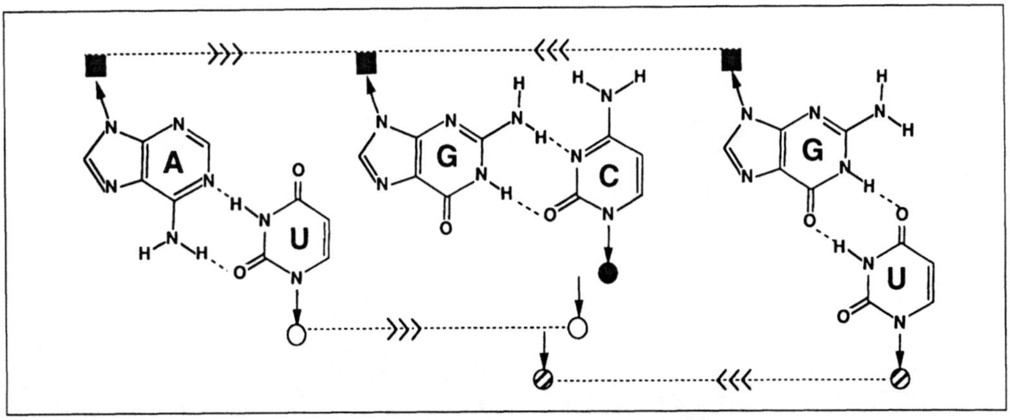

Fig. 4 Relative geometries of three *trans* Watson–Crick base pairs. Squares and circles represent the positions of the C1' atoms of the ribose attached to the purine or pyrimidine bases, respectively. Superposition of the squares would yield the indicated circle positions. Notice that the *trans* A–C base pair (not shown) is isosteric to the *trans* G–U base pair. From Jaeger *et al.* (8).

the deep groove, the interaction involves Hoogsteen pairs, while in the shallow groove the hydrogen-bonding contacts are mainly made to N2 and N3 of guanines and to O2' of riboses. A regularly shaped 3' end single strand leaving one helix can only interact in the deep groove of another co-axially and right-handedly stacked helix (in *Figure 5*, the 3' end of one strand of helix H1 interacts with the deep

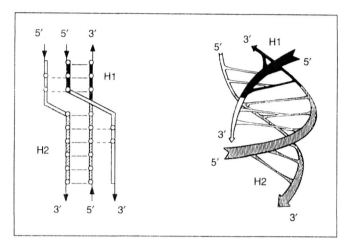

Fig. 5 2-D diagram and 3-D view illustrating how a triple helix is related to a pseudoknot. Instead of crossing the deep groove, the single-stranded 3' end (white) of helix H1 (black) interacts with the deep groove of helix H2 (grey). Similarly, instead of crossing the shallow groove, the single-stranded 5' end (white) of helix H2 (black) interacts with the shallow groove of helix H1 (black). In catalytic introns of group I, such a double triple-helix motif has been modelled on the basis of sequence comparisons and mutational data (17, 24). Recently, NMR evidence has been put forward supporting some of the suggested interactions (85). Adapted from Westhof and Jaeger (19).

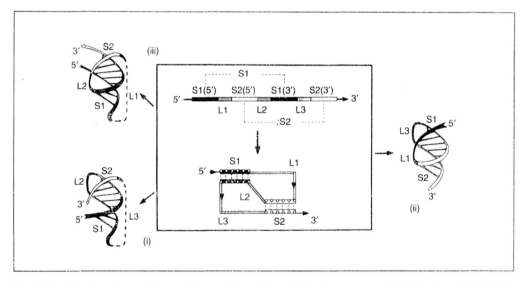

Fig. 6 Classification of pseudoknots with co-axial stacking. The helices are noted S1 (black) and S2 (white) and the single-stranded stretches L1, L2, and L3 (grey). At the top of the inset, the arrangement of the strands along the sequence is shown, stressing the interdigitation of the 5' and 3' strands constituting the two helices (see section 2.1). Below this is shown a planar view of the pairing in which the 3-D co-axiality is lost. The absence of the connecting segment L2 leads to the most frequent pseudoknot (ii). The other two pseudoknots are less frequent and are obtained when segments L1 or L3 are missing, leading to pseudoknots (i) and (iii), respectively. It is geometrically easy to cross the 'major' groove because of the depth and narrowness of RNA helices. Pseudoknotting is thus rooted in the particular geometry of double-stranded right-handed RNA helices. Adapted from Jaeger et al. (18) and Westhof and Jaeger (19).

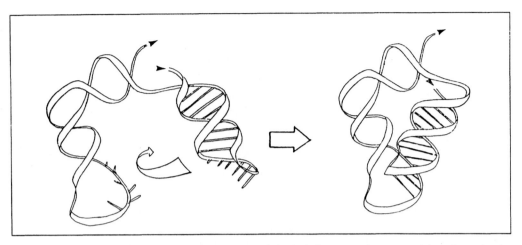

Fig. 7 Diagram illustrating how the preformed loop of the hairpin can nucleate pseudoknot formation by interacting with a single-stranded stretch. Arrowheads indicate the 5' and 3' ends. Adapted with permission of Luc Jaeger (92).

groove of helix H2). Similarly, a regularly shaped 5' end single strand entering at the 5' end of one helix will interact with the shallow groove of another co-axially and right-handedly stacked helix (in *Figure 5*, the 5' end of one strand of helix H2 interacts with the shallow groove of helix H1). In the available instances (tRNA structure; 17, 22–24), the number of base triples is less than two or three and they need not be regular.

3.2.2 Single-stranded stretch with loop: pseudoknots

In triple helices, the third strands form hydrogen-bonded pairs to the double helix, whereas in pseudoknots the third strands are chemically linked by 3'–5' phosphodiester linkages to the helix. The 'classic' pseudoknot structures, in which the two helical stems are co-axially arranged, can be divided into three types (19, 25). In type (ii), the best described and the most frequent one, the first loop in the 5' to 3' direction crosses the deep groove and the second one the shallow groove (L1 and L3, respectively, in structure (ii) of *Figure 6*). In type (i), there is no loop crossing the deep groove and, in type (iii), there is no loop crossing the shallow groove but, in those two types, there is a segment (which may be quite long) connecting the ends of the two stacked helical stems (L3 and L1, respectively, in structures (i) and (iii) of *Figure 6*).

3.3 Final 3-D anchors: motifs without co-axiality

Those motifs are usually formed from secondary structure elements which are at a distance in the 2-D structure. Motifs involving bases which belong to a loop do not lead to co-axial stacking of two adjacent helices but can still maintain some parallelism between the helical axes of the supporting helices.

3.3.1 Loop with helix: binding of G–C and A–U pairs by GNRA loops

As will be detailed below, phylogenetic sequence comparisons can be used not only to identify secondary structure elements but also to uncover tertiary interactions. Thus, sequence comparisons of group I introns identified covariations between some terminal GNRA loops and Watson–Crick base pairs in helices (17) which was interpreted and modelled as base pairing in the shallow groove of helical G–C and A–U base pairs (see below). Recently, Jaeger et al. (26) have provided strong evidence in support of that previous proposal (17). The ability of GNRA tetraloops to behave as anchors linking separate sections of RNA scaffolds must contribute to their remarkable frequency in structured RNAs (20).

3.3.2 Loop–loop interactions

The residues of two terminal loops can hydrogen bond to form an additional base-paired helix. The archetypal example is the anticodon–anticodon base pairing between two tRNAs as seen in the crystal structure of yeast tRNA^{Asp} (15) (*Figure 8*). Complexes between complementary anticodons are characterized by unexpectedly large free energies of formation (27). Antisense RNA control of plasmid replication also involves loop–loop interactions between sense and antisense RNAs, leading to 'kissing complexes' (28, 29). In all these examples, the loops involved contain between six and eight nucleotides and could adopt a conformation very similar to that of the anticodon loop with a tendency to favour stacking on the 3' side of the loop in continuity with the 3' strand of the helix. Interestingly, once an initial

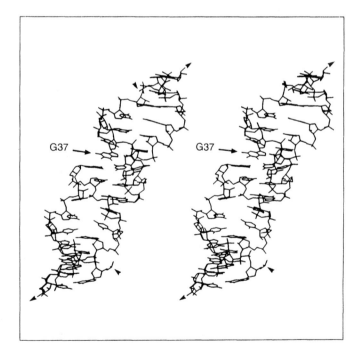

Fig. 8 Stereo drawings of the two anticodon stems related by a 2-fold symmetry axis parallel to the middle base of the anticodon triple (GUC) in the crystal structure of yeast tRNA^{Asp}. The view illustrates the fact that the helical axes are not co-linear, yet they are parallel to each other. Notice how the G37 (in fact a 1-methyl-G, shown by an arrow for the top anticodon loop) stabilizes the mini-helix made of three base pairs by stacking at the 3' end of the C–G base pairs at both extremities of that mini-helix. The stabilizing role of 3' dangling residues has been stressed (27). Arrowheads indicate the 5' and 3' ends. From Westhof et al. (86).

complex is formed with three to five base pairs, the newly formed helix grows at the expense of the helices closing the loops, thus destroying secondary structure (in the CopA–CopT inhibition complex of plasmid R1, the final loop–loop interaction comprises eight base pairs). Interactions with possible stacking on the 3' end are frequent and are found, for example in the group II introns (30).

4. Identification of interacting partners

4.1 Secondary structure prediction from energy minimization

This approach (reviewed in Turner et al. (31) and Zuker (32)) attempts to generate thermodynamically optimal, complex secondary structures from sets of thermodynamic parameters pertaining to elementary building blocks, such as two neighbouring base pairs in a double-stranded helix (the underlying assumption, which, as already discussed, is partly supported by experimental data, is that there exists sufficient hierarchy in RNA folding for secondary structure to be dealt with independently from tertiary folding). This is a formidable problem: for a random sequence of N nucleotides, taking only contiguous, classical base pairs into account, the number of possible pairings grows according to N^2, but the number of potential combinations of these pairings has been estimated to grow according to 1.8^N (33). Also, computed minimal energy structures may not be biologically relevant. The problem does not lie merely in the incompleteness of parameter sets, the naïveté of simple additive models, or the fact that input thermodynamic values were derived under conditions that may not truly mimic *in vivo* situations. The ultimate difficulty is rather that many natural RNAs are likely to require helpers (proteins or other RNAs) in order to fold into biologically active forms; even among catalytic RNAs, many group I introns and most group II introns are inactive *in vitro* (for the role of proteins in the excision of group I introns, see Saldanha et al. (34)). Using only molecules which are known to adopt an active structure *in vitro* and then, only planar, unknotted structures (the ones explored by dynamic programming), 70–90% of the helices deduced by comparative analysis turn out to be present in secondary structure models calculated by current energy minimization programs (35), which provides a fair estimate of the effectiveness of this approach.

4.2 Comparative sequence analysis

Exploring potential secondary structure by energy minimization can be of great help, but the choice method for biologists, whenever several related sequences are available, is comparative sequence analysis, which takes advantage of the fact that the structures of functional RNAs tend to be better conserved by evolution rather than their sequences. Comparative sequence analysis provides information about biologically relevant structures, but as pointed out in a recent review by Woese and Pace (36), it took some time for the power of this approach to be generally recognized by hard-core biochemists. Just 12 years ago, a manuscript (5) which offered what turned out to be the correct secondary structures of group I and group II self-splicing introns, established by comparative analysis, was rejected by *Nucleic Acids Research* on the grounds that the work was mere 'paper biochemistry'.

4.2.1 Prediction of classical base pairing

Prediction of Watson–Crick base pairs by comparative sequence analysis has been well reviewed (for example, Woese and Pace (36) and James *et al.* (37)). Its principle is straightforward: given an alignment of two or more sequences, look for sites where classical pairing is maintained despite nucleotide replacements. If in sufficient number (and if exceptions are few enough), compensatory base changes at two sites constitute a statistical constraint, which, in turn, betrays the existence of some selection pressure. This selection pressure can be either of natural origin, when dealing with biological sequences or human-imposed, in the case of *in vitro* selection or evolution experiments, which have been successfully used to explore RNA structure (38–40). In practice, the units for statistical reasoning are double-stranded helices, that is, extended pairings comprising several contiguous base pairs, rather than pairs of individual sites. A putative helix is considered proven when a sufficient number (typically two or three) of distinct compensatory changes have been observed and this allows the very efficient sifting of the thousands of potential helices generated by a mere 500+ nucleotide (nt) sequence. This is how secondary structure models of (among others) tRNA, 5S rRNA (41), 16S rRNA (42), 23S rRNA (43), RNaseP M1 RNA (44), and group I (*Figure 2a*) and group II self-splicing introns (5, 45) were generated: all those models have essentially withstood the test of time.

It is important to recall that, in principle at least, comparative sequence analysis does not say anything about the functional basis of the selective pressures it uncovers. At the time the secondary structures of group I and group II introns were established by Michel *et al.* (5) and Davies *et al.* (45), it was not realized that the molecules under scrutiny would turn out to have an active role in splicing. Also, even though the output of a search for classical base pairs is commonly organized into a 'secondary structure model', there is no reason why all deduced pairings should co-exist in the real molecule. In actual fact, incompatible base pairings are rare or non-existent in molecules which are known to have well-defined and stable structures (rRNA, self-splicing introns), whereas they may predominate when successive rearrangements are the essence of function (in the case of attenuators or for some of the snRNA components of the spliceosome).

4.2.2 Prediction of other types of interaction

Even within helices, there are sites where U–G pairs freely replace Watson–Crick pairs, whereas at others, U–Gs tend to persist over evolutionary time or alternate preferentially with C–A, thus betraying some selective pressure for non-standard geometry (46, 47). Comparative sequence analysis has indeed been used on a number of occasions to search for non-canonical interactions (4, 48). In the case of group I self-splicing introns, a systematic search for non-Watson–Crick, long-range interactions that we conducted (17) eventually led to a 3-D model of the catalytic core of these molecules (*Figure 2c*). A serious problem in the detection of non-canonical pairings, compared to Watson–Crick base pairs, is the lack of pre-defined search patterns. Among the 25 or so non-classical base pairs with two hydrogen bonds, there are few strictly isosteric series and none includes more than two members (see Tinoco (49) and *Figure 4*). Of course, some measure of geometrical tolerance is expected to exist in nature, such as for U–G pairs in double-stranded helices. In practice, this means that one should look for anything that deviates from expectations of randomness (see below and the following subsection). Some of the statistical constraints thus uncovered may result from structural constraints other than base–base pairing (for example, interactions between sugar hydroxyls O2' and phosphates or bases). For instance, group I introns with two consecutive U-G pairs in their P1 'substrate' (5' splice site) helix have a distinctive J4/5 (*Figure 2*) ribozyme core sequence and this covariation has been assumed to reflect contacts between hydroxyl groups of the P1 sugar–phosphate backbone and the J4/5 bases (*Table 3*; 17). Even those co-variations that reflect direct base–base contact may not lend themselves to straightforward geometrical interpretation. For instance, the C–A pair, although strictly isosteric with the U–G wobble pair only when both are in the *trans*-orientation, can form a protonated C–A$^+$ *cis* pair isosteric with the *cis* U–G pair (such a protonation occurs at pH 7; 50).

An additional benefit resulting from the modelling of the catalytic core of group I introns was the identification of novel, recurrent 3-D motifs (16, 17). For example, analysis of co-variations between GNRA tetraloops and base pairs in distant helices indicated that the two purine residues of the tetraloop could hydrogen bond in the shallow groove with either two G–C base pairs (for example, N6A . . . N3G and

Table 3 Co-variation between the nucleotide at position J4/5–3 of group I introns and presence versus absence of two consecutive U–G pairs in the substrate helix[a]

Nucleotide at J4/5–3	One U–G pair on each side of the splice site in P1[b]	Other pair on each side of the splice site in P1[b]
A	1	48
G	1	9
U	1	5
C	9	3

[a] Adapted from Michel and Westhof (17).
[b] Numbers correspond to occurrences of each of the specified situations in available group I sequences. The 5'-splice site of group I introns is specified in part by the presence of a U–G pair in the P1 helix (*Figure 2a*): cleavage occurs immediately 3' of the U. In a minority of introns, the G of that U–G pair is preceded by a second G, which can pair with another U (either the first nucleotide of the intron or the first nucleotide of the 3' exon; see Michel and Westhof (17)). In this subgroup of sequences, the distribution of bases at J4/5–3 is highly atypical (compare columns at left and right) and this was interpreted as evidence of a direct contact between the J4/5 loop and the backbone of the P1 helix in the vicinity of the 5'-splice site.

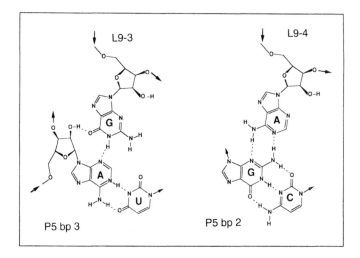

Fig. 9 Proposed hydrogen bonding contacts (dotted lines) between nucleotides G3 and A4 of the L9 GUGA tetraloop and base pairs 3 and 2 of helix P5 in the autocatalytic intron *sunY* of bacteriophage T4. From Jaeger et al. (26).

N1A ... N2G) or one A–U and one G–C base pair (see *Figure 9* and *Tables 4* and *5*). The proposed interactions were modelled in two instances:

(1) between the L2 terminal loop and helix P8 with one A ... G–C base triple;

(2) between the L9 terminal loop and helix P5 with one A ... G–C and one G ... A–U base triple.

Recently, evidence was provided showing that the L9 GUGA loop changes state during the cooperative disruption of the tertiary architecture of the intron. Further, the postulated interaction was successfully replaced by the classical base pairs of a pseudoknot, in agreement with binding of the GNRA loop into the shallow groove (26).

The need to be unprejudiced about patterns of covariation when searching for non-canonical interactions leads in turn to a major difficulty, since not only

Table 4 Co-variations between P8 bp 4 and presence versus absence of a GNRA loop at the tip of the P2 helix in group I introns[a]

P8 base pair 4	L2 terminal loop: GNRA	L2 terminal loop: other or absent
C–G	25	8
Other	3	50

[a] Adapted from Michel and Westhof (17).
Among 58 group I introns with either no P2 hairpin or one that does not end with a GNRA loop, there are no more C–G pairs at P8 bp 4 than expected from a random assortment of nucleotides. In contrast, 25 out of 28 sequences with a GNRA loop at the tip of P2 have C–G as P8 bp 4, which suggests that this base pair is directly contacted by one of the two conserved nucleotides of the loop.

Table 5 Covariation between P8 bp 5 and loop nucleotide 3 of the P2 hairpin in group I introns with a GNRA loop at the tip of P2 [a]

P8 base pair 5	L2 terminal loop: GNAA	L2 terminal loop: GNGA
C–G	14	0
U–A	1	10
Other	3	

[a] Adapted from Michel and Westhof (17).
Among 28 group I introns with a GNRA loop at the tip of helix P2, there is a marked preference for either an A as loop nucleotide 3 and a C–G pair at P8–5 or a G as loop nucleotide 3 and an A–U pair at P8–5. The remaining combinations are rare or absent. This was interpreted as evidence that the base at position 3 of the P2 loop interacts in the minor groove of the P8 helix, with bp 5 (see *Figure 9*).

structural and/or biological necessity, but also historical accidents generate non-random patterns, as exemplified in *Figure 10*. In practice, only those sites that have undergone multiple changes (and are therefore discarded as 'homoplasic' when building phylogenetic trees under the assumption of parsimony, because they are considered to result from convergence or reversion processes) can be used for structural inferences. This is why vast data sets (many sequences, with all degrees of divergence) and means of counting substitution events are essential when searching for tertiary contacts (for Watson–Crick base pairs, observation of the four isosteric combinations ensures that there have been at least three underlying events). Otherwise, the same caveats as mentioned for secondary structure models (that is, lack of information about possible temporal co-existence and functional significance of interactions) apply to the use of comparative sequence analysis to build 3-D models. In order to distinguish coordinated changes that result from authentic molecular constraints from the ones that arise from historical contingencies, we have applied the following method to 87 sequences of group I introns (17). The sequences were first classified and grouped on the basis of sequence similarity without explicit reference to phylogeny. This classification was based on a sequence distance which was simply determined by counting the number of nucleotide changes between two sequences at chosen positions of the aligned secondary structure consensus of the catalytic core. These pairwise comparisons eventually yielded a distance matrix and automatic clustering methods (for example, Delorme and Hénaut (51)) allowed the identification of sequence classes. The parsing of covariations among those classes finally allowed a decision concerning the statistical and, hence, structural significance of uncovered associations to be made: those covariations that recur in distinct classes and subclasses are the ones that can safely be ascribed to structural and functional constraints which, ideally, should be time independent and set by the physics and chemistry of biological processes.

4.2.3 Automation of comparative sequence analysis
There have been a number of attempts to automate the search for both 2-D and 3-D interactions by comparative sequence analysis. Use of an 'expected mutual infor-

Fig. 10 Two possible, extreme interpretations of a covariation. (a) Phylogenetic tree requiring only two events to generate sequences sq1–sq8. (b) At least eight changes are required. In the latter case, the contents of sites 1 and 2 can be concluded to be constrained by one another, while no such inference can be drawn from the situation depicted in (a).

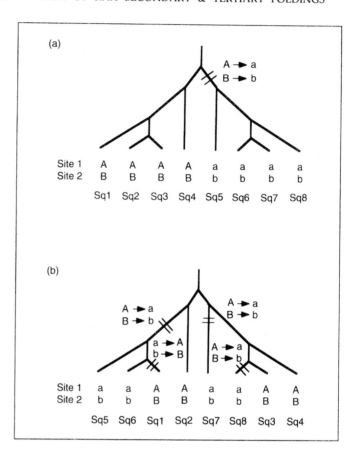

mation' measure (52) makes it possible to detect any constraint between two sites but does not solve the difficulty of sorting out the relative roles of history and necessity in generating biases. Winker *et al.* (53) attempted to tackle this latter problem by counting the number of times each combination of bases at two sites arose independently in a phylogenetic tree. These approaches assume that the sequences of interest have already been aligned by independent procedures. Depending on data sets, practical sequence alignment may be trivial (in which case the search for secondary, if not tertiary structure, is also trivial) or require that potential secondary structure (and even potential 3-D motifs) be taken into account (see Sankoff (54) for an attempt to solve the problem of iterative alignment): alignment of highly divergent RNA sequences is an occupation in which human expertise has not yet been superseded.

4.3 Experimental identification of interacting partners

We review here experimental approaches other than NMR and X-ray diffraction analysis. We will not describe either the current protocols for chemical and enzymatic probing of RNA, since reviews exist (55, 56) and these methods do not generally allow the identification of partners without systematic site-directed mutagenesis or the help of secondary structure predictions.

4.3.1 Crosslinking and directed cleavage

Cross-linking provides an efficient way of establishing vicinity relationships in complex RNAs, particularly when cross-linked forms can be shown to retain activity. The introduction of highly cross-linkable nucleotides at specific positions (57–59) makes it possible to explore the neighbourhood of the sites of particular interest in an organized fashion. A related approach consists of grafting a metal chelator to a nucleotide in order to direct the cleavage of surrounding backbone sites (60).

4.3.2 Search for compensatory effects

In nature, compensatory base changes are selected because of the need to preserve biological activity. Most man-made experiments also monitor some form of activity to assess the consequences of multiple changes, but projects have been designed that sought restoration of renaturability (61), thermal stability (8, 26), or ability to bind a substrate (62, 63). Among possible strategies for generating compensatory effects, combination of single nucleotide changes in a haphazard way seldom leads to compensation and even when successful, is unlikely to lead to straightforward structural interpretations. Reproducing those nucleotide combinations that prevail (or are missing) in biological sequences under the assumption that they should be the most favourable (respectively, deleterious) ones, will at best confirm that the sites under scrutiny are subject to no constraints other than those that could be deduced by analysing available datasets (and, also, that chosen experimental conditions mimic reasonably well those environmental pressures under which the molecules of interest evolved). A more rewarding approach, which may sometimes be carried out entirely *in vitro* (64), consists in directly selecting molecules with second-site 'suppressor' mutations. Alternatively, preconceived structural models can be tested by generating nucleotide combinations that have not yet been observed and this approach need not be confined to proving Watson–Crick base pairing. Thus, suspicions of selective pressure for non-standard helical geometry at the 5' splice site of group I introns were confirmed by the finding that C–A is the best possible substitute for the U–G pair that is normally found at that site in nature (65). Still more fruitful may be the use of unnatural nucleotides: a model according to which the N1-H position of the guanosine co-factor of a group I self-splicing intron is bound to the O6 group of a conserved G–C pair in the ribozyme was verified by restoring the activity of an A–U mutant molecule with 2-aminopurine ribonucleoside (62). *In vitro* experiments in which the coordination of magnesium ions to specific phosphate groups is checked by replacing the latter by phosphorothioates and the former with manganese belong to the same category (66, 67).

5. Modelling tertiary structure

Even with the simplified representations presently used in molecular mechanics and molecular dynamics simulations, detailed treatments at the atomic level are, for the time being, ruled out for handling the global folding of large RNAs (>100 nt). In programs based on force field calculations, the main problems reside in proper handling of electrostatics and solvation. Indeed, in all forms of nucleic acids, water molecules should be considered as an integral part of the structure, since intra- and interresidue water bridges fulfil the hydrogen-bonding capacity of the polar atoms, forming strings, spines, or filaments in which water molecules have enough reorientational mobility for additional screening of the phosphate charges (68).

Thus, the present intractability of modelling such an overwhelming amount of mutually coupled interactions led us to favour interactive graphics techniques for modelling large RNAs, despite the unavoidable heavy reliance on human judgements for selecting local conformations leading to global compactness. The heuristic power of manipulating and visualizing 3-D structures should not be underestimated. The visualization suggests potential contacts or covariations which can then be confirmed by sequence analysis and other experimental data. The proposal of a consensus 3-D structure results therefore from a constant back and forth movement between experimental data and computer modelling (see section 5.2). Although a more global technique, distance geometry, was recently introduced in RNA modelling (69), it has led to improbable tangled or knotted structures that the algorithm cannot remove from the set of solutions. Malhotra *et al.* (70) have proposed an automatic folding procedure in which a nucleotide is represented as a pseudoatom located at the phosphate atom. Such an approach ignores all the finely grained interactions between helices or between loops and helices which govern and stabilize the 3-D folding (16). A technique based on a constraint satisfaction algorithm has been put forward (71). For small systems, this approach could be very useful for delineating possible chain paths.

5.1 The assembly of 3-D architecture from secondary structure motifs

The model building approach follows the hierarchical view of folding described above. It involves, first, the recognition of the elementary motifs constituting the secondary structure, followed either by the construction of the appropriate motif, using NAHELIX or PSEUDOKNOT (72) or by its extraction from a 3-D structure bank (using FRAGMENT) of the 3-D structures of crystallographically determined nucleic acid structures, as well as of already modelled structures. With the program FRAGMENT, one can put any base sequence on any structural motif existing in the structure bank (72). The separate fragments are then hooked together in a 2-D/3-D structure (in which some links are not properly made) using ADDFRAG (72, 73).

Each fragment can be assembled separately from its constituent motifs interactively on the graphics system with a program such as FRODO (74). Each subfragment and the structure as a whole are subjected to geometrical and stereochemical refinement in order to ensure proper geometry and to prevent bad contacts using the restrained least-squares refinement program NUCLIN-NUCLSQ (15).

5.2 Accuracy and limits of molecular modelling

It should be understood that the use of atomic detail does not imply that the constructions are valid at atomic resolution in the crystallographic sense. Being meticulous about distances, angles, contacts, etc., ensures that at least the structural model is precise. The accuracy of a model can only be assessed either by X-ray crystallography or *a posteriori* by the incentives and new ideas it gives rise to, since energetic and stereochemical considerations alone cannot be taken as proof of correctness. With modelling conceived as a tool producing 3-D hypotheses for devising new experiments, several validity criteria can be envisaged. The sites of interaction with other macromolecules obtained from footprinting experiments could give useful indications as to the correctness of a model. The generalization of a model to other molecules of the same class also constitutes a convincing criterion for its accuracy, especially if the 3-D fold is able to accommodate insertions or deletions of fragments without major rearrangements of the core structure. Invariant or semi-invariant residues are of no great help for the construction itself. However, a rationalization of their occurrence through their involvement in the maintenance of the 3-D fold or of the active site is a strong *a posteriori* argument for the validity of a model.

6. Cooperativity of folding

In the 1970s, thermodynamic and kinetic melting studies on tRNAs had already provided evidence of

(1) cooperativity in denaturation and renaturation processes;

(2) the very large activation energies involved ($\sim$60 kcal mol^{-1});

(3) melting of tertiary structure and the weakest secondary structure pairings independently from the rest of the secondary structure (9–11).

These observations bear out the hierarchy of folding whereby 3-D architecture results from the compaction of separate, pre-existing, and stable secondary structure elements. Small autonomous functional mini-assemblies have been described (for example, hammerhead ribozymes (75) or the Arg-binding site of HIV TAR RNA (76)). In several instances, it can also be shown that large catalytic RNAs such as self-splicing introns can be reconstructed by bringing together in *trans* what could be considered separately folded domains (61, 77, 78). The M1 RNA of RNase P, a ribozyme that is a true enzyme, can also be reconstituted from inactive fragments to give functional complexes (79). Recently, Murphy and Cech (80) have characterized such an independently folding domain within the *Tetrahymena* ribozyme. They showed that a subdomain structure must first be in place before the domain forms, emphasizing not only the modular organization of structured RNAs but also their assembly from smaller tertiary units.

By monitoring either UV absorbance or self-splicing reaction kinetics as a function of temperature, Jaeger *et al.* (8) have shown that transcripts of the *sunY* group I intron of bacteriophage T4 undergo highly cooperative unfolding/inactivation upon heating. These two methods provide similar estimates of the corresponding thermodynamic parameters (see *Table 1*). Further, by comparing the sensitivity to heat inactivation of various mutants, it is possible to assess the energetic contributions of specific interactions to the overall 3-D fold while monitoring the nature of the contacts (*Figure 11*; 8, 26). Several of the interactions under scrutiny must also exist in the transition state for the folding process, since their disruption was found to affect refolding rates to approximately the same extent as the ratio of folded to unfolded molecules at equilibrium (see the discussions in Matouschek *et al.* (81)).

All but the last five residues of the *sunY* intron are required for its active form to be favoured over the unfolded state under seemingly physiological conditions. This and the fact that the final folding process is highly cooperative should ensure control of catalytic power and optimal coupling between the two steps of the splicing reaction. Another reason why formation of long-range tertiary interactions

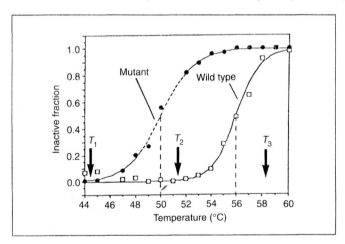

Fig. 11 Diagram illustrating the assessment of thermodynamic parameters corresponding to specific tertiary interactions. The curves represent the extent of inactivation, as a function of temperature, of autocatalytic ribozymes (wild type and mutant). Assuming an all-or-none transition, analysis of these curves in a van't Hoff representation yields equilibrium constants, as well as free energy, enthalpy and entropy variations associated with the process under scrutiny. Additionally, chemical probing of the RNA at temperatures T_1 (wild type and mutant folded in an active state), T_2 (only wild type folded in an active state), and T_3 (both wild type and mutant denatured), at which conformational states are well defined, should allow identification of those contacts that change state upon denaturation (8, 26).

in complex RNA molecules should be delayed until synthesis has been completed is the need for free intertwining of helix strands during the build-up of the secondary structure elements. Both these types of requirement could also be fulfilled by having 3-D folding dependent on recognition of completed (or partial) transcripts by other molecules: this could be one of the reasons why most contemporary RNAs, including many of the surviving ribozymes, depend on protein helpers to achieve active forms *in vivo*.

Acknowledgement

Some of the results and conclusions presented in this review would not have been obtained without the patient and dedicated work of Luc Jaeger during his PhD years in our two groups.

References

1. Doudna, J. A., Grosshans, C., Gooding, A., and Kundrot, C. E. (1993) Crystallization of ribozymes and small RNA motifs by a sparse matrix approach. *Proc. Natl Acad. Sci. USA*, **90**, 7829.
2. Watson, J. D. and Crick, F. H. C. (1953) A structure for deoxyribose nucleic acid. *Nature*, **171**, 737.
3. Fuller, W. and Hodgson, A. (1967) Conformation of the anticodon loop in tRNA. *Nature*, **215**, 817.
4. Levitt, M. (1969) Detailed molecular model for transfer ribonucleic acid. *Nature*, **224**, 759.
5. Michel, F., Jacquier, A., and Dujon, B. (1982) Comparison of fungal mitochondrial introns reveals extensive homologies in RNA secondary structure. *Biochimie*, **64**, 867.
6. Burke, J. M., Belfort, M., Cech, T. R., Davies, R. W., Schweyen, R. J., Shub, D. A., Szostak, J. W., and Tabak, H. F. (1987) Structural conventions for group I introns. *Nucleic Acids Res.*, **15**, 7217.
7. Banerjee, A. R., Jaeger, J. A, and Turner, D. H. (1993) Thermal melting of a group I ribozyme: the low temperature transition is primarily disruption of tertiary structure. *Biochemistry*, **32**, 153.
8. Jaeger, L., Westhof, E., and Michel, F. (1993) Monitoring of the cooperative unfolding of the *sunY* group I intron of bacteriophage T4. The active form of the *sunY* ribozyme is stabilized by multiple interations with 3' terminal intron components. *J. Mol. Biol.*, **234**, 331.
9. Coutts, S. M., Gangloff, J., and Dirheimer, G. (1974) Conformational transitions in tRNA^{Asp} (brewer's yeast). Thermodynamic, kinetic, and enzymatic measurements on oligonucleotide fragments and the intact molecule *Biochemistry*, **13**, 3938.
10. Crothers, D. M., Cole, P. E., Hilbers, C. W., and Shulman, R. G. (1974) The molecular mechanism of thermal unfolding of *Escherichia coli* formylmethionine transfer RNA. *J. Mol. Biol.*, **87**, 63.
11. Hawkins, E. R., Chang, S. H., and Mattice, W. L. (1977) Kinetics of the renaturation of yeast tRNA$_3^{Leu}$. *Biopolymers*, **16**, 1557.
12. Sussman, J. L., Holbrook, S. R., Wade Warrant, R., Church, G. M., and Kim, S.-H.

(1978) Crystal structure of yeast phenylalanine tRNA. I. Crystallographic refinement. *J. Mol. Biol.*, **123**, 607.
13. Hingerty, B., Brown, R. S., and Jack, A. (1978) Further refinement of the structure of yeast tRNA^{Phe}. *J. Mol. Biol.*, **124**, 523.
14. Westhof, E. and Sundaralingam, M. (1986) Restrained refinement of the monoclinic form of yeast phenylalanine transfer RNA. *Biochemistry*, **25**, 4868.
15. Westhof, E., Dumas, P., and Moras, D. (1985) Crystallographic refinement of yeast aspartic acid transfer RNA. *J. Mol. Biol.*, **184**, 119.
16. Westhof, E. and Michel, F. (1992) Some tertiary motifs of RNA foldings. In *Structural Tools for the Analysis of Protein–Nucleic Acid Complexes*. Lilley D. M. J., Heumann, H., and Suck, D. (eds). Birkhäuser Verlag, Basel, p. 255.
17. Michel, F. and Westhof, E. (1990) Modelling of the three-dimensional architecture of group I introns based on comparative sequence analysis. *J. Mol. Biol.*, **216**, 585.
18. Jaeger, L., Westhof, E., and Michel, F. (1991) Function of P11, a tertiary base pairing in self-splicing introns of subgroup IA. *J. Mol. Biol.*, **221**, 1153.
19. Westhof, E. and Jaeger, L. (1992) RNA pseudoknots. *Curr. Opin. Struct. Biol.*, **2**, 327.
20. Woese, C. R., Winker, S., and Gutell, R. R. (1990) Architecture of ribosomal RNA: constraints on the sequence of 'tetra-loops'. *Proc. Natl Acad. Sci. USA*, **87**, 8467.
21. Tuerk, C., Gauss, P., Thermes, C., Groebe, D. R., Gayle, M., Guild, N., Stormo, G., d'Aubenton-Carafa, Y., Uhlenbeck, O. C., Tinoco, I., Brody, E., and Gold, L. (1988) CUUCGG hairpins: Extraordinarily stable RNA secondary structures associated with various biochemical processes. *Proc. Natl Acad. Sci. USA*, **85**, 1364.
22. Westhof, E., Romby, P., Romaniuk, P. J., Ebel, J.-P., Ehresmann, C., and Ehresmann, B. (1989) Computer modeling from solution data of spinach chloroplast and of *Xenopus laevis* somatic and oocyte 5S rRNAs. *J. Mol. Biol.*, **207**, 417.
23. Brunel, C., Romby, P., Westhof, E., Ehresmann, C., and Ehresmann, B. (1991) Three-dimensional model of *Escherichia coli* ribosomal 5S RNA as deduced from structure probing in solution and computer modelling. *J. Mol. Biol.*, **221**, 293.
24. Michel, F., Ellington, A. D., Couture, S., and Szostak, J. W. (1990) Phylogenetic and genetic evidence for base-triples in the catalytic domain of group I introns. *Nature*, **347**, 578.
25. Pleij, C. W. A. (1990) Pseudoknots: a new motif in the RNA game. *TIBS*, **15**, 143.
26. Jaeger, L., Michel, F., and Westhof, E. (1994) Involvement of a GNRA tetraloop in long-range RNA tertiary interactions. *J. Mol. Biol.*, **236**, 1271.
27. Grosjean, H., Söll, D. G., and Crothers, D. (1976) Studies of the complex between tRNA with complementary anticodons. I. Origins of enhanced affinity between complementary triplets. *J. Mol. Biol.*, **103**, 499.
28. Persson, C., Wagner, E. G. H., and Nordström, K. (1990) Control of replication of plasmid R1: formation of an initial transient complex is for anti-sense RNA/target RNA pairing. *EMBO J.*, **9**, 3777.
29. Tomizawa, J. I. (1993) Evolution of functional structures of RNA. In *The RNA World*. Gesteland, R. F. and Atkins, J. F. (eds). Cold Spring Harbor Laboratory Press, Cold Spring Harbor, NY, p. 419.
30. Michel, F., Umesono, K., and Ozeki, H. (1989) Comparative and functional anatomy of group II catalytic introns. *Gene*, **82**, 5.
31. Turner, D. H., Sugimoto, N., and Freier, S. M. (1988) RNA structure prediction. *Ann. Rev. Biophys. Chem.*, **17**, 167.
32. Zuker, M. (1989) Computer prediction of RNA structure. *Methods Enzymol.*, **180**, 262.

33. Zuker, M. and Sankoff, D. (1984) RNA secondary structures and their prediction. *Bull. Math. Biol.*, **46**, 591.
34. Saldanha, R., Mohr, G., Belfort, M., and Lambowitz, A. M. (1993) Group I and group II introns. *FASEB J.*, **7**, 15.
35. Jaeger, J. A., Turner, D. H., and Zuker, M. (1989) Improved predictions of secondary structures for RNA. *Proc. Natl Acad. Sci. USA*, **86**, 7706.
36. Woese, C. R. and Pace, N. R. (1993) Probing RNA structure, function, and history by comparative analysis. In *The RNA World*. Gesteland, R. F. and Atkins, J. F. (eds). Cold Spring Harbor Laboratory Press, Cold Spring Harbor, NY, p. 91.
37. James, B. D., Olsen, G. J., and Pace, N. R. (1989) Phylogenetic comparative analysis of RNA secondary structure. *Methods Enzymol.*, **180**, 227.
38. Green, R., Ellington, A. D., and Szostak, J. W. (1990) *In vitro* genetic analysis of the *Tetrahymena* self-splicing intron. *Nature*, **347**, 406.
39. Bartel, D. P., Zapp, M. L., Green, M. R., and Szostak, J. W. (1991) HIV-1 rev regulation involves recognition of non-Watson–Crick base pairs in viral RNA. *Cell*, **67**, 529.
40. Green, R. and Szostak, J. W. (1992) Selection of a ribozyme that functions as a superior template in a self-copying reaction. *Science*, **258**, 1910.
41. Fox, G. E. and Woese, C. R. (1975) 5S RNA secondary structure. *Nature*, **256**, 505.
42. Woese, C., Magrum, L. J., Gupta, R., Siegel, R. B., Stahl, D. A., Kop, J., Crawford, N., Brosius, J., Gutell, R., Hogan, J. J., and Noller, H. F. (1980) Secondary structure model for bacterial 16S ribosomal RNA: phylogenetic, enzymatic and chemical evidence. *Nucleic Acids Res.*, **8**, 2275.
43. Noller, H. F., Kop, J., Wheaton, V., Brosius, J., Gutell, R., Kopylov, A. M., Dohme, F., Herr, W., Stahl, D. A., Gupta, R., and Woese, C. R. (1981) Secondary structure model for 23S ribosomal RNA. *Nucleic Acids Res.*, **9**, 6167.
44. James, B. D., Olsen, G. J., Liu, J., and Pace, N. R. (1988) The secondary structure of ribonuclease P RNA, the catalytic element of a ribonucleoprotein enzyme. *Cell*, **52**, 19.
45. Davies, R. W., Waring, R. B., Ray, J. A., Brown, T. A., and Scazzocchio, C. (1982) Making ends meet: a model for RNA splicing in fungal mitochondria. *Nature*, **300**, 719.
46. Rousset, F., Pelandakis, M., and Solignac, M. (1991) Evolution of compensatory substitutions through G–U intermediate state in *Drosophila* rRNA. *Proc. Natl Acad. Sci. USA*, **88**, 10 032.
47. Gutell, R. R., Larsen, N., and Woese, C. R. (1993) Lessons from an evolving ribosomal RNA: 16S and 23S rRNA structure from a comparative perspective. In *Ribosomal RNA: Structure, Evolution, Gene Expression and Function in Protein Synthesis*. Zimmerman, R. A. and Dahlberg, A. E. (eds). Telford Press, Caldwell, NJ.
48. Gutell, R. R. and Woese, C. R. (1990) Higher order structural elements in ribosomal RNAs: pseudo-knots and the use of noncanonical pairs. *Proc. Natl Acad. Sci. USA*, **87**, 663.
49. Tinoco, I., Jr (1993) Structures of base pairs involving at least two hydrogen bonds. In *The RNA World*. Gesteland, R. F. and Atkins, J. F. (eds). Cold Spring Harbor Laboratory Press, Cold Spring Harbor, NY, p. 603.
50. Puglisi, J. D., Wyatt, J. R., and Tinoco, I., Jr (1990) Solution conformation of an RNA hairpin loop. *Biochemistry*, **29**, 4215.
51. Delorme, M. O. and Hénaut, A. (1988) Merging of distance matrices and classification by dynamic clustering. *CABIOS*, **4**, 453.
52. Chiu, D. K. Y. and Kolodziejczak, T. (1991) Inferring consensus structure from nucleic acid sequences. *CABIOS*, **7**, 347.

53. Winker, S., Overbeek, R., Woese, C. R., Olsen, G. J., and Pfluger, N. (1990) Structure detection through automated covariance search. *CABIOS*, **6**, 365.
54. Sankoff, D. (1985) Simultaneous solution of the RNA folding, alignment and protosequence problems. *SIAM J. Appl. Math.*, **45**, 810.
55. Krol, A. and Carbon, P. (1989) A guide for probing native small nuclear RNA and ribonucleoprotein structures. *Methods Enzymol.*, **180**, 212.
56. Ehresmann, C., Baudin, F., Mougel, M., Romby, P., Ebel, J. P. L., and Ehresmann, B. (1987) Probing the structure of RNAs in solution. *Nucleic Acids Res.*, **15**, 9109.
57. Burgin, A. and Pace, N. R. (1990) Mapping the active site of ribonuclease P RNA using a substrate containing a photoaffinity agent. *EMBO J.*, **9**, 411.
58. Wyatt, J. R., Sontheimer, E. J., and Steitz, J. A. (1992) Site-specific crosslinking of mammalian U5 snRNP to the 5' splice site prior to the first step of premessenger RNA, splicing. *Genes Devel.*, **6**, 2542.
59. Wang, J.-F., Downs, W. D., and Cech, T. R. (1993) Movement of the guide sequence during RNA catalysis for a group I ribozyme. *Science*, **260**, 504.
60. Wang, J.-F. and Cech, T. R. (1992) Tertiary structure around the guanosine-binding site of the *Tetrahymena* ribozyme. *Science*, **256**, 526.
61. Michel, F., Jaeger, L., Westhof, E., Kuras, R., Tihy, F., Xu, M.-Q., and Shub, D. A. (1992). Activation of the catalytic core of a group I intron by a remote 3' splice junction. *Genes Devel.*, **6**, 1373.
62. Michel, F., Hanna, M., Green, R., Bartel, D. P., and Szostak, J. W. (1989) The guanosine binding site of the *Tetrahymena* ribozyme. *Nature*, **342**, 391.
63. Pyle, A. M., Murphy, F. L., and Cech, T. R. (1992) RNA substrate binding site in the catalytic core of the *Tetrahymena* ribozyme. *Nature*, **358**, 123.
64. Szostak, J. W. (1992) *In vitro* genetics. *TIBS*, **17**, 89.
65. Doudna, J., Cormack, B. P., and Szostak, J. W. (1989) RNA structure, not sequence, determines the 5' splice-site specificity of a group I intron. *Proc. Natl Acad. Sci. USA*, **86**, 7402.
66. Dahm, S. and Uhlenbeck, O. C. (1991) Role of divalent metal ions in the hammerhead RNA cleavage reaction. *Biochemistry*, **30**, 9464.
67. Piccirilli, J. A., Vyle, J. S., Caruthers, M. H., and Cech, T. R. (1993) Metal ion catalysis in the *Tetrahymena* ribozyme reaction. *Nature*, **361**, 85.
68. Westhof, E. (1988) Water: an integral part of nucleic acid structure. *Ann. Rev. Biophys. Biophys. Chem.*, **17**, 125.
69. Hubbard, J. M. and Hearst, J. E. (1991) Predicting the three-dimensional folding of transfer RNA with a computer modeling method. *Biochemistry*, **30**, 5458.
70. Malhotra, A., Tan, R. K. Z., and Harvey, S. (1990) Prediction of the three-dimensional structure of *Escherichia coli* 30S ribosomal subunit: a molecular mechanics approach. *Proc. Natl Acad. Sci. USA*, **87**, 1950.
71. Major, F., Turcotte, M., Gautheret, D., Lapalme, G., Fillion, E., and Cedergren, R. (1991) The combination of symbolic and numerical computation for three-dimensional modeling of RNA. *Science*, **253**, 1255.
72. Westhof, E., Romby, P., Ehresmann, C., and Ehresmann, B. (1990) Computer-aided structural biochemistry of ribonucleic acids. In *Theoretical Biochemistry, and Molecular Biophysics*. Beveridge, D. L. and Lavery, R. (eds). Adenine Press, New York, Vol. 1, p. 399.
73. Westhof, E. (1993) Modelling the three-dimensional structure of ribonucleic acids. *J. Mol. Struct. (Theochem)* **286**, 203.

74. Jones, T. A. (1978) A graphic model building and refinement system for macromolecules. *J. Appl. Crystal.*, **11**, 268.
75. Uhlenbeck, O. C. (1987) A small catalytic oligoribonucleotide. *Nature*, **328**, 596.
76. Puglisi, J. D., Tan, R., Calnan, B. J., Frankel, A. D., and Williamson, J. R. (1992) Conformation of the TAR RNA–arginine complex by NMR spectroscopy. *Science*, **257**, 76.
77. Van der Horst, G., Christian, A., and Inoue, T. (1991) Reconstruction of a group I intron self-splicing reaction with an activator RNA. *Proc. Natl Acad. Sci. USA*, **88**, 184.
78. Jarrell, K. A., Dietrich, R. C., and Perlman, P. S. (1988) Group II intron domain 5 facilitates a *trans*-splicing reaction. *Mol. Cell. Biol.*, **8**, 2361.
79. Guerrier-Takada, C. and Altman, S. (1992) Reconstitution of enzymatic activity from fragments of M1 RNA. *Proc. Natl Acad. Sci. USA*, **89**, 1266.
80. Murphy, F. L. and Cech, T. R. (1993). An independently folding domain of RNA tertiary structure within the *Tetrahymena* ribozyme. *Biochemistry*, **32**, 5291.
81. Matouschek, A., Kellis, J. T., Serrano, L., and Fersht, A. R. (1989) Mapping the transition state and pathway of protein folding by protein engineering. *Nature*, **340**, 122.
82. Michel, F. and Dujon, B. (1983) Conservation of RNA secondary structure in two intron families including mitochondrial-, chloroplast- and nuclear-encoded members. *EMBO J.*, **2**, 33.
83. Westhof, E. (1992) Westhof's rule. *Nature*, **358**, 459.
84. Lavery, R., Zakrzewska, K., Sun, J.-S., and Harvey, S. C. (1992) A comprehensive classification of nucleic acid structural families based on strand direction and base pairing. *Nucleic Acid Res.*, **20**, 5011.
85. Chastain, M. and Tinoco, I., Jr (1992) A base-triple structural domain in RNA. *Biochemistry*, **31**, 12733.
86. Westhof, E., Dumas, P., and Moras, D. (1983) Loop stereochemistry and dynamics in transfer RNA. *J. Biomol. Struct. Dyn.*, **1**, 337.
87. SantaLucia, J., Kierzek, R., and Turner, D. H. (1992). Context dependence of hydrogen bond free energy revealed by substitutions in an RNA hairpin. *Science*, **256**, 217.
88. Antao, V. P., Lai, S. Y., and Tinoco, I., Jr (1991). A thermodynamic study of unusually stable RNA and DNA hairpins. *Nucleic Acids Res.*, **19**, 5901.
89. Dock-Bregeon, A. C., Westhof, E., Giegé, R., and Moras, D. (1989) Solution structure of a tRNA with a large variable region: yeast tRNA$^{Ser.}$ *J. Mol. Biol.*, **206**, 707.
90. Heus, H. A. and Pardi, A. (1991) Structural features that give rise to the unusual stability of RNA hairpins containing GNRA loops. *Science*, **253**, 191.
91. Cheong, C., Varani, G., and Tinoco, I., Jr (1990) Solution structure of an unusually stable RNA hairpin, 5'GGAC(UUCG)GUCC. *Nature*, **346**, 680.
92. Jaeger, L. (1993) Les introns auto-catalytiques de groupe I comme modèle d'étude du repliement des acides ribonucléiques. PhD Thesis, Strasbourg.

3 | Aminoacyl-tRNA synthetase–tRNA recognition

JOHN G. ARNEZ and DINO MORAS

1. Introduction

Aminoacylation is the first and key step in the translation of genetic messages into proteins. The reaction brings together amino acids and their cognate tRNAs and is catalysed by aminoacyl-tRNA synthetases (aaRSs). In the reaction the amino acid is esterified to one of the hydroxyl groups of the 3'-terminal adenosine; the energy in the overall reaction is supplied by the hydrolysis of ATP. aaRSs catalyse the reaction via a two-step mechanism through an aminoacyl-adenylate intermediate,

$$\text{amino acid} + \text{ATP} \rightleftharpoons \text{aminoacyl-AMP} + PP_i$$

$$\text{aminoacyl-AMP} + \text{tRNA} \rightleftharpoons \text{aminoacyl-tRNA} + \text{AMP}$$

where the slower second step determines the overall reaction rate (1). Normally, the substrates bind in the same order as they react. However, arginyl (ArgRS), glutaminyl (GlnRS), and glutamyl (GluRS)-tRNA synthetases require the presence of enzyme-bound tRNA for the formation of aminoacyl-adenylate (2). There are three substrates in each aminoacylation: the tRNA, an amino acid, and ATP. Often there are several tRNA species (isoacceptors) specifying each amino acid, but there is generally one aaRS for each amino acid (1, 3). Each aaRS is specific for the amino acid and its cognate isoacceptor tRNAs.

The fidelity of translation depends on the accuracy of the actual decoding of the genetic messages by the ribosomal mechinery and on the specificity of amino-acylation. Overall, fewer than 1 in 10^4 amino acids are misincorporated into proteins. During aminoacylation, errors due to tRNA misselection are <1 in 10^7, whereas amino acid misselection, mainly the case with similar amino acids, occurs at a rate of ~5 in 10^5, which approaches the overall error rate of translation. Thus, amino acid selection limits the precision of aminoacylation (4). The focus of this chapter, however, is the interactions between the tRNA and the cognate aaRS, from the tRNA point of view.

The specificity of charging of cognate tRNAs by an aaRS is achieved through binding and catalysis. However, the differences in binding affinities for cognate and non-cognate tRNAs are only approximately two orders of magnitude in favour of the cognate species, so differentiation at the binding level is weak. However, the rate of catalysis is three to four orders of magnitude higher for the cognate species. Aminoacylation thus relies primarily on the rate of catalysis to discriminate between cognate and non-cognate tRNAs (5).

tRNA recognition by aaRSs has been proposed to constitute the second genetic code; the hypothesis states that the code is imprinted into the structure of an aaRS which matches with the structural features of its cognate tRNAs (6). However, there is no direct link between the anticodon loop and the site of aminoacylation (7). The assignment of amino acids to particular tRNAs and associated anticodons arose fortuitously, as a result of varied specificities for amino acids and tRNAs by primordial aaRS. The original set of specificities was limited, but it diversified as the enzymes evolved (8). The genetic code is based on tRNA recognition by aaRS and manifested in the process of translation of messages into proteins (9).

2. Transfer RNAs

Different tRNA sequences can all assume the same basic clover-leaf secondary structure (10) (*Figure 1a*); their lengths range from 75 to 90 nucleotides (nt). In addition to the four standard nucleotides they contain modified residues. The crystal structures of several tRNAs have been solved, for example tRNA^{Phe} (11, 12) (*Figure 1b*), tRNA^{Asp} (13, 14), and initiator tRNA$_i^{Met}$(15). All of them share the same L-shaped conformation with pseudodyad symmetry. Base pairing and base stacking are the main types of interactions stabilizing the tertiary structure; 95% of the bases in tRNA are stacked, although only half are involved in a helical structure. In addition to Watson–Crick base pairing found in the helical regions, other types of base-pairing interactions were found (*Figure 2*): purine–purine (*Figure 2c,e,f,* and *h*), reverse Watson–Crick in parallel chains (*Figure 2a*), Hoogsteen and reverse Hoogsteen (*Figure 2d* and *g*), and base triples (*Figure 2e,f,* and *h*). Most of these interactions are tertiary features involving conserved bases in the elbow of the structure. The bases can form hydrogen bonds with the sugar–phosphate backbone atoms as well (*Figure 2b* and *e–g*). Metal ions bind to tRNA cooperatively and stabilize the structure (10).

tRNAs from all organisms possess modified nucleotides, which may influence the efficiency and fidelity of translation as well as maintaining the three-dimensional (3-D) structure of tRNAs (16, 17). Modifications may not affect recognition of a tRNA species by its cognate aaRS but may act as antideterminants by preventing a non-cognate aaRS from recognizing a tRNA (17). Unmodified tRNA^{Asp}, for example, is charged with Asp as efficiently as its fully modified counterpart, while it can be mischarged with arginine (18). Free unmodified tRNA is less stable than but structurally similar to fully modified tRNA. Comparison by nuclear magnetic resonance (NMR) of unmodified yeast tRNA^{Phe} and its fully modified counterpart suggested that tRNA^{Phe} produced *in vitro* folded normally

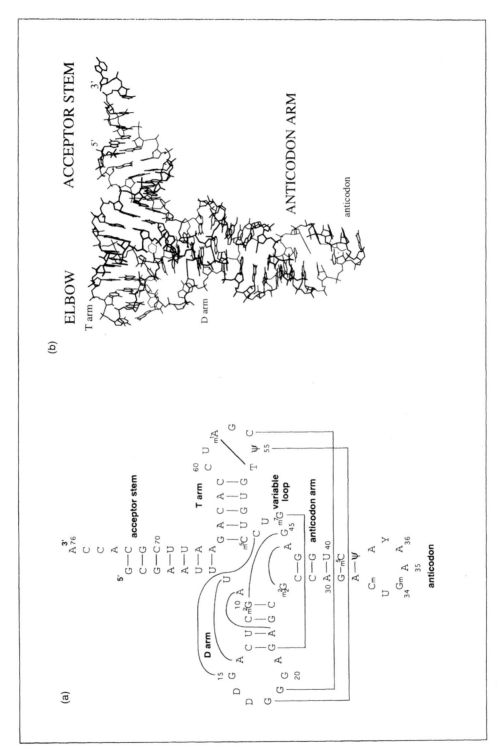

Fig. 1 tRNA^{Phe} (a) secondary cloverleaf structure showing tertiary interactions found in (b) three-dimensional structure as elucidated by X-ray crystallography (11, 12). This and all figures of three-dimensional structures were made by MOLSCRIPT (89).

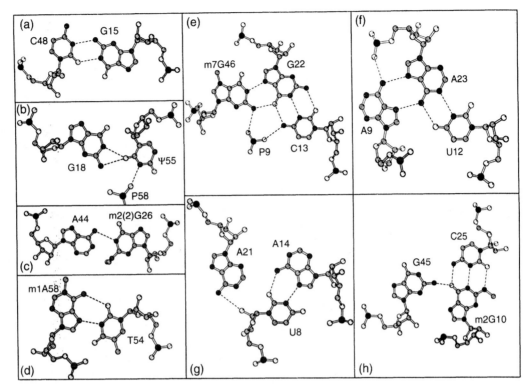

Fig. 2 Tertiary interactions in tRNA^{Phe} (11, 12). The pair G19–C56, a standard Watson–Crick base pair, is not shown. Atoms P–N–C–O are grey-coded in this and subsequent figures with P being the darkest and O the lightest grey.

and structurally resembled fully modified tRNA^{Phe} (19), however, melting profile analyses indicated that the unmodified species has a less stable structure (20). Unmodified tRNA^{Gln} may be less stable than its modified counterpart, but when it is complexed with GlnRS it may be stabilized through interactions with the protein. It has the same structure as modified tRNA^{Gln} (J. G. Arnez and T. A. Steitz, in press).

tRNAs are structurally similar so their precursors can first be processed by common tRNA-processing enzymes and their aminoacylated derivatives are able to interact with the protein synthesis apparatus. However, tRNAs need to be distinct so they can be differentiated by aaRSs and, thus, be aminoacylated correctly (9). The features necessary for tRNA identification and aminoacylation by the cognate aaRSs are called the recognition elements; if they are altered the tRNA is aminoacylated at a slower rate. The set of necessary and sufficient elements for recognition by the cognate aaRSs and rejection by non-cognate aaRSs constitutes the identity for a given tRNA. Hence, recognition elements are a subset of the identity. Each cognate set of tRNAs thus has a set of identity elements and these vary among different sets of tRNAs. The identity determinants for a number of

Table 1 Major identity determinants of selected aminoacyl-tRNA synthetase systems

aaRS System	Acceptor Stem	Elbow			Anticodon		References
		D Arm	T Arm	V Loop	Stem	Loop	
Ala (E. coli)	G3–U70						36
Arg (E. coli)		C16 U17 A20	A59			C35 G36	28 32 33
Asp (yeast)	G73	G10–C25				G34 U35 C36	48 79 82
Gln (E. coli)	G73 U1–A72 G2 G3	G10 C16				U/C34 U35 G36	66
Met (E. coli)						C34 A35 U36	29
Phe (yeast)	A73	G20				G34 A35 A36	26
Phe (E. coli)	A73	U20	U59 U60	G44 U45	G27–C43 G28–C42	ND^a	32
Ser (E. coli)	G73 C72 G2–C71 A3–U70	C11–G24		long			35

^{a} Suppressor tRNA was used in the analysis.

tRNAs have been elucidated through genetic and *in vitro* experiments (21) (*Table 1*). In spite of the universality of the genetic code, tRNA identity may not be phylogenetically conserved. For example, an *Escherichia coli* tRNA^{Tyr} is a leucine-specific tRNA in yeast; *E. coli* tRNA^{Tyr} and yeast tRNA^{Leu} are structurally related (22). There appears to be no single universal recognition mechanism but rather idiosyncratic recognition of tRNA by aaRSs (23).

Two approaches have been used to analyse tRNA identity. Firstly, the introduction of the fewest alterations into a tRNA to shift its aaRS recognition, that is, to change its identity (also called identity swapping), which shows and proves the elements constitute the identity of the new system. Secondly, the generation of variants of a particular tRNA whose properties as substrates for the aaRS involved are then analysed, primarily *in vitro* (24) and sometimes used to find the elements and to gauge their relative importance.

A comparison of 67 tRNA sequences suggested that distinguishing features are located primarily in the anticodon loop, the acceptor arm, and a few base pairs in the T- and D-stems (25). The anticodon bases are crucial for the recognition of many tRNAs by their cognate aaRSs, such as tRNA^{Phe} (26, 27), tRNA^{Arg} (28), tRNA^{Met} (29), tRNA^{Glu} (30), tRNA^{Gln} (30) tRNA^{Ile} (27), and tRNA^{Val} (27). For example, exchanging the anticodons between tRNA^{Met} and tRNA^{Val} swaps their aminoacylation specificities (31). Transferring the anticodon from tRNA^{Arg} to elongator tRNA$_m^{Met}$ sufficed to confer recognition by ArgRS on the latter (28).

In the crystal structure of yeast tRNA^{Phe} the variable bases 16, 17, 20, 59 and 60 were found to form a pocket, called the variable pocket, in the elbow region, that is, where the D-loop interacts with the T-loop, suggesting a potential discrimination site (12). In the amber suppressor derived from tRNA^{Phe} from E. coli, features contributing to its identity are located in this pocket and also in the part of the anticodon stem adjacent to the variable loop (32). When the bases in the variable pocket are changed to those of tRNA^{Arg}, the resulting tRNA inserts Arg (33). Thus, part of the tRNA^{Arg} identity also resides in the variable pocket.

In some cases, tRNA identity attributes lie outside the anticodon arm entirely. Normanly et al. (34) have transformed a suppressor derived from tRNA^{Leu} into a Ser-specific species by changing 12 nucleotides; subsequently, the tRNA^{Ser} identity was found to consist of eight bases, a subset of the 12 (35). They are all located in the acceptor and D-arms (34, 35). The G3–U70 base pair in the acceptor stem is a major determinant of tRNA^{Ala} identity (36); a transfer of this element to, for example, suppressors derived from tRNA^{Cys} and tRNA^{Phe}, both of which are substantially different from tRNA^{Ala}, resulted in alanine acceptance by the two variants (36, 37). The G3–U70 base pair in tRNA^{Ala} from E. coli is a wobble pair; replacing it with another wobble pair does not affect its acceptor identity. Translocating the wobble pair to another site in the acceptor helix preserves some of the original alanine acceptance. Moreover, introducing any wobble pair into the corresponding site in tRNA^{Lys} conferred alanine acceptance upon this tRNA. Thus, identity in this case may be associated with a distortion in the acceptor stem helix due to a wobble base pair (38). A mini-helix corresponding to the acceptor stem and the TψC-arm of tRNA is able to function as a substrate for alanyl-tRNA synthetase (AlaRS) provided it contains a single G–U base pair (39).

An interesting case of tRNA identity is suppressor tRNAs. Changing the anticodon of tRNA^{fMet} from E. coli to an amber suppressor (CUA) confers glutamine acceptance on to this tRNA in vitro (40), as the amber suppressor anticodon differs only in position 36 from the glutamine-specific anticodon CUG. The amber suppressor derived from tRNA^{Trp} can be charged with both tryptophan and glutamine in approximately equimolar amounts in vitro (41). The amber suppressor derived from tRNA^{Tyr}, which normally inserts tyrosine, can be transformed into a glutamine acceptor by introducing any of the following changes: A73→G (42, 43), C71→U, G1-C72→A-U, and C72→U (42, 43), C72→A (43), G2→A (43), and G1→A (42). A suppressor tRNA does not necessarily have the same ability to serve as a substrate of its cognate aaRS as its wild-type counterpart. The amber suppressor derived from tRNA^{Gln} is a 1000-fold poorer substrate of GlnRS in vivo than the wild-type cognate tRNA although it is a functional amber suppressor in vitro (44).

A drawback of using suppressor tRNAs in identity studies is that the contribution of the anticodon to the identity cannot be analysed (21, 24).

Aminoacylation specificity may depend on the competition for a tRNA by different aaRSs. Parallel cognate tRNA–aaRS systems in the cytoplasm enhance selectivity of each aaRS. The amounts of aaRSs and tRNAs are approximately equimolar and interactions between tRNAs and their cognate aaRSs make tRNAs less available to non-cognate aaRs. In this competition, cognate associations are favoured and weaker non-cognate interactions disfavoured (3). Thus, correct ratios of aaRS and tRNA concentrations are essential for the maintenance of aminoacylation specificity. When GlnRS is overproduced in *E. coli* it acylates the amber suppressor derived from tRNA^{Tyr} with glutamine. However, when the level of tRNA^{Glu} is simultaneously increased as well mischarging is abolished (45).

Efficient aminoacylation depends not only on the presence of the set of identity elements but also on the conformation of the tRNA. Hence, transplantation of identity elements into other tRNA contexts works, but the resulting tRNAs are not optimized as the presence of negative features at least partially hinders interactions (46).

3. Aminoacyl-tRNA synthetases

Aminoacyl-tRNA synthetases are a diverse group of enzymes. Their quaternary structures range from monomeric to dimeric to tetrameric (1, 47) (*Table 2*). However, they can be grouped into two classes of ten members each that represent two structural solutions to a problem (47, 48). Class I includes enzymes with signature amino acid sequences, KMSKS (Lys–Met–Ser–Lys–Ser) and HIGH (His–Ile–Gly–His) and whose structures contain the Rossmann fold nucleotide-binding motif (47) (*Figure 3a*). The classical Rossmann fold consists of an alternating α–β structure with a central parallel β-sheet (49). Class II comprises aaRSs that have three concatenated homologous motifs, 1, 2, and 3 and do not contain the Rossmann fold in their structures but an antiparallel β-sheet (47, 50). Motif 1 is involved in the dimer interface, as all class II aaRSs are dimers, while motifs 2 and 3 form the catalytic site (50) (*Figure 3b*). In addition, there is a functional correlation; aaRSs belonging to class I attach the amino acid to the 2'-OH of the 3'-terminal ribose, while class II aaRSs attach it to the 3'-OH. Phenylalanyl-tRNA synthetase (PheRS), a class II enzyme, acylates the 2'-OH and is thus the only exception to the rule (7, 47, 51, 52).

The basic design of aaRSs is modular. Each aaRS has a requisite core of ~300–400 residues — the active-site domain. The diversity of sizes and tRNA specificities is generated by additions or insertions of extra peptides, that is, modules (50, 53). Cysteinyl-tRNA synthetase (CysRS) from *E. coli* is the smallest known monomeric aaRS; the protein consists of 461 amino acid residues and may represent the minimal structure necessary for tRNA binding and charging (54). Tryptophanyl-tRNA synthetase (TrpRS) from *E. coli* is an α_2-dimer of two 334-residue, 37 kDa monomers, which are the smallest aaRS subunits known (55). AlaRS from *E. coli* is

Table 2 Classification of aminoacyl-tRNA synthetases on the basis of their structural and functional organization (1,7,47,51)

Class	Characteristic motifs	Group	aaRS	4° Str.	OH Acyl.	Species
I	**H**I**G**H	a	CysRS	α	?	Eco
	KM**S**K**S**		IleRs	α	2'	Eco, SC
			LeuRS	α	2'	Eco, Bst
				α₂		SC
			MetRS	α₂	2'	Eco, Bst
			ValRS	α	2'	Eco, Bst, SC
		b	TrpRS	α₂	2'	Eco, Bst
			TyrRS	α₂	2'	Eco, Bst, SC
		c	ArgRS	α	2'	Eco, Bst
			GlnRS	α	2'	Eco
			GluRS	α	2'	Eco
				αβ		Eco
II	(1) ... P ...	a	HisRS	α₂	3'	Eco, Sty
	(2) ... FRXE ...		ProRS	α₂	3'	Eco
	(3) ... (GX)₃ER ...		SerRS	α₂	3'	Eco, Tth, SC
			ThrRS	α₂	3'	Eco
		b	AsnRS	α₂	3'	SC
			AspRS	α₂	3'	Eco, Tth, SC
			LysRS	α₂	3'	Eco, SC
		c	AlaRS	α₄	3'	Eco
				α		SC
			GlyRS	(αβ)₂	3'	Eco, BM
			PheRS	(αβ)₂	2'	Eco, SC

The species names are abbreviated as follows: Eco, *Escherichia coli*; Bst, *Bacillus stearothermophilus*; SC, yeast (*Saccharomyces cerevisiae*); Tth, *Thermus thermophilus*; Sty, *Salmonella typhimurium*; BM, *Bombyx mori*.

a large enzyme; it is an α_4-tetramer of 875 amino acid residue monomers. However, much of the polypeptide is dispensable for activity. A core of 461 residues of AlaRS shows complete activity and substrate specificity; this core does not tetramerize, which demonstrates that tetramerization is not essential for function (53). However, recent sequence analysis has led to the discovery of motif 1 in AlaRS, implying dimeric structure (56). Similarly, the sequence of glycyl-tRNA synthetase (GlyRS) from *Bombyx mori* contains motif 1 and, thus, the dimer interface (57). Finally, PheRS from *Thermus thermophilus* is a dimer of dimers, as shown by X-ray crystallographic studies (58).

In addition to the active site homologies that define the two classes, aaRSs share additional similarities that result in the partition of each class into subgroups (7). For example, aspartyl (AspRS)-, asparaginyl (AsnRS)- and lysyl (LysRS)-tRNA synthetases contain consensus sequences at the N-termini and prolyl (ProRS)-, threonyl (ThrRS)-, and histidyl (HisRS)-tRNA synthetases contain consensus sequences at the C-termini. The subgroups tend to comprise aaRSs that charge similar amino acid types, be it hydrophobic, charged, large, or small. In class I, group *a* aaRSs charge hydrophobic and large amino acids, group *b* charge large

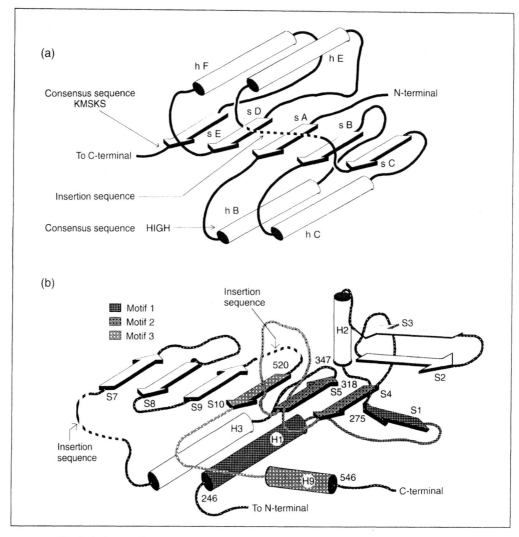

Fig. 3 Active site folds in (a) Class I and (b) Class II aminoacyl-tRNA synthetases (50).

and aromatic amino acids, and group c charge charged amino acids (50); members of the latter group require the presence of tRNA for the formation of aminoacyl-adenylate (2). In class II, group a aaRSs tend to charge small and polar amino acids, group b charge the smaller set of charged amino acids, while group c aaRSs are distinct in their quaternary structures and charge the smallest two amino acids and an aromatic amino acid (50) (*Table 2*).

Examples of both class I (methionyl (MetRS)- (59) and tyrosyl (TyrRS)- (60) tRNA synthetases alone, GlnRS complexed with tRNA^{Gln} (61)), and class II aaRSs (AspRS complexed with tRNA^{Asp} (48), seryl-tRNA synthetase (SerRS) alone (62), and complexed with tRNA^{Ser} (63, 64)) have been studied at the structural level. These structures are discussed below.

4. tRNA recognition by class I aminoacyl-tRNA synthetases

4.1 Glutaminyl-tRNA synthetase

Glutaminyl-tRNA synthetase (GlnRS) from *E. coli*, a class I aaRS (47), is a monomer of 553 amino acid residues and has a molecular weight of 63.4 kDa (65). It requires the presence of enzyme-bound tRNA for the formation of glutaminyl-adenylate (2). The crystal structure of the complex of GlnRS with tRNA^{Gln} was solved at 2.8 Å resolution (60) and refined at 2.5 Å resolution (66) (*Figure 4*). The enzyme is an elongated protein consisting of two major domains. The active-site domain comprises the Rossmann fold into which is inserted the acceptor-binding subdomain (61); the size of the entire combined domain is similar to that of the three-motif domain found in AspRS (48). The anticodon-binding domain comprises two β-barrels.

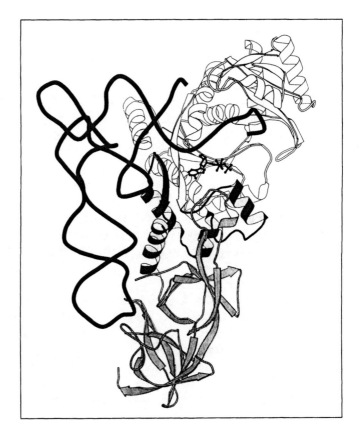

Fig. 4 Schematic drawing of the crystal structure of GlnRS complexed with tRNA^{Gln} (61, 66). The active site domain is light grey, the anticodon binding domain grey and the connecting domain black.

tRNA^{Gln} binds to GlnRS with the variable loop facing the solvent and interacts with the active-site domain on the minor groove side (61). The interface between the tRNA and GlnRS is extensive — the buried surface is 2700 Å^2 (66) — and exemplifies an induced fit: the terminal base pair is denatured as the 3'-terminal CCA bends and dips into the active site and the anticodon undergoes a conformational change (61). Many nucleotides of the tRNA are involved in specific recognition by GlnRS (*Figure 5*). Nucleotides whose bases interact directly with the protein are in the acceptor stem, the D-arm, and the anticodon loop. Furthermore, many additional bases in these regions form water-mediated contacts with the protein. Some residues are not directly involved in protein–RNA interactions; they enable the tRNA to adopt the conformation that facilitates its binding to GlnRS. These residues are G73 and the U1–A72 base pair in the acceptor stem and residues 32, 33, 37, and 38 in the anticodon loop. U1 is very flexible as it is disordered in the electron density map. GlnRS forms extensive contacts with the ribose–phosphate backbone mainly in the acceptor arm, part of the D-arm, and the anticodon loop (66).

The acceptor stem of tRNA^{Gln} contains the 3'-terminal CCA, which accepts the amino acid. Two loops and an α-helix in GlnRS interact directly with the three terminal base pairs in the acceptor stem of tRNA^{Glu} (*Figure 6a*). The first loop, tipped with Leu136, forms a wedge which breaks the U1–A72 base pair and, thus, facilitates the bending of the 3'-terminal CCA into the active site. The exocyclic amino group of G73 interacts with the phosphate moiety of residue 72, stabilizing the bend. The second loop (residues 179–184) contacts the exocyclic amino group of G2 through the backbone carbonyl oxygen of Pro181 and forms a water-mediated contact with C72 through the peptide nitrogen of Ile183. The α-helix interacts with the G3–C70 base pair and extends into the active site. Residue Asp235 on this helix forms a direct contact with G3 and a water-mediated contact with C70 (61). The active site is a parallel β-sheet nucleotide-binding fold comprising the two motifs characteristic of class I aaRSs, HIGH and MSK. The motifs interact with each other, forming a surface that binds the ATP molecule. The α-phosphate of ATP and the 2'-OH of tRNA^{Gln} are within hydrogen-bonding distance. Glutamine has been positioned in an adjacent pocket by model building; its reactive carboxylate is close to the α-phosphate of ATP while its side chain makes specific contacts with the enzyme, explaining the discrimination against Glu (67).

The anticodon loop in tRNA^{Gln} provides essential recognition elements for GlnRS (*Figure 6b*). The loop undergoes a conformational change as the three anticodon bases (C34, U35, and G36) are splayed out so that each one of them binds snugly into a complementary pocket in the protein; these pockets are specific for tRNA^{Gln} anticodon bases. The anticodon stem is extended by two base pairs (32–38 and 33–37) which are not of the Watson–Crick type; these base pairs are not present in the tRNAs whose structures are known. The C-terminal double β-barrel domain of GlnRS interacts with the anticodon and forms a tight interface. The C34 binding pocket can also accommodate the 2-thio-U34 of the isoacceptor tRNA. The U35–

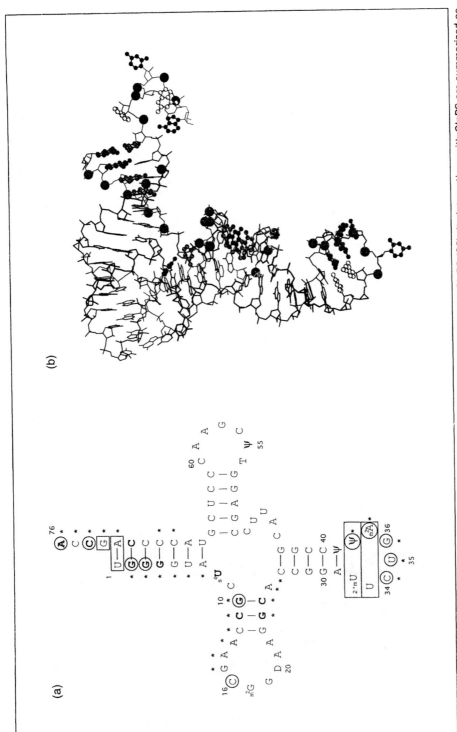

Fig. 5 tRNA^{Gln} (a) secondary cloverleaf structure and (b) as it appears in 3-D in complex with GlnRS (66). Its interactions with GlnRS are summarized as follows: bases interacting directly with the protein are circled in (a) and solid black in (b) and bases whose contacts with the protein are mediated by water molecules are bold face in (a) and dark grey in (b). Nucleotides implicated in enabling (in a base-specific manner), the tRNA to assume the conformation necessary for its binding to the enzyme are boxed in (a) and light grey in (b). Residues whose backbone moieties are in contact with the protein are indicated by '*' in (a) and large grey spheres in P atom positions in (b).

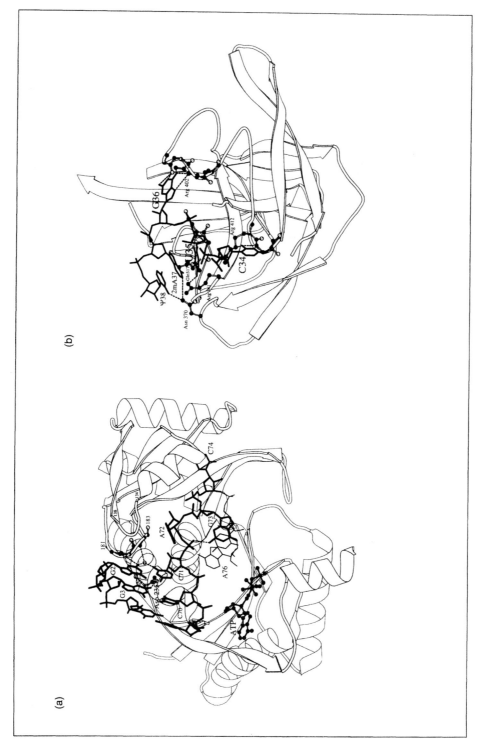

Fig. 6 Details of the interactions between GlnRS and tRNA^{Gln} (61, 66) in (a) the acceptor region (shown are residues 2 to 4 and 69 to 76 of the tRNA and the active site domain of GlnRS. The first nucleotide is not shown because it is disordered in the crystal structure) and (b) the anticodon region (shown are residues 34 to 38 of the tRNA and the anticodon-binding double β-barrel domain of GlnRS). For clarity only the main base-specific interactions are shown.

specific cleft is the tightest binding pocket. The G36-binding pocket is very specific for G. The pockets are structurally very similar: each consists of a polypeptide segment of five or six residues; at least one positively charged side chain makes a salt bridge with the adjacent phosphate while the aliphatic part of this side chain packs against either the base or ribose. The bases are recognized by direct hydrogen bonding between the side chains or the backbone of the peptide and the hydrogen-bonding groups of bases. At the junction of the β-barrels is an insertion of a long two-stranded β-ribbon that reaches the active-site HIGH and MSK region (*Figure 4*); the ribbon was suggested to form an allosteric link between the anticodon-binding region and the active centre (66).

Three mischarging mutant proteins of GlnRS that were able to misacylate the amber suppressor derived from tRNA^{Tyr} with glutamine were isolated using an *in vivo* suppression screen (68). Two of the mutations were at residue 235 (*Figure 6a*), Asp235 → Asn (GlnRS7) and Asp235 → Gly (GlnRS10), while the third had a change in the acceptor stem-binding domain, Ile129 → Thr (GlnRS15) (61, 68, 69). GlnRS15 has the weakest mischarging activity and GlnRS10 mischarges less efficiently than GlnRS7 (68). All these mutant enzymes retain their specificity for tRNA^{Gln} and they misacylate only a small set of non-cognate suppressor tRNAs and wild-type non-cognate species with glutamine at low levels *in vitro*; they do not exhibit a general broadening of specificity (69). The mischarging efficiency of GlnRS7, the highest among the mutant GlnRS, is at best 100-fold less (and on average 1000-fold less) than its ability to glutaminylate tRNA^{Gln}. Wild-type GlnRS, when present in high concentrations, is capable of misacylating the same set of non-cognate tRNAs as the mutants; its mischarging efficiency is at best 10 000-fold lower than its ability to glutaminylate tRNA^{Gln} (45).

In addition to the study of the crystal structure of GlnRS complexed with tRNA^{Gln} and ATP (61, 66), the identity of tRNA^{Gln} has been analysed biochemically by introducing mutations into the tRNA (44) and examining glutamine acceptance of the amber suppressors derived from tRNA$_1^{Ser}$ (70), tRNA^{Trp} (41), and tRNA^{Tyr} (42) that had some or all of the postulated tRNA^{Gln} elements. All these studies led to the conclusion that the identity of tRNA^{Gln} lies primarily in the acceptor stem and the anticodon (*Table 1*).

GlnRS bears considerable structural homology with the other two class I enzymes whose structures are known, TyrRS and MetRS (61).

4.2 Methionyl-tRNA synthetase

Methionyl-tRNA synthetase (MetRS) from *E. coli* is an $α_2$-dimer of total molecular weight 172 kDa. The crystal structure of the 64 kDa tryptic monomer fragment of MetRS that retains specificity for methionine and tRNA^{Met} has been solved at 2.5 Å resolution (59). The protein is an elongated molecule which contains the Rossmann fold. MetRS is unique in that it recognizes both initiator tRNA^{fMet} and elongator tRNA^{Met} (59). The structure of MetRS is very similar to that of GlnRS not only in its nucleotide-binding fold but also in the domain which inserts therein. MetRS also possesses a long α-helix which corresponds to the long α-helix in GlnRS that extends along the D- and anticodon stems of the tRNA. This helix probably orients the tRNA with respect to the enzyme. The analogous helix in MetRS suggests that tRNA^{Met} binds to MetRS in a similar way. These structural similarities with GlnRS, whose structure has been solved in a complex with tRNA^{Gln}, have given indications of the way in which tRNA^{Met} may bind to the enzyme (71). Positioning of tRNA^{Met} on MetRS (71) was aided by genetic studies of suppressor mutants of MetRS that were selected for their ability to recognize tRNA^{Met} with the CUA anticodon, while retaining specificity for the CAU anticodon (72). *Figure 10(a)* shows the structure of MetRS with tRNA^{Gln}; the binding of tRNA^{Gln} was reconstructed from the figures in Perona *et al.* (71).

The 3-D structure of MetRS from *E. coli* is remarkably similar to that of TyrRS from *Bacillus stearothermophilus* over 140 amino acid residues comprising the nucleotide-binding fold, a five-stranded β-sheet. The two enzymes also have common structural features such as conserved Cys and His residues which occupy identical positions in the nucleotide-binding fold. In TyrRS, these Cys and His residues interact with tyrosyl-adenylate (Tyr-AMP) (73).

4.3 Tyrosyl-tRNA synthetase

Tyrosyl-tRNA synthetase (TyrRS) from *B. stearothermophilus* is an $α_2$-dimer of two 47 kDa subunits. The crystal structure of the dimer alone was solved at 2.7 Å (60) and refined at 2.3 Å resolution (74). TyrRS is an elongated dimer of two globular monomers that harbour the Rossman fold. The structures of TyrRS alone and in complex with either Tyr-AMP or an inhibitory analogue of Tyr-AMP have been refined. The ligands bind in similar conformations in a deep cleft that has a narrow inner pocket for the tyrosine moiety (74).

The crystal structure of TyrRS complexed with tRNA^{Tyr} is not available. Site-directed mutagenesis of the enzyme and tRNA^{Tyr} have aided in the docking of tRNA^{Phe} to TyrRS; the model placed the 3' end of the tRNA in the active site near Tyr-AMP (75). However, the 3'-end CCA in this model is straight and more akin to the conformation found in class II aaRSs described below. An alternative model is shown in *Figure 10(b)* whereby tRNA^{Gln} from the complex with GlnRS was docked to the TyrRS dimer according to the principles of the class I tRNA-binding mode. The 3' end of tRNA^{Tyr} may not bend as dramatically as that of tRNA^{Gln} since the terminal base pair is a G–C. The TyrRS dimer is asymmetric in solution and binds one tRNA (76). Combined results of photo cross-linking studies (77, 78) suggest that the enzyme binds the acceptor stem, the anticodon arm, part of the D-arm, and the variable loop of tRNA^{Tyr}. It is conceivable that the C-terminal domain of TyrRS, which is disordered in the crystal structure (60), would bind to the D-arm and the variable loop of the tRNA.

5. tRNA recognition by class II aminoacyl-tRNA synthetases

5.1 Aspartyl-tRNA synthetase

Aspartyl-tRNA synthetase (AspRS) from yeast is an α_2-dimer of two 557-residue, 63 kDa monomers and it is a member of the class II family of aaRSs. The crystal structure of AspRS complexed with tRNA^{Asp} was solved at 3.0 Å resolution (48) (*Figure 7*). The enzyme is a rather flat and diamond-shaped dimer of two elongated subunits. Each AspRS monomer consists of two domains connected by a hinge. The N-terminal domain of one subunit interacts primarily with the C-terminal domain of the other. Most of the dimer interface is between the C-terminal domains. Each monomer is complexed to a molecule of tRNA^{Asp}. The buried surface of 2500 Å^2, which represents 20% of the solvent accessible surface of tRNA^{Asp}, is similar to the extent of the protein-tRNA interface found in the GlnRS-tRNA^{Gln} complex (66, 79). The anticodon undergoes a substantial conformational change, that is a protein-induced fit (48). Each subunit contacts the tRNA in three major areas, each involving at least one putative identity element of tRNA^{Asp}. The dimer binds tRNAs symmetrically, although some differences at the active site have been observed (79). The C-terminal domain contains the catalytic site; it is composed of an antiparallel β-sheet flanked by α-helices. This structure does not constitute and is larger than the Rossmann fold. The tRNA interacts with AspRS on the variable loop side and the protein contacts the bases of the tRNA in the major groove (48), all of which constitutes the mirror image of the mode of tRNA^{Gln} binding to GlnRS (compare *Figures 4* and *7b*) (61). A barrel-like N-terminal domain binds to the anticodon loop of the tRNA (48).

Many nucleotides of tRNA^{Asp} take part in recognition by AspRS (*Figure 8*). Nucleotides whose bases interact directly with the protein are in the acceptor stem and the anticodon loop. Some residues are not directly involved in protein-RNA interactions; they enable the tRNA to adopt the conformation that facilitates its binding to AspRS. These residues are G37 in the anticodon loop and the G10-C25 base pair in the D-loop. AspRS contacts the ribose-phosphate backbone mainly in the acceptor stem, the D-stem, and the anticodon loop (48).

The C-terminal domain is the largest of the two domains and contains the active site, that is the three amino acid sequence motifs conserved among class II aaRSs (*Figures 3* and *9a*). Motif 1 and part of motif 2 form the dimer interface. Motifs 2 and 3 interact with the 3'-terminal CCA of tRNA^{Asp} and ATP, which binds in a manner characteristic of class II aaRSs (79). Aspartic acid binds in a pocket adjacent to the α-phosphate of AMP and the 3'-OH of the terminal adenosine. The side chain inserts itself into a pre-existing network of hydrogen bonds and salt bridges and forms specific contacts with the enzyme. This environment accepts only Asp and discriminates against Asn (80). Motifs 1 and 2 position the acceptor stem of the tRNA. Motif 2 interacts with G73 and the first base pair of tRNA^{Asp}. The 3'-terminal GCCA of the tRNA is in a helical conformation (48), unlike that of tRNA^{Gln} in complex

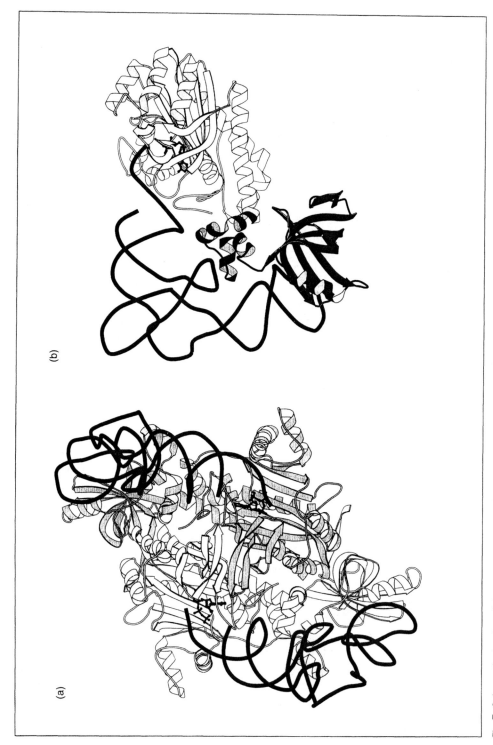

Fig. 7 Schematic drawing of the crystal structure of AspRS complexed with tRNA^{Asp} (48) (a) the dimer and (b), one of the monomers. The domains in (b) are grey-coded as described in Fig. 4.

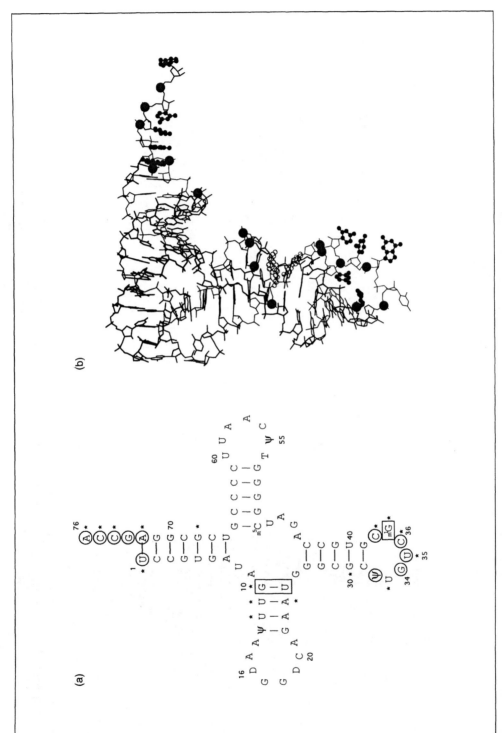

Fig. 8 tRNA^Asp (a) secondary cloverleaf structure and (b) as it appears in complex with AspRS (48). Its interactions with AspRS are coded as in Fig. 5.

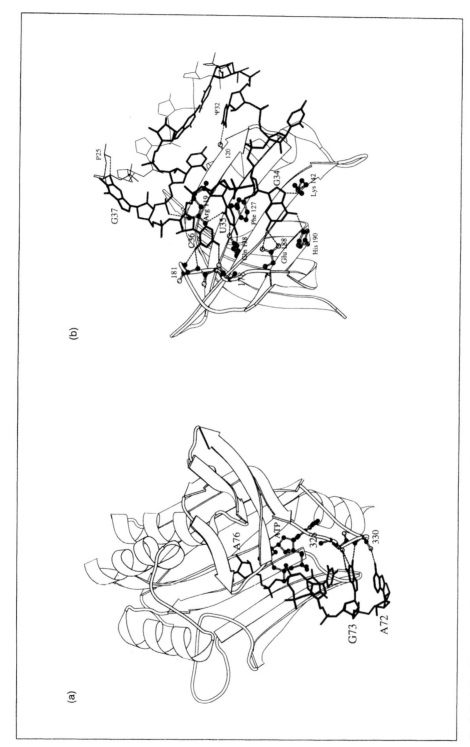

Fig. 9 Details of the interactions between AspRS and tRNA^Asp (48) in (a) the acceptor stem (shown are residues 72 to 76 of the tRNA and the active site domain of AspRS. U1 has been omitted for clarity) and (b) the anticodon region (shown are residues 31 to 37 of the tRNA (thick lines), the backbone portion of the tRNA from residues 25 to 30 (thin lines) and the anticodon-binding domain of AspRS. Note that G37 does not interact with the protein but contacts the phosphate moiety of residue 25). For clarity only the main base-specific interactions are shown.

with GlnRS (61) and the first base pair (U1–A72) is undisrupted. The GCCA interacts directly with the helices and loops of the protein that form part of the active-site pocket (*Figure 9a*). The base of G73 interacts directly with a backbone oxygen and nitrogen atom of the variable loop (residues 327–334) of motif 2 (48). Two other loops contact C75 and A76. The ribose of the terminal adenosine is positioned in such a way that the 3'-OH can accept Asp from Asp-AMP (80). The conformational differences in this region of the tRNAs in the AspRS and GlnRS systems are due to different interactions with the cognate enzymes, as the tRNA sequences are similar in this region. All but one direct contact involve the same subunit; only the phosphate of U1 interacts with Lys293 of the other subunit (79). The principles involved in acceptor binding can be generalized to class II aaRSs.

The anticodon loop of tRNA^{Asp} contains essential recognition elements for AspRS (*Figure 9b*). It binds to the N-terminal domain, which is a five-stranded β-sheet and interacts with it on the major groove side. The anticodon arm undergoes a protein-induced conformational change, which results in the bulging out of residue 37 and the shortening and bending of the anticodon stem–loop. The three anticodon bases are unstacked and protrude out to maximize contacts with the protein. The resulting conformation is stabilized in part by intramolecular hydrogen bonding between the exocyclic amino group of mG37 and the phosphate of residue 25. The anticodon bases are recognized by direct hydrogen bonding between the side chains or backbone segments of the enzyme and the hydrogen-bonding groups of the bases (48, 79). These interactions are AspRS specific, but the principles can be extended to the remaining two aaRS of subgroup b (79).

The hinge region connecting the N- and C-terminal domains (*Figure 7b*) is a globular module; it interacts with the ribose and phosphate moieties of residues 11 and 12, which are near the G10–U25 base pair, whose integrity is important for the structural stability of the tRNA and its ability to bind to AspRS.

AspRS has been shown to tolerate some conformational variability in tRNA^{Asp} at the level of the D-loop, the variable region, and the T-arm, while integrity of the anticodon arm of the tRNA is crucial for aspartylation (81). The three anticodon bases and G73 were found to be the main identity determinants and the G10–C25 base pair is an accessory element. When these elements were transplanted on to a yeast tRNA^{Phe} the resulting derivative was aspartylated, which showed that the set indeed constituted the tRNA^{Asp} identity. However, a G–C in position 1–72 was as good as a U–A in tRNA^{Asp} (82), which is different from E. coli tRNA^{Gln} (44). Chemical footprinting of the interaction of wild-type and mutant yeast tRNA^{Asp} transcripts with yeast AspRS resulted in patterns consistent with direct binding of the determinants and some other residues by AspRS, whereas additional protections in the variable loop and the T- and D-arms resulted from conformational changes upon binding to the enzyme. tRNA^{Phe} with transplanted tRNA^{Asp} identity had protection patterns similar to those of wild-type tRNA^{Asp} (83).

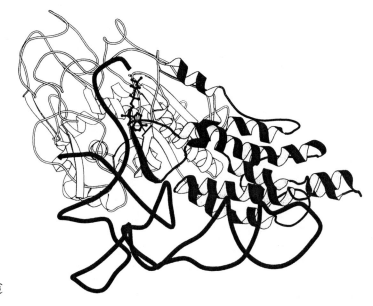

(b)

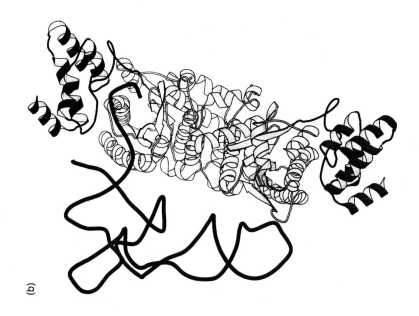

(a)

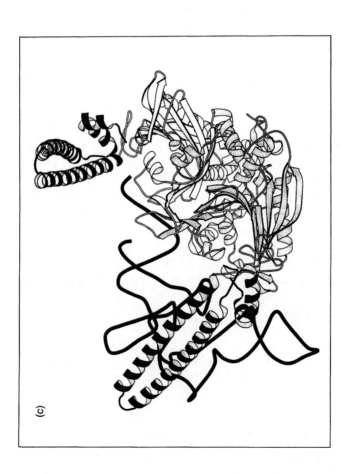

Fig. 10 Schematic drawings of the crystal structures of (a) methionyl-tRNA synthetase monomeric fragment from *E. coli* (59), (b) tyrosyl-tRNA synthetase from *B. stearothermophilus* (60, 74), and (c) seryl-tRNA synthetase from *E. coli* (62) showing their modular organization. The domains are grey-coded as in Fig. 4. The tRNA molecules were docked as rigid bodies, that is without any alterations, by computer graphics using FRODO (90) as described in the text.

5.2 Asparaginyl- and lysyl-tRNA synthetases

Asparaginyl (AsnRS)- and lysyl (LysRS)-tRNA synthetases are very homologous with AspRS (84). Therefore, all the conclusions pertaining to the AspRS system can probably be extended to these two systems.

5.3 Seryl-tRNA synthetase

Seryl-tRNA synthetase (SerRS) from *E. coli* is an α_2 dimer of 48.4 kDa subunits (*Figure 10c*); its crystal structure has been solved at 2.5 Å resolution (61). The monomer consists of two domains. The first 100 N-terminal residues form a 60 Å antiparallel coiled coil of two α-helices. The rest of the molecule is globular and is made of a seven-stranded mostly antiparallel β-sheet surrounded by α-helices (62). The topology of the sheet is similar to that in AspRS (48) and is characteristic of class II aaRSs.

Chemical footprinting and enzymatic probes of tRNA^{Ser} have shown that the 3' end of the acceptor stem, part of the anticodon stem at the base of the variable loop, part of the TψC loop, and the base-paired portion of the long variable arm interact with SerRS. The anticodon is free in solution (85, 86), which agrees with the fact that the tRNAs charged by the enzyme, five tRNA^{Ser} isoacceptors, and the tRNA^{SeCys} (selenocysteinyl tRNA), possess a variety of anticodon sequences (62).

The structures of SerRS complexed with tRNA^{Ser} from *E. coli* (64) and *T. thermophilus* (63) have been solved; similar features have been observed in both complexes. The tRNA binds across both subunits of the dimer. The terminal part of the acceptor end contacts the active site of one subunit, while the rest of the tRNA is bound to the other subunit. The anticodon arm forms no contacts with SerRS, while the long variable loop and TψC loop interact with the long helical arm of the protein, which in turn changes its conformation upon tRNA binding. The enzyme seems to recognize the unique shape of tRNA^{Ser} rather than its sequence (63).

In *Figure 10(c)*, the model of tRNA^{Ser} (87) was docked to the structure of the SerRS dimer (62) according to the principles of the class II tRNA-binding mode and guided by the footprinting and preliminary structural information described above.

5.4 Phenylalanyl-tRNA synthetase

Phenylalanyl-tRNA synthetase (PheRS) from *T. thermophilus* has an $\alpha_2\beta_2$ quaternary structure, which can be construed as being composed of two ($\alpha\beta$) class II characteristic dimeric entities (58). Recently, the enzyme has been co-crystallized with tRNA^{Phe} (88).

6. Conformational changes in tRNA

To summarize, both GlnRS and AspRS induce dramatic conformational changes (*Figure 11*) in their cognate tRNA substrates to fit their binding surfaces. When

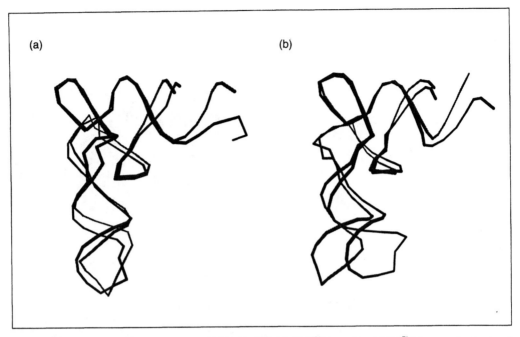

Fig. 11 Comparison of free (thin) and bound (thick) tRNA (a) tRNA^{Gln} bound and tRNA^{Phe} free as standard and (b) tRNA^{Asp} bound and free.

tRNA^{Gln} binds to GlnRS the anticodon loop bends inward, unstacking the anticodon bases and maximizing their interactions with the protein. Two new non-Watson–Crick base pairs form in the loop, extending the helical stem. The 3′-terminal CCA of the acceptor stem bends into the active site, which is facilitated by the melting of the terminal base pair (U1–A72) and the resulting conformation is stabilized by the base-specific intramolecular hydrogen bond between G73 and the backbone (61). In contrast, tRNA^{Asp} does not undergo any substantial conformational change in the acceptor arm; the CCA-terminus remains helical as it binds to the active site. The nucleotide bases, however, are unstacked as they form contacts with the protein. The conformational change in this case is concentrated in the anticodon loop and is very dramatic. The base of mG37 bulges out, shortening the anticodon stem and bending the loop. The bases are unstacked in the process and make specific interactions with the protein. The new conformation is stabilized by inter- and intramolecular hydrogen bonding (48, 79).

7. Conclusion

Aminoacyl tRNA synthetases are modular enzymes, built around two types of active-site platforms, which define the two classes of these enzymes (*Figure 12*). Class I aaRSs share the Rossmann fold nucleotide-binding domain, which has also been observed in many other nucleotide-binding enzymes and is built around a

76 | AMINOACYL-tRNA SYNTHETASE–tRNA RECOGNITION

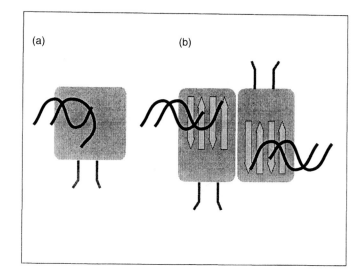

Fig. 12 General scheme of organization of (a) Class I aaRS and (b) Class II aaRS.

parallel β-sheet (*Figure 3*) (49). Class II enzymes possess a new type of nucleotide-binding domain, observed only in aaRSs. It is based on an antiparallel β-sheet (47). Enzymes of each class share sequence motifs which are related to structural motifs. In addition, the site of amino acid attachment on the tRNA is class specific, with one exception. To this central platform are attached system-specific insertions and additional domains that confer tRNA and amino acid specificity upon each enzyme.

The crystal structures of two aaRS–tRNA complexes, one for each class, have been examined. The main class I characteristics, as exemplified by GlnRS, are minor groove recognition and a distorted CCA-terminus of the tRNA, while the principal features characterizing class II enzymes, exemplified by AspRS, are dimeric organization and major groove recognition of the tRNA at the acceptor end. tRNA specificity in both classes is the result of direct base pair–protein interactions.

References

1. Schimmel, P. R. and Söll, D. (1979) Aminoacyl-tRNA synthetases: general features and recognition of transfer RNAs. *Ann. Rev. Biochem.*, **48**, 601.
2. Ravel, J. M., Wang, S. F., Heinemeyer, C., and Shive, W. (1965) Glutamyl and glutaminyl ribonucleic acid synthetases of *Escherichia coli* W. Separation, properties and stimulation of adenosine triphosphate exchange by acceptor ribonucleic acid. *J. Biol. Chem.*, **240**, 432.
3. Yarus, M (1972) Intrinsic precision of aminoacyl-tRNA synthesis enhanced through parallel system of ligands. *Nature New Biol.*, **239**, 106.
4. Yarus, M. (1980) The accuracy of translation. *Prog. Nucleic Acids Res. Mol. Biol.*, **23**, 195.
5. Ebel, J. P., Giegé, R., Bonnet, J., Kern, D., Befort, N., Bollack, C., Fasiolo, F., Gangloff,

J., and Dirheimer, G. (1973) Factors determining the specificity of the tRNA aminoacylation reaction. *Biochimie*, **55**, 547.
6. de Duve, C. (1988) The second genetic code. *Nature*, **333**, 117.
7. Moras, D. (1992) Structural and functional relationships between aminoacyl-tRNA synthetases. *Trends Biochem. Sci.*, **17**, 159.
8. Weiner, A. M. and Maizels, N. (1987) tRNA-like structures tag the 3′ ends of genomic RNA molecules for replication: implications for the origin of protein synthesis. *Proc. Natl Acad. Sci. USA*, **84**, 7383.
9. Schimmel, P. (1989) Parameters for the molecular recognition of transfer RNA. *Biochemistry*, **28**, 2747.
10. Saenger, W. (1982) *Principles of Nucleic Acid Structure*. Springer-Verlag, New York.
11. Suddath, F. L., Quigley, G. J., McPherson, A., Sneden, D., Kim, J. J., Kim, S. H., and Rich, A. (1974) Three-dimensional structure of yeast phenylalanine transfer RNA at 3.0 Å resolution. *Nature*, **248**, 20.
12. Ladner, J. E., Jack, A., Robertus, J. D., Brown, R. S., Rhodes, D., Clark, B. F. C., and Klug, A. (1975) Structure of yeast phenylalanine transfer RNA at 2.5 Å resolution. *Proc. Natl Acad. Sci. USA*, **72**, 4414
13. Moras, D., Comarmond, M. B., Fischer, J., Weiss, R., Thierry, J. C., Ebel, J. P., and Giegé, R. (1980) Crystal structure of yeast tRNA^{Asp}. *Nature*, **288**, 669.
14. Westhof, E., Dumas, P., and Moras, D. (1985) Crystallographic refinement of yeast aspartic acid transfer RNA. *J. Mol. Biol.*, **184**, 119.
15. Basavappa, R. and Sigler, P. B. (1991) The 3 Å crystal structure of yeast initiator tRNA: functional implications in initiator/elongator discrimination. *EMBO J.*, **10**, 3105.
16. Björk, G. R., Ericson, J. U., Gustafsson, C., Hagervall, T. G., Jönsson, Y. H., and Wikstrom, P. M. (1987) Transfer RNA modification. *Ann. Rev. Biochem.*, **56**, 263.
17. Muramatsu, T., Nishikawa, K., Nemoto, F., Kuchino, Y., Nishimura, S., Miyazawa, T., and Yokoyama, S. (1988) Codon and amino acid specificities of a transfer RNA are both converted by a single post-transcriptional modification. *Nature*, **336**, 179.
18. Perret, V., Garcia, A., Grosjean, H., Ebel, J. P., Florentz, C., and Giegé, R. (1990) Relaxation of a transfer RNA specificity by removal of modified nucleotides. *Nature*, **344**, 787.
19. Hall, K. B., Sampson, J. R., Uhlenbeck, O. C., and Redfield, A. G. (1989) Structure of an unmodified tRNA molecule. *Biochemistry*, **28**, 5794.
20. Sampson, J. R. and Uhlenbeck, O. C. (1988) Biochemical and physical characterization of an unmodified yeast phenylalanine transfer RNA transcribed *in vitro*. *Proc. Natl Acad. Sci. USA*, **85**, 1033.
21. Schulman, L. H. (1991) Recognition of tRNAs by aminoacyl-tRNA synthetases. *Prog. Nucleic Acids Res. Mol. Biol.*, **41**, 23.
22. Edwards, H., Trézéguet, V., and Schimmel, P. (1991) An *Escherichia coli* tyrosine transfer RNA is a leucine specific transfer RNA in the yeast *Saccharomyces cerevisiae*. *Proc. Natl Acad. Sci. USA*, **88**, 1153.
23. Crothers, D. M., Seno, T., and Soll, D. (1972) Is there a discriminator site in transfer RNA? *Proc. Natl Acad. Sci. USA*, **69**, 3063.
24. Normanly, J. and Abelson, J. (1989) tRNA identity. *Ann. Rev. Biochem.*, **58**, 1029.
25. McClain, W. H. and Nicholas, H. B., Jr (1987) Differences between transfer RNA molecules. *J. Mol. Biol.*, **194**, 635.
26. Sampson, J. R., DiRenzo, A. B., Behlen, L. S., and Uhlenbeck, O. C. (1989) Nucleotides in yeast tRNA^{Phe} required for specific recognition by its cognate synthetase. *Science*, **243**, 1363.

27. Pallanck, R. and Schulman, L. H. (1991) Anticodon-dependent aminoacylation of a noncognate tRNA with isoleucine, valine, and phenylalanine *in vivo*. *Proc. Natl Acad. Sci. USA*, **88**, 3872.
28. Schulman, L. H. and Pelka, H. (1989) The anticodon contains a major element of the identity of arginine transfer RNAs. *Science*, **246**, 1595.
29. Schulman, L. H. and Pelka, H. (1983) Anticodon loop size and sequence requirements for recognition of formylmethionine tRNA by methionyl-tRNA synthetase. *Proc. Natl Acad. Sci. USA*, **80**, 6755.
30. Seno, T., Agris, P. F., and Söll, D. (1974) Involvement of the anticodon region of *Escherichia coli* tRNA^{Gln} and tRNA^{Glu} in the specific interaction with cognate aminoacyl-tRNA synthetase: alteration of the 2-thiouridine derivatives in the anticodon of the tRNAs by BrCN or sulfur deprivation. *Biochim. Biophys. Acta*, **349**, 328.
31. Schulman, L. H. and Pelka, H. (1988) Anticodon switching changes the identity of methionine and valine tRNAs. *Science*, **242**, 765.
32. McClain, W. H. and Foss, K. (1988) Nucleotides that contribute to the identity of *Escherichia coli* tRNA^{Phe}. *J. Mol. Biol.*, **202**, 697.
33. McClain, W. H. and Foss, K. (1988) Changing the acceptor identity of a transfer RNA by altering nucleotides in a 'variable pocket'. *Science*, **241**, 1804.
34. Normanly, J., Ogden, R. C., Horvath, S. J., and Abelson, J. (1986) Changing the identity of a transfer RNA. *Nature*, **321**, 213.
35. Normanly, J., Ollick, T., and Abelson, J. (1992) Eight base changes are sufficient to convert a leucine-inserting tRNA into a serine-inserting tRNA. *Proc. Natl Acad. Sci. USA*, **89**, 5680.
36. Hou, Y. M. and Schimmel, P. R. (1988) A simple structural feature is a major determinant of the identity of a transfer RNA. *Nature*, **333**, 140.
37. McClain, W. H. and Foss, K. (1988) Changing the identity of a tRNA by introducing a G–U wobble pair near the 3' acceptor end. *Science*, **240**, 793.
38. McClain, W. H., Chen, Y. M., Foss, K., and Schneider, J. (1988) Transfer RNA acceptor identity associated with a helical irregularity. *Science*, **242**, 1681.
39. Francklyn, C. and Schimmel, P. (1989) Aminoacylation of RNA minihelices with alanine. *Nature*, **337**, 478.
40. Schulman, L. H. and Pelka, H. (1985) *In vitro* conversion of a methionine to a glutamine acceptor tRNA. *Biochemistry*, **24**, 7309.
41. Knowlton, R. G., Soll, L., and Yarus, M. (1980) Dual specificity of su^+7 tRNA. Evidence for translational discrimination. *J. Mol. Biol.*, **139**, 705.
42. Ghysen, A. and Celis, J. E. (1974) Mischarging single and double mutants of *Escherichia coli sup3* tyrosine transfer RNA. *J. Mol. Biol.*, **83**, 333.
43. Hooper, J. L., Russel, R. L., and Smith, J. D. (1972) Mischarging in mutant tyrosine transfer RNAs. *FEBS Lett.*, **22**, 149.
44. Jahn, M., Rogers, M. J., and Söll, D. (1991) Anticodon and acceptor stem nucleotides in tRNA^{Gln} are major recognition elements for *E. coli* glutaminyl-tRNA synthetase. *Nature*, **352**, 258.
45. Swanson, R., Hoben, P., Sumner-Smith, M., Uemura, H., Watson, L., and Söll, D. (1988) Accuracy of *in vivo* aminoacylation requires proper balance of tRNA and aminoacyl-tRNA synthetase. *Science*, **242**, 1548.
46. Perret, V., Florentz, C., Puglisi, J. D., and Giegé, R. (1992) Effect of conformational features on the aminoacylation of tRNAs and consequences on the permutation of tRNA specificities. *J. Mol. Biol.*, **226**, 323
47. Eriani, G., Delarue, M., Poch, O., Gangloff, J., and Moras, D. (1990) Partition of tRNA

synthetases into two classes based on mutually exclusive sets of sequence motifs. *Nature*, **347**, 203.

48. Ruff, M., Krishnaswamy, S., Boeglin, M., Poterszman, A., Mitschler, A., Podjarny, A., Rees, B., Thierry, J. C., and Moras, D. (1991) Class II aminoacyl-transfer RNA synthetase: crystal structure of yeast aspartyl-tRNA synthetase complexed with tRNA^{Asp}. *Science*, **252**, 1682.

49. Rossmann, M. G., Moras, D., and Olsen, K. W. (1974) Chemical and biological evolution of a nucleotide-binding protein. *Nature*, **250**, 194.

50. Delarue, M. and Moras, D. (1993) The aminoacyl-tRNA synthetase family: modules at work. *BioEssays*, **15**, 1.

51. Fraser, T. H. and Rich, A. (1975) Amino acids are not all initially attached to the same position on transfer RNA molecules. *Proc. Natl Acad. Sci. USA*, **72**, 3044.

52. Sprinzl, M. and Cramer, F. (1975) Site of aminoacylation of tRNAs from *Escherichia coli* with respect to the 2′- or 3′-hydroxyl group of the terminal residue. *Proc. Natl Acad. Sci. USA*, **72**, 3049.

53. Jasin, M., Regan, L., and Schimmel, P. (1983) Modular arrangement of functional domains along the sequence of an aminoacyl-tRNA synthetase. *Nature*, **306**, 441.

54. Hou, Y. M., Shiba, K., Mottes, C., and Schimmel, P. (1991) Sequence determination and modeling of structural motifs for the smallest monomeric aminoacyl-tRNA synthetase. *Proc. Natl Acad. Sci. USA*, **88**, 976.

55. Hall, C. V., vanCleemput, M., Muench, K. H., and Yanofsky, C. (1982) The nucleotide sequence of the structural gene for *Escherichia coli* tryptophanyl-tRNA synthetase. *J. Biol. Chem.*, **257**, 6132.

56. Ribas de Pouplana, L., Buechter, D. D., Davis, M. W., and Schimmel, P. R. (1993) Idiographic representation of conserved domains of a Class II aminoacyl-tRNA synthetase of unknown structure. *Protein Sci.*, **2**, 2259.

57. Nada, S., Chang, P. K., and Dignam, J. D. (1993) Primary structure of the gene for glycyl-tRNA synthetase from *Bombyx mori*. *J. Biol. Chem.*, **268**, 7660.

58. Reshetnikova, L., Chernaya, M., Ankilova, V., Lavrik, O., Delarue, M., Thierry, J. C., Moras, D., and Safro, M. (1992) Three-dimensional structure of phenylalanyl-transfer RNA synthetase from *Thermus thermophilus* HB8 at 0.6-nm resolution. *Eur. J. Biochem.*, **208**, 411.

59. Zelwer, C., Risler, J. L., and Brunie, S. (1982) Crystal structure of *E. coli* methionyl-tRNA synthetase at 2.5 Å resolution. *J. Mol. Biol.*, **155**, 63.

60. Bhat, T. N., Blow, D. M., Brick, P., and Nyborg, J. (1982) Tyrosyl-tRNA synthetase forms a mononucleotide binding fold. *J. Mol. Biol.*, **158**, 699.

61. Rould, M. A., Perona, J. J., Söll, D., and Steitz, T. A. (1989) Structure of *E. coli* glutaminyl-tRNA synthetase complexed with tRNA^{Gln} and ATP at 2.8 Å resolution. *Science*, **246**, 1135.

62. Cusack, S., Berthet-Colominas, C., Härtlein, M., Nassar, N., and Leberman, R. (1990) A second class of synthetase structure revealed by X-ray analysis of *Escherichia coli* seryl-tRNA synthetase at 2.5 Å. *Nature*, **347**, 249.

63. Biou, V., Berthet-Colominas, C., and Cusack, S. (1992) Structure of the complex of seryl-tRNA synthetase and tRNA^{Ser} from *T. thermophilus* at 2.9 Å resolution. In *Research Reports 1992*. The European Molecular Biology Laboratory, Heidelberg, p. 232.

64. Cusack, S., Berthet-Colominas, C., Biou, V., Price, S., and Leberman, R. (1992) Crystal structure of the *E. coli* seryl-tRNA synthetase–tRNA$_2^{Ser}$ complex. In *Research Reports 1992*. The European Molecular Biology Laboratory, Heidelberg, p. 232.

65. Hoben, P., Royal, N., Cheung, A., Yamao, F., Biemann, K., and Söll, D. (1982)

Escherichia coli glutaminyl-tRNA synthetase. II. Characterization of the *glnS* gene product. *J. Biol. Chem.*, **257**, 11 644.

66. Rould, M. A., Perona, J. J., and Steitz, T. A. (1991) Structural basis of anticodon loop recognition by glutaminyl-tRNA synthetase. *Nature*, **352**, 213.
67. Perona, J. J., Rould, M. A., and Steitz, T. A. (1993) Structural basis for transfer RNA aminoacylation by *Escherichia coli* glutaminyl-tRNA synthetase. *Biochemistry*, **32**, 8758.
68. Perona, J. J., Swanson, R. N., Rould, M. A., Steitz, T. A., and Söll, D. (1989) Structural basis for misaminoacylation by mutant *E. coli* glutaminyl-tRNA synthetase enzymes. *Science*, **246**, 1152.
69. Inokuchi, H., Hoben, P., Yamao, F., Ozeki, H. and Söll, D. (1984) Transfer RNA mischarging mediated by a mutant *E. coli* glutaminyl-tRNA synthetase. *Proc. Natl Acad. Sci. USA*, **81**, 5076.
70. Rogers, M. J. and Söll, D. (1988) Discrimination between glutaminyl-tRNA synthetase and seryl-tRNA synthetase involves nucleotides in the acceptor helix of tRNA. *Proc. Natl Acad. Sci. USA*, **85**, 6627.
71. Perona, J. J., Rould, M. A., Steitz, T. A., Risler, J. L., Zelwer, C., and Brunie, S. (1991) Structural similarities in glutaminyl- and methionyl-tRNA synthetases suggest a common overall orientation of tRNA binding. *Proc. Natl Acad. Sci. USA*, **88**, 2903.
72. Meinnel, T., Mechulam, Y., Le Corre, D., Panvert, M., Blanquet, S., and Fayat, G. (1991) Selection of suppressor methionyl-tRNA synthetases: mapping the tRNA anticodon binding site. *Proc. Natl Acad. Sci. USA*, **88**, 291.
73. Blow, D. M., Bhat, T. N., Metcalfe, J. L., Risler, J. L., Brunie, S., and Zelwer, C. (1983) Structural homology in the amino-terminal domains of two aminoacyl-tRNA synthetases. *J. Mol. Biol.*, **171**, 571.
74. Brick, P., Bhat, T. N., and Blow, D. M. (1989) Structure of tyrosyl-tRNA synthetase refined at 2.3 Å resolution. *J. Mol. Biol.*, **208**, 83.
75. Bedouelle, H. and Winter, G. (1986) A model of synthetase/transfer RNA interaction as deduced by protein engineering. *Nature*, **320**, 371.
76. Ward, W. H. J. and Fersht, A. R. (1988) Asymmetry of tyrosyl-tRNA synthetase in solution. *Biochemistry*, **27**, 1041.
77. Ackerman, E. J., Joachimiak, A., Klinghofer, V., and Sigler, P. B. (1985) Directly photocrosslinked nucleotides joining transfer RNA to aminoacyl-tRNA synthetase in methionine and tyrosine systems. *J. Mol. Biol.*, **181**, 93.
78. Schoemaker, H. J. P. and Schimmel, P. R. (1974) Photo-induced joining of a transfer RNA with its cognate aminoacyl-tRNA synthetase. *J. Mol. Biol.*, **84**, 503.
79. Cavarelli, J., Rees, B., Ruff, M., Thierry, J. C., and Moras, D. (1993) Yeast tRNA^{Asp} recognition by its cognate class II aminoacyl-tRNA synthetase. *Nature*, **362**, 181.
80. Cavarelli, J., Eriani, G., Rees, B., Ruff, M., Boeglin, M., Mitschler, A., Martin, F., Gangloff, J., Thierry, J. C., and Moras, D. (1994). The active site of yeast aspartyl-tRNA synthetase: structural and functional aspects of the aminoacylation reaction. *EMBO J.*, **13**, 327.
81. Giegé, R., Florentz, C., Garcia, A., Grosjean, H., Perret, V., Puglisi, J., Théobald-Dietrich, A., and Ebel, J. P. (1990) Exploring the aminoacylation function of transfer RNA by macromolecular engineering approaches. Involvement of conformational features in the charging process of yeast tRNA^{Asp}. *Biochimie*, **72**, 453.
82. Pütz, J., Puglisi, J. D., Florentz, C., and Giegé, R. (1991) Identity elements for specific aminoacylation of yeast tRNA^{Asp} by cognate aspartyl-tRNA synthetase. *Science*, **252**, 1696.

83. Rudinger, J., Puglisi, J. D., Pütz, J., Schatz, D., Eckstein, F., Florentz, C., and Giegé, R. (1992) Determinant nucleotides of yeast tRNA^{Asp} interact directly with aspartyl-tRNA synthetase. *Proc. Natl Acad. Sci. USA*, **89**, 5882.
84. Eriani, G., Dirheimer, G., and Gangloff, J. (1990) Aspartyl-tRNA synthetase from *Escherichia coli*: cloning and characterisation of the gene, homologies of its translated amino acid sequence with asparaginyl- and lysyl-tRNA synthetases. *Nucleic Acids Res.*, **18**, 7109.
85. Dock-Bregeon, A. C., Garcia, A., Giegé, R., and Moras, D. (1990) The contacts of yeast tRNA^{Ser} with seryl-tRNA synthetase studied by footprinting experiments. *Eur. J. Biochem.*, **188**, 283.
86. Schatz, D., Leberman, R., and Eckstein, F. (1991) Interaction of *Escherichia coli* tRNA^{Ser} with its cognate aminoacyl-tRNA synthetase as determined by footprinting with phosphorothioate-containing tRNA transcripts. *Proc. Natl Acad. Sci. USA*, **88**, 6132.
87. Dock-Bregeon, A. C., Westhof, E., Giegé, R., and Moras, D. (1989) Solution structure of a tRNA with a large variable region: yeast tRNA^{Ser}. *J. Mol. Biol.*, **206**, 707.
88. Reshetnikova, L., Khodyreva, S., Lavrik, O., Ankilova, V., Frolow, F., and Safro, M. (1993) Crystals of the phenylalanyl-tRNA synthetase from *Thermus thermophilus* HB8 complexed with tRNA^{Phe}. *J. Mol. Biol.*, **231**, 927.
89. Kraulis, P. J. (1991) 'MOLSCRIPT: a program to produce both detailed and schematic plots of protein structures.' *J. Appl. Crystal.*, **24**, 946.
90. Jones, T. A. (1982) 'FRODO.' In *Computational Crystallography*. Sayre, D. (ed.). Clarendon Press, Oxford, p. 303.

4 | RNA–protein interactions in ribosomes

DAVID E. DRAPER

1. Introduction

In recent years, the functional role of ribosomal RNAs in protein synthesis has received a great deal of attention, as domains of highly conserved structure have been identified with the decoding site, the peptidyl transferase active centre, binding sites for the elongation factors, and other functional centres of the ribosome (1). Now that a large database of ribosomal protein sequences is available, it is also apparent that many of the ribosomal proteins are as highly conserved as the rRNAs, including many of the proteins which bind directly and independently to the RNA (2–4). The basic ribosomal machinery is probably an RNA–protein complex which traces its ancestry to the beginnings of biotic evolution.

The ribosome was the first RNP to be seriously investigated in the early days of molecular and structural biology. Although experimental tools available 20 years ago may now seem limited, many of the basic concepts and methodologies currently in use for protein–RNA studies have their origins in these first studies. This chapter will first discuss the main experimental methods available for defining the structure and properties of a protein–RNA complex and then review several of the better understood complexes derived from the ribosome.

2. Nomenclature

The ribosomal proteins are given an L or S designation indicating association with the large or small subunit, followed by a number which originally indicated the order of elution from an ion exchange column (S4, L23, etc.). The set of ribosomal proteins from *Escherichia coli* has been completely sequenced and characterized and similar enterprises are under way for archaebacterial thermophile, and rat ribosomes (2–5). Although many of the proteins are homologous between kingdoms, the nomenclature does not necessarily imply a relationship, for example the *E. coli* L23 is homologous to yeast L25. In this review, the name will refer to the *E. coli* protein unless otherwise specified.

There are two large ribosomal RNAs. The small subunit contains a single species, usually in the range of 1500–1800 nucleotides (nt) and is referred to as small subunit rRNA or 16S-like rRNA. Likewise the large RNA in the large subunit is called large subunit rRNA or 23S-like rRNA and is approximately twice the size of the small subunit rRNA. Databases of hundreds of sequences of each large rRNA are available (6–8). The large subunit also contains the 5S rRNA (~120 nt). Eukaryotes contain a 5.8S rRNA, which is homologous to the 5' 160 nt of prokaryotic 23S rRNA.

Both the small and large subunits of *E. coli* ribosomes have been assembled *in vitro* from the component proteins and rRNAs (9–10). In each subunit, approximately one-third of the ribosomal proteins bind independently to the rRNA and are referred to as primary-binding proteins. The remainder depend on the presence of one or more of the primary-binding proteins. The primary-binding proteins may organize the rRNA structure in a way that facilitates interactions of the remaining proteins and some recent work strongly suggests that RNA conformational changes are an important aspect of assembly (11). Protein–protein interactions are probably also important (12). This review will be concerned with the better characterized primary-binding proteins.

3. Thermodynamic measurements

The first questions to ask about any protein–RNA complex are whether the interaction is specific, what are the optimum conditions for binding, and what is the magnitude of the binding affinity. Answers to these questions require methods for detecting and quantitating protein–RNA complexes. There are a limited number of options, all of which depend on physical separation of complexes from free RNA and protein. (Spectroscopic methods for detecting RNA–protein complexes have not been developed.) The filter-binding assay, which depends on nitrocellulose strongly binding protein but not RNA, is probably the most consistently applicable technique: protein and radiolabelled RNA are incubated together, the solution pulled through a nitrocellulose filter by gentle suction, the filter rinsed with buffer to remove non-specifically retained RNA and the remaining RNA on the filter determined by scintillation counting (13). Though artefacts can arise from rinsing the filter and the range of solvent and temperature conditions over which the assay operates is sometimes limited, reliable equilibrium constants and stoichiometries are generally obtained. With ribosomal proteins, it has been found that complexes with the large ribosomal RNAs are usually not retained on the filter, even though labelled protein binds quantitatively to the filter membranes. The large RNA must prevent the protein from contacting the membrane efficiently. The filter-binding assay can still be put to use in these cases, by labelling the protein and measuring the reduction in bound protein upon titration with rRNA (14). Fragments of the ribosomal RNAs, even of relatively large size, are usually retained on the filters by ribosomal proteins (15).

The gel-shift assay was first developed to detect protein–DNA complexes (16), but can also be applied to protein–RNA binding. Complexed and uncomplexed RNAs are separated by electrophoretic mobility: the larger size and (usually) reduced negative charge of the protein–RNA complex slows migration in a standard polyacrylamide gel. Titrations of labelled RNA with protein in this assay can determine equilibrium constants and stoichiometries, though the potential for artefacts is greater than in the filter-binding assay. This is because the time taken for separation of the complex is much longer (minutes to hours, rather than a second or less) and exceeds the dissociation half-time for all but the strongest RNA–protein complexes. Many ribosomal proteins have a relatively weak affinity for RNA ($\sim 10^7$ M^{-1}) and do not show a gel-shift effect or only show non-specific binding in the assay (13). Binding affinities measured by gel shift have been compared to those obtained by other techniques in only a few cases; in one example, the binding constant obtained by filter binding was three orders of magnitude stronger than the gel shift-derived constant (17), suggesting that substantial dissociation of the complex takes place during electrophoresis. A better gel method for measuring weaker protein–RNA binding constants is zone-interference electrophoresis, in which zones of increasing RNA concentration are electrophoresed through a band of protein (18). The complex can re-equilibrate continuously and the distance of protein migration is proportional to the fraction of time it spends bound to the RNA. This is a rapid and powerful method, but has not been applied to ribosomal proteins.

The third method for separating complexes, which has not been used much in recent years, is sedimentation or gel chromatography of labelled protein with increasing concentrations of RNA (19, 20). This was of use in studying protein association with rapidly-sedimenting large rRNAs. It suffers from the same potential artefact as gel-shift assays, since the complex has sufficient time to dissociate during the sedimentation experiment, thus underestimating binding affinity and stoichiometry. Corrections for dissociation can be made to obtain true equilibrium constants (21).

Equilibrium association constants for ribosomal proteins with RNA are in the range 10^6–10^8 M^{-1} (22). Non-specific binding of basic proteins to RNA can approach this magnitude of association constant in low-salt buffers, so it is necessary to demonstrate that an interaction is in fact specific. A first prerequisite is a 1:1 stoichiometry of RNA and protein in the complex. A second criterion is that the affinity for the putative binding site is stronger than for other RNAs, preferably of approximately the same size and structural complexity. For example, a protein recognizing the 16S rRNA should bind more weakly to the 23S rRNA and vice versa. Homopolymers are sometimes much less efficient at non-specific binding than tRNA or other rRNA fragments (15), so it is important to try a series of different RNAs. Convincing evidence for specificity is a large decrease in protein affinity when one or a few RNA bases are mutated.

Since RNA generally requires significant concentrations of salt to maintain its structure, most ribosomal proteins are assayed in a minimum of ~ 100 mM monovalent ions. (Protein aggregation is also a frequent problem and is usually reduced in the higher salt concentrations.) Some, but not all, ribosomal proteins require a minimum concentration of Mg^{2+} or other multivalent ions for optimum binding

affinity. Since K⁺ is the most prevalent intracellular cation, studies of ribosome assembly and protein–RNA binding have been traditionally done in K⁺ salts. It is also known that intact ribosomes have optimum activity in NH_4^+ or K⁺ salts and are inactivated by Na⁺ or Li⁺ salts (23, 24). (For reasons rooted in historical accident,
work on tRNA and most physical studies of RNA structure and thermodynamics have been carried out in Na⁺.) Recently the association constant of a ribosomal protein with a small rRNA fragment has been shown to decrease by an order of magnitude in the series $NH_4^+ > K^+ > Na^+$ (25). Whether this is a more general phenomenon for ribosomal or other RNA-binding proteins has not yet been checked.

4. Methods for studying RNA–protein complexes
4.1 Definition of the RNA-binding site

The first question to ask of a protein–RNA complex is what is the minimum RNA capable of sustaining protein recognition: all primary-binding ribosomal proteins can bind rRNA fragments smaller than the intact rRNA species. The original method was the so-called 'bind and chew' experiment: the complex was formed, digested mildly with a ribonuclease, and the protected RNA fragment isolated (26). For a control, it is necessary to show that the RNA fragment can still rebind the protein. Now that sequencing gels are available, a more elegant version of this experiment is filter selection: a partial alkaline hydrolysate of 3' or 5' end-labelled RNA is incubated with protein, passed through a nitrocellulose filter and the RNA lengths retained on the filter analysed by electrophoresis (27, 28). Both the 5' and 3' 'edges' of the protein-binding site can be determined this way. Using the polymerase chain reaction (PCR) to amplify specific DNA sequences for T7 RNA polymerase transcription, generating a series of RNA fragments of defined length to test the suggested fragment size becomes straightforward.

4.2 Localization of RNA contacts

Once an RNA fragment containing the protein recognition site has been defined, the next step is to define the region actually contacted by the protein. This may be only a small portion of the fragment. In an extreme case, the S4-binding site on 16S rRNA can be narrowed no further than 460 nt (15), essentially the 5' third of the rRNA; within this structure the 24 kD protein can only be in contact with a small fraction of the nucleotides. Mapping these specific contacts has turned out to be much more difficult than for DNA-binding proteins. The basic approach is to assess either the effect of the protein on an RNA modification reaction or (vice versa) the effect of an RNA modification on the protein binding. Several variations of this approach have been described in the literature and applied to many proteins; the difficulty comes in the interpretation of the results.

The most common method used for mapping RNA–protein contacts is a 'footprint', first developed for detecting protein-binding sites on DNA (29). The RNA complexes are formed and reacted with a battery of reagents that attack the bases (for example, DMS, DEPC, kethoxal), phosphodiester (single- and double-strand-specific nucleases, phosphate alkylating reagents), and sugar (hydroxyl radical). Methods for carrying out most of these reactions have been extensively reviewed (30, 31). Comparison of the reaction rates in the presence and absence of the binding protein delineates a subset of nucleotides to which the protein limits access of reagents. The resolution of the method is limited by the number of sites which are initially reactive in the absence of protein; for a highly structured RNA, this may be a small fraction of the bases.

The experiment can also be set up as in the filter-selection method for defining the RNA site: the RNA is first reacted with a modifying reagent to less than one modification per molecule and the RNAs retaining binding activity isolated by either a gel-shift or nitrocellulose filter-binding assay with the protein (32). This 'damage selection' or 'interference' technique gives a complementary picture to that of the 'footprint'. Because the initial reaction is done on unfolded RNA, a greater number of reaction sites can be examined.

A third variation of the modification experiment is to change the sequence of the RNA, to see which bases are required for protein recognition. For a small enough RNA, this can be done by generating individual mutations and measuring the binding constant of each (33). (Since the secondary structures of the ribosomal RNAs are known from phylogenetic studies, compensatory mutations can be made in helices, reducing the total number of mutations which need to be examined.) Selection/amplification methods can speed up this procedure for larger RNAs: mutations are introduced into the RNA by transcription from DNA either synthesized from 'dirty' (contaminated) monomers or amplified by error-prone PCR. The set of RNAs still capable of binding protein are then selected by filter binding. To eliminate sequences binding non-specifically, the protocol is repeated until a set of RNAs with wild-type binding constants is obtained (34).

Cross-linking experiments have also been used to define RNA domains interacting with proteins. A number of reagents have been designed and used (reviewed in Brimacombe *et al.* (35)). The easiest cross-links to interpret are those which require the closest proximity of the RNA and protein; the incorporation of 4-thiouridine into RNA for making UV-sensitized cross-links has been a useful development in this regard (36). Identification of cross-linking sites is fairly laborious, which has probably limited the application of this approach, though a good deal of high-quality cross-linking data has been accumulated and used to position rRNA and proteins in current models of the ribosomal subunits (37).

4.3 Interpretation of footprint-type experiments

It is tempting to interpret protection and RNA modification experiments in terms of direct protein–RNA contacts; indeed the term 'footprint' suggests that the protein location on the RNA is revealed directly. The situation is much more complicated than this, as *Figure 1* suggests. There are at least three structural levels

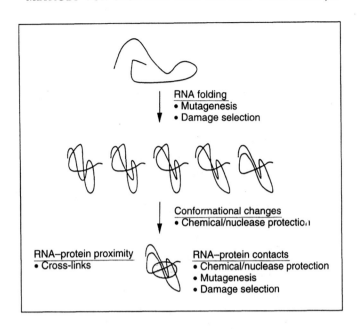

Fig. 1 Methods for studying RNA–protein complexes. Steps in the formation of an RNA–protein complex are shown, along with experimental methods sensitive to each step. Starting from the top of the figure, an RNA transcript or denatured RNA must first fold, but may adopt multiple conformations under a given set of renaturation and solution conditions. A protein may then bind to only a subset of the available conformations or may recognize an otherwise unstable conformation. See the text for discussion of the methods.

at which these experiments probe RNA–protein complexes. First, the RNA must fold into a secondary and tertiary structure. Damage selection and mutagenesis experiments may affect the kinetics of folding or the stability of the folded RNA, without actually changing any of the RNA features contacted by the protein. Second, the folded RNA may have a number of alternative conformations available, only some of which are able to interact with the protein; the most stable RNA conformation may not be the one recognized by the protein. Another way of saying this is that the protein may 'induce' a conformational change in the RNA. Any conformational change (or change in the populations of RNA conformations) will show up in footprint experiments as enhanced and repressed reactivities of the residues; these changes can be outside of the RNA region actually contacted. Thirdly, all three kinds of modification experiments (footprint, damage selection, and mutagenesis), as well as cross-linking, may detect nucleotides which are actually contacted by the protein.

How frequently do RNA conformational changes bias the results of footprint-type experiments? The available evidence suggests that substantial conformational changes are more the rule than the exception. The best understood protein–RNA interactions are the complexes between tRNAs and their cognate synthetases; although tRNAs are very stable and fairly rigid, the synthetases dramatically distort the tRNA structure, going so far as to disrupt a Watson–Crick pair and create new hydrogen bonds in one instance (38, 39). Other RNAs do not have nearly as stable and well-defined tertiary structure as tRNA and present even more opportunity for large-scale structural rearrangements upon binding protein. For example, recent nuclear magnetic resonance (NMR) studies show a drastic change in a bulge loop structure upon binding of the simple ligand arginine (40). Many ribosomal proteins probably have similar effects. Enhancements of RNA reactivity are frequently seen in footprint experiments (41) and in the case of protein L11, the set of footprinted bases does not include any of the bases at which mutations have strong effects on binding (33, 42). It thus seems best to be extremely cautious when building detailed models of protein–RNA contacts based on these kinds of experiments.

The problems inherent in the footprinting types of experiments have been apparent for some time and there have been attempts to establish principles for their interpretation. The most extensive set of protein–rRNA footprints comes from the Noller (41) laboratory, which looked at all 20 proteins assembling with the 16S rRNA. They argued that conformational changes are as likely to enhance chemical or enzymatic reactivities as protection; therefore regions showing only protection are likely to be protein-contact sites. This was a reasonable argument at the time, but data obtained since then have shown it to be weak. For instance, a hairpin loop showing only protection by the protein L11 can be changed to any sequence without affecting the protein-binding affinity (33). Many RNA-binding proteins are fairly basic, and upon binding will allow a large RNA to adopt a more compact conformation by simply reducing electrostatic repulsion between phosphates. Conformational changes probably more often lead to protection rather than enhancement of reactivity.

A more stringent criterion for determining the protein-contact site is suggested by *Figure 1*: only those nucleotides that are identified as essential by at least two different experimental techniques should be considered direct contacts. (Damage selection and mutagenesis are closely related and should not count as independent techniques.) The stringency with which this criterion is applied depends on the application. If one only needs to know what secondary structural domain within a large ribosomal RNA a protein is binding, a footprint aided by a cross-link or a few mutants may be sufficient; detailed models at the nucleotide level require more extensive results. The reliability with which RNA-contact sites are known will be discussed for individual ribosomal proteins in the following sections.

5. Ribosomal protein–RNA complexes

More than a dozen ribosomal proteins bind directly and independently to rRNA in *E. coli* and homologues of most of these are found in archaebacteria and eukaryotes as well. They vary greatly in the size and complexity of the RNA-binding site which is recognized. I have somewhat arbitrarily divided the proteins into three classes: ones recognizing relatively short, irregular helices, those that bind larger (approximately tRNA size) RNA structures likely to contain some tertiary structure, and two proteins which recognize very large domains of hundreds of nucleotides. The best studied examples of each are discussed below.

5.1 S8 and other proteins recognizing an irregular helix

The S8-binding site was localized to the central third of the 16S rRNA in early 'bind and chew' experiments and these experiments were later refined to give a single

Fig. 2 16S rRNA-binding site for S8; *E. coli* sequence and numbering are shown. Phosphates in bold face type are protected from ethylnitrosourea by S8 (98) and bases in bold face are also present in an mRNA fragment recognizing S4 (48, 49).

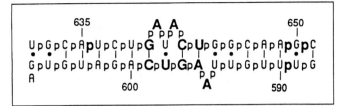

irregular hairpin, 583–605/624–653 (*Figure 2*) (43). Three cross-links have been identified between S8 and this region (44). S8 also protected several nucleotides in this helix from base-specific reagents and phosphate alkylation (45). Similar experiments carried out in another laboratory found that S8 protected nucleotides in a different region of the rRNA central domain, in addition to those within the irregular helix shown in *Figure 2* (46). Recently the S8-binding site has been formed by annealing two synthetic RNA strands to form the *Figure 2* structure; the S8-binding affinity for this fragment is comparable to its affinity for the intact 16S rRNA (47). The additional region footprinted in the Svensson *et al.* (46) experiments is therefore most likely irrelevant to S8 recognition. The different footprint results of the two groups are possibly a result of the different procedures used to extract and renature 16S rRNA.

S8 represses translation of itself and other ribosomal proteins in the *spc* operon, which it accomplishes by specifically binding to the messenger RNA. The mRNA site has been identified by deletion analysis and site-directed mutagenesis (48) and nuclease protection experiments (49). Sequences of the homologous mRNA from different bacteria (48) help define a secondary structure for the binding site which is strikingly similar to the rRNA site (*Figure 2*). The conservation of the three bulged A-residues, which may stack with each other and affect the overall helical twist of this region (45), is noteworthy.

The L25 protein also recognizes a simple, helix-like structure in the 5S rRNA. Protection studies localized the L25-binding site to a purine-rich internal loop in one arm of the rRNA, the so-called loop E (50, 51). An extended hairpin structure derived from 5S rRNA and containing this loop bound L25 (52). NMR studies of this loop suggest that there is non-canonical hydrogen bonding (53) which is perturbed (but not disrupted) by the binding of L25 (54). Gel-mobility experiments have shown that the loop has nearly the same helical twist as a standard RNA duplex, though it is bent significantly; the overall conformation is only slightly distorted by protein binding (55). The related loop sequence from eukaryotes, which has a bulged base on one strand, has also been shown to be an approximately A-form helix (56) (see Chapter 1).

Hairpin loops seem to be infrequently used as recognition features by ribosomal proteins, despite the numerous structures available. Protein S20 does protect some nucleotides in the loop and stem of a short hairpin (316–336) (57) and sequence changes within the loop and a conserved G–A pair adjacent to the base pair closing the loop dramatically reduce S20-binding affinity (58). This hairpin is a good candidate for direct recognition by S20, although mutations and footprints suggest that another stem may also be contacted (58).

5.2 L11 and other proteins requiring rRNA tertiary structures

The L11 protein recognizes a conserved structure in the large subunit rRNA which is a step up in complexity from the S8 recognition site; three helices form a 'junction' structure (*Figure 3*). An rRNA fragment binding L11, *E. coli* sequence 1052–1112, was first isolated in RNase protection experiments (59). Filter selection using L11 and partial alkaline hydrolysates of a 94 nt RNA fragment gave essentially the same result; if the helix with G1051 or U1108 is reduced in length by one or two base pairs the binding affinity is severely reduced (28). The sequence requirement could be simply explained by the protein recognizing the short helix defined by 1051–1108, but another possibility is that the junction itself is an important component of the recognition site and unravels when the 1051–1108 helix becomes too short. The latter explanation turns out to be correct, as discussed below.

The L11 protein and its recognition site have been highly conserved during evolution. Both the *E. coli* and yeast homologues will recognize mouse rRNA, for instance and the *E. coli* protein recognizes other eubacterial and archaebacterial rRNAs (60, 62). The rRNA domain from *E. coli* has been substituted for the corresponding domain in yeast (63) and *vice versa* (64), a total of 20 base changes. In both cases the ribosomes are active. Alignment of homologous L11 sequences from all three kingdoms shows several very highly conserved regions through the

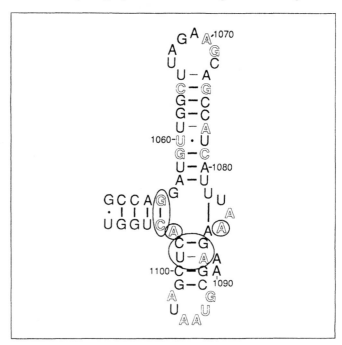

Fig. 3 23S-rRNA binding site for L11; *E. coli* sequence and numbering are shown. Open face type indicates universally conserved nucleotides (excepting mitochondrial sequences). Circled nucleotides or sets of nucleotides cannot be altered without reducing the L11-binding affinity 5-fold or more (33).

molecule, including the unusual sequence Pro–X–Pro–Pro–hph–Gly–Pro–X–Leu–Gly (X is any amino acid and hph is a hydrophobic side chain) (4). The RNA sequence also contains many universally conserved bases, distributed throughout the domain (*Figure 3*).

The highly conserved bases are prime candidates for protein recognition. To test this, a set of mutations collectively changing every base within the 1051–1108 sequence was prepared; double mutations were made where required to maintain the conserved secondary structure. L11 binding was weakened more than 5-fold for only five of the mutations (*Figure 3*); four of them are clustered around the junction and involve universally conserved bases or base pairs (33). The existence of an unusual lone base pair within the junction loop, 1082–1086, has been suggested by phylogenetic studies (41) and was confirmed as necessary for L11 recognition, though the identity of the pair is not crucial (65). It appears that the junction itself is an important component of the recognition site.

As discussed in reference to mutagenesis experiments, there are two different ways the junction could be important for protein recognition: L11 could contact the junction region directly or the junction may fold into a specific tertiary structure which maintains a needed structure elsewhere in the RNA (for instance, a specific orientation between two helices contacting the protein). If the latter is true, then the mutants disrupting protein binding should also destabilize the RNA structure. An early melting transition in the 1051–1108 fragment has been assigned to the unfolding of tertiary structure in this RNA and all four of the junction region mutations affecting L11 binding also destabilize this structure drastically, as does disruption of the 1082/1086 base pair (M. Lu and D. E. Draper, unpublished observations). From these experiments, it seems certain that the L11-recognition site is defined by tertiary structure in the junction region. Whether some of the same bases required to form the tertiary structure also contact L11 cannot be decided for sure.

Extensive footprinting experiments have also been carried out with the L11 protein. Given the mutagenesis results, it is surprising that protections are seen scattered through the RNA, particularly concentrated in the 1066–1072 hairpin loop and either side of both U1081 and the 1089–1090 bulge (41). There is virtually no overlap between the mutagenesis and footprint results. Some of the protections could be due to stabilization of otherwise weak structures (such as the base pairs adjacent to the 1089–1090 bulge, which are accessible to single-strand-specific reagents in the absence of protein) or other protein-induced conformational changes. Many of the protections are from nucleases and may be caused by protein contacts with the backbone. A recent footprinting study with hydroxyl radicals has identified protections centred on the minor groove of the 1057–1060/1078–1081 helix, as well as the vicinity of 1085 (67). It may be that L11 sits primarily on one of the helices leading into the junction and either contacts the junction as well or bridges over it in some way.

RNA structures can specifically coordinate counter-ions such as Mg^{2+} and these may potentially play a role in protein recognition. The L11 tertiary structure is stabilized by a single Mg^{2+} ion; since ions as similar as Ca^{2+} are much less effective at stabilizing the structure, the Mg^{2+} is presumably coordinating to RNA ligands in a specific way (68). A single NH_4^+ ion is also specifically bound to the tertiary structure (25). A single Mg^{2+} ion is taken up into the complex when L11 binds (28) and NH_4^+ stimulates protein binding as much as 10-fold compared to other monovalent ions (25). These ions are presumably stabilizing the RNA tertiary structure required by L11, but it is also possible that the protein coordinates or hydrogen bonds directly to the ion embedded in the RNA structure. It is well known that zinc stabilizes the structures of some RNA-binding proteins (69) and metal ions could also be specifically coordinated at the protein–RNA interface as part of the recognition mechanism.

Several other proteins also recognize junction-type structures with unusual conformations. The *E. coli* L23 and yeast L25 proteins are homologous (70) and each of the proteins protects similar structures in both the *E. coli* and yeast rRNAs (71, 72). The RNA secondary structure is sketched in *Figure 4*. Phylogenetic comparisons have suggested hydrogen bonding between two pairs of bases; the pairing would cross-link two internal loops as indicated by the heavy line in *Figure 4* (73). Compensatory base changes have been made and tested for L25-binding affinity; the experiment demonstrates that the pairing is needed for yeast L25 recognition (74). It is interesting to note that several of the bases protected from single-strand-specific chemical reagents (75) are part of phylogenetically-conserved base pairs within the junction loops (*Figure 4*). It may be that these pairings are weak and depend on bound protein for stabilization.

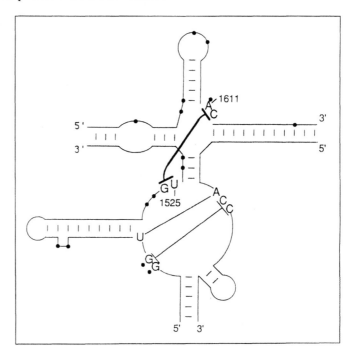

Fig. 4 rRNA-binding site for L23 (*E. coli*) or L25 (yeast). The secondary structure depicted is conserved between *E. coli* and yeast in domain III of the large subunit RNA. Additional pairings within and between junction loops, as suggested by phylogenetic comparisons, are shown by lines connecting bases or pairs of bases (6). Approximately this domain is protected from nuclease digestion by L23 (99) and bases protected from single-strand specific reagents (*E. coli* sequence) are indicated by dots (75). The pair of tertiary interactions that has been experimentally tested is indicated by heavier lines (the numbering is from the yeast sequence).

5.3 S4 recognition of a large rRNA domain

In the large subunit, L24 recognizes an unusually large rRNA domain near the 5'-terminus (76) and is probably the first protein to bind during 50S ribosome assembly (77). A small subunit protein, S4, is also the first protein to bind during 30S subunit assembly *in vivo* (78) and recognizes the rRNA 5' domain (*Figure 5*). S4 has been the better characterized of the two proteins. In RNase T_1 protection experiments, S4 was found in a complex with ~550 nt comprising the 5' third of the 16S rRNA (79); a few nicks and missing fragments were contained within the domain. Digestion experiments with carrier-bound RNase A showed that the 5' domain could be obtained as a large fragment from mild digestion of 16S rRNA alone and that S4 conferred no further protection under these conditions. This indicated that the 5' domain is already a compactly folded structure, which S4 further organizes to

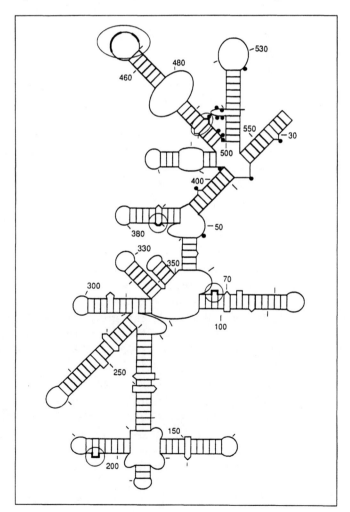

Fig. 5 16S rRNA-binding site for S4; *E. coli* sequence and numbering are shown. Dots are next to bases protected by S4 from reaction with chemical reagents (99). Circled loops are positions where deletion of bulged bases or single-base changes in a hairpin loop and a mismatch reduce S4-binding affinity by 5–10-fold (83).

confer protection from T_1 nuclease. Since S4 is only 23.3 kD, it is unlikely that it actually protects this entire domain (~150 kD) from nuclease digestion, but only protects some key points. A major role for S4 in organizing the 16S rRNA structure is suggested from its key position in the subunit assembly map (9, 78); S4 binding may well stimulate the binding of other proteins indirectly, by virtue of its effects on the rRNA structure.

The nuclease protection results were essentially duplicated by measurements of S4-binding affinity to RNA fragments with successively larger 5' and 3' deletions (15). As long as a fragment contained nucleotides 39–500, its affinity for S4 was approximately the same as that of intact 16S rRNA; smaller fragments bound at least 10-fold more weakly. The only surprise in these results is that the long range helix G27–U37/A547–C556, which defines the boundary of the 5' domain, is not essential for S4 recognition, neither is an adjacent hairpin within the 5' domain, G500–C545. These two helices are part of a five-helix junction structure; presumably the remaining three helices are able to maintain the needed RNA conformation in the absence of the other two.

Extensive 'footprint' experiments with S4 and intact 16S rRNA have been carried out and show a cluster of protections within the five-helix junction structure (57, 80). The sites are shown in *Figure 5*. This result seems to fit nicely with the nuclease protection data, which suggested that S4 should stabilize a small but critical region for folding of the RNA domain. Some caution is needed in drawing conclusions, though; some of the protections are within the G500–C545 hairpin that has been deleted without affecting S4 binding. Enhancements of chemical reactivity are also seen within the loop of this hairpin, indicating that S4 affects RNA structure outside its contact site.

If the S4-binding site is really localized to the five-helix junction region, then might a substantial internal deletion of the remaining RNA be possible? Earlier studies in which the S4-binding site could be reconstituted from two non-contiguous RNA fragments showed that at least a small deletion can be made (81). Attempts to delete large portions of the 'lower' half of the domain (*Figure 5*), using the phylogenetically conserved secondary structure as a guide, came up empty handed (15). A survey of smaller hairpin deletions within the 5' domain turned up only two which had no effect on S4 binding: the G500–C545 hairpin already mentioned and the C316–G337 hairpin (82). Further studies with mutants have identified five bulge or hairpin loops in which changes have 5–10-fold effects on S4-binding affinity (83); these are scattered throughout the 5' domain (*Figure 5*). Only one of these positions overlaps with the 'footprint' site.

How can the differing results from protection and mutagenesis experiments be reconciled? I suggest that the entire 5' domain folds or packs into a loose but fairly specific tertiary structure. Small sequence changes may alter the packing in significant ways: deletion of the two-base bulge GA205, for instance, could alter the bend and twist of the helix (84) and, thus, the ability of the whole hairpin to fit into the rest of the structure or the bases could be hydrogen bonded to a loop elsewhere in the structure. Thus, mutagenesis picks up bases distributed throughout the structure which are required for the proper folding of the RNA. The footprint results detect conformational changes induced by the protein as well as direct contacts; the protections are localized in the five-helix junctions but are scattered elsewhere. The one candidate identified in these experiments as a specific S4 contact is the A441–A493 mismatch: S4 protects it from chemical reagents and mutation to an A–U base pair weakens S4 binding 10-fold.

One implication of these studies on the S4 protein is that the 16S 5′ rRNA domain is highly organized in the absence of protein. Although the protein probably stabilizes and further organizes the structure (as suggested by its stimulation of other ribosomal-protein binding), the rRNA alone must have a specific tertiary structure which is not evident in the standard phylogenetic representations of secondary structure.

6. Protein structure

One of the most promising recent developments in ribosome structural studies has been the application of X-ray crystallography to the proteins. The *E. coli* ribosomal proteins have long been legendary for their reluctance to crystallize; only L30 and a fragment of the L7/L12 protein have yielded their structures (85, 86). More recent studies have shown proteins from thermophilic bacteria to be much better behaved, as crystals have been obtained of several *Bacillus stearothermophilus* and *Thermus thermophilus* proteins (87, 88).

A helpful development for structural studies has been plasmid expression vectors based on T7 promoters, which allow ribosomal proteins to be overexpressed in good yields in *E. coli* and sometimes purified without strong denaturants (87). This has greatly simplified the purification of large quantities of high-purity protein and the generation of isotopically labelled proteins needed for structural determination by NMR.

The *B. stearothermophilus* proteins S5 (89) and L6 (90) have been solved to ~2.7 Å resolution by X-ray crystallography. Neither of these interacts independently with ribosomal RNA, though incorporation of S5 into an assembling subunit does protect specific 16S rRNA residues (91) and L6 assembles with L3 onto an ~300 nt domain of 23S rRNA (92) where it can also be cross-linked to the RNA (93). The S5 protein folds in two domains with a deep cleft between them; a loop at the end of one domain has six conserved lysine and arginine residues and is a candidate for rRNA interaction. L6 also folds into two domains, but the two appear almost independent and form an elongated L shape nearly 60 Å across. This is suggestive of some cross-linking function, but more functional information about the protein is needed to begin speculations.

More recently the structure of a ribosomal protein that binds independently to ribosomal RNA, S17, has been solved by NMR (94). This protein protects a single extended hairpin in 16S rRNA (57). Although binding studies with rRNA fragments have not been done yet, the recognition site may be small enough for detailed NMR studies.

On evolutionary grounds, one might expect that the RNA-binding structures of ribosomal proteins would have been appropriated by many other RNA-binding proteins in the cell. The reasoning runs as follows: conservation of most primary rRNA-binding proteins across kingdom lines indicates that they were present in the first cells and were probably the first proteins to figure out how to bind RNA specifically; ergo later developing RNA-binding proteins would be able to draw on a store of RNA recognition motifs among the ribosomal proteins. A simple search for sequence motifs among RNA-binding proteins does not support this argument at all: there are certainly sequence motifs among the non-ribosomal RNA-binding proteins, such as the RNA recognition motif (RRM) and zinc finger proteins (95), but none of these is reproduced within the primary-binding ribosomal proteins. (There is a report on zinc finger ribosomal proteins (96), but they have not been conserved and are not found among the proteins independently binding RNA.) It is of course possible that the relationships are distant enough that they will become evident only when protein tertiary structures are compared and there are arguments that the RRM motif is a basic protein fold reproduced in L7/L12 and L30 (97). It is also possible that the RNA-binding capacity of ribosomal proteins would not be that useful outside of the ribosome: RNA-binding affinities of individual ribosomal proteins are usually weak (since cooperativity with other ribosomal proteins ensures irreversible binding) and they tend to bind idiosyncratic and complex structures. Until structures of more RNA-binding proteins and RNA–protein complexes are known, only speculation about potential relationships can be offered.

Acknowledgement

I am grateful to NIH grant GM29048 for support of ribosomal protein research in my laboratory.

References

1. Noller, H. F. (1991) Ribosomal RNA and translation. *Ann. Rev. Biochem.*, **60**, 191.
2. Wool, I. G., Endo, Y., Chan, Y.-L, and Glück, A. (1990) Structure, function, and evolution of mammalian ribosomes. In *The Ribosome: Structure, Function, and Evolution*. Hill, W. E., Dahlberg, A., Garrett, R. A., Moore, P. B., Schlessinger, D., and Warner, J. R. (eds). American Society for Microbiology, Washington, DC, p. 203.
3. Liljas, A. (1991) Comparative biochemistry and biophysics of ribosomal proteins. *Int. Rev. Cytology*, **124**, 103.
4. Draper, D. E. (1993) Ribosomal protein–RNA interactions. In *Ribosomal RNA: Structure, Evolution, Processing and Function in Protein Synthesis*. Dahlberg, A. and Zimmermann, R. (eds). CRC Press, Caldwell, NJ, in press.
5. Wittmann-Liebold, B., Köpke, A. K. E., Arndt, E., Krömer, W., Hatakeyama, T., and Wittmann, H.-G. (1990) Sequence comparison and evolution of ribosomal proteins and their genes. In *The Ribosome: Structure, Function, and Evolution*. Hill, W. E., Dahlberg, A., Garrett, R. A., Moore, P. B., Schlessinger, D., and Warner, J. R. (eds). American Society for Microbiology, Washington, DC, p. 598.

6. Gutell, R. R., Gray, M. W., and Schnare, M. N. (1993) A compilation of large subunit (23S and 23S-like) ribosomal RNA structures: 1993. *Nucleic Acids Res.*, **21**, 3055.
7. Neefs, J.-M., Van de Peer, Y., De Rijk, P., Chapelle, S., and De Wachter, R. (1993) Compilation of small ribosomal subunit RNA structures. *Nucleic Acids Res.*, **21**, 3025.
8. Gutell, R. R. (1993) Collection of small subunit (16S- and 16S-like) ribosomal RNA structures. *Nucleic Acids Res.*, **21**, 3051.
9. Nomura, M. and Held, W. A. (1974). Reconstitution of ribosomes: studies of ribosome structure, function and assembly. In *Ribosomes*. Nomura, M., Tissières, A., and Lengyel, P. (eds). Cold Spring Harbor Laboratory, Cold Spring Harbor, NY, p. 193.
10. Dohme, F. and Nierhaus, K. H. (1976) Total reconstitution and assembly of 50S subunits from *Escherichia coli* ribosomes *in vitro*. *J. Mol. Biol.*, **107**, 585.
11. Powers, T., Daubresse, G., and Noller, H. F. (1993) Dynamics of *in vitro* assembly of 16S rRNA into 30S ribosomal subunits. *J. Mol. Biol.*, **232**, 362.
12. Tindall, S. H. and Aune, K. (1981) Assessment by sedimentation equilibrium analysis of a heterologous macromolecular interaction in the presence of self-association: interaction of S5 with S8. *Biochemistry*, **20**, 4861.
13. Draper, D. E., Deckman, I. C., and Vartikar, J. V. (1988) Physical studies of ribosomal protein–RNA interactions. *Methods Enzymol.*, **164**, 203.
14. Schwarzbauer, J. and Craven, G. R. (1981) Apparent association constants for *E. coli* ribosomal proteins S4, S7, S8, S15, and S20 binding to 16S RNA. *Nucleic Acids Res.*, **9**, 2223.
15. Vartikar, J. V. and Draper, D. E. (1989) S4–16S ribosomal RNA complex: binding constant measurements and specific recognition of a 460 nucleotide region. *J. Mol. Biol.*, **209**, 221.
16. Revzin, A., Ceglarek, J. A., and Garner, M. M. (1986) Comparison of nucleic acid–protein interactions in solution and in polyacrylamide gels. *Anal. Biochem.*, **153**, 172.
17. Hall, K. B. and Stump, W. T. (1992) Interaction of N-terminal domain of U1A protein with an RNA stem/loop. *Nucleic Acids Res.*, **20**, 4283.
18. Abrahams, J. P., Kraal, B., and Bosch, L. (1988) Zone-interference gel electrophoresis: a new method for studying weak protein–nucleic acid complexes under native equilibrium conditions. *Nucleic Acids Res.*, **16**, 10 999.
19. Muto, A. and Zimmermann, R. A. (1974) RNA–protein interactions in the ribosome. III. Differences in the stability of ribosomal protein binding sites in the 16S RNA. *J. Mol. Biol.*, **121**, 17.
20. Schulte, C. and Garrett, R. A. (1972) Optimal conditions for the interaction of ribosomal protein S8 and 16S RNA and studies on the reaction mechanism. *Mol. Gen. Genet.*, **119**, 345.
21. Draper, D. E. and von Hippel, P. H. (1979) Measurement of macromolecular equilibrium binding constants by a sucrose gradient band sedimentation method: application to protein–nucleic acid interactions. *Biochemistry*, **18**, 753.
22. Draper, D. E. (1990) Structure and function of ribosomal protein–RNA complexes: thermodynamic studies. In *The Structure, Function, and Evolution of Ribosomes*. Hill, W., Dahlberg, A., Garrett, R. A., Moore, P. B., Schlessinger, D., and Warner, J. (eds). American Society for Microbiology, Washington, DC, p. 160.
23. Zamir, A., Miskin, R., and Elson, D. (1971) Inactivation and reactivation of ribosomal subunits: amino acyl-transfer RNA binding activity of the 30 s subunit of *Escherichia coli*. *J. Mol. Biol.*, **60**, 347.
24. Miskin, R., Zamir, A., and Elson, D. (1970) Inactivation and reactivation of ribosomal

subunits: the peptidyl transferase activity of the 50 s subunit of *Escherichia coli*. *J. Mol. Biol.*, **54**, 355.
25. Wang, Y.-X., Lu, M., and Draper, D. E. (1993) Specific ammonium ion requirement for functional ribosomal RNA tertiary structure. *Biochemistry*, **32**, 12 279.
26. Branlant, C., Krol, A., Waidada, J. S., and Fellner, P. (1973) The identification of the RNA binding site for a 50 S ribosomal protein by a new technique. *FEBS Lett.*, **35**, 265.
27. Carey, J., Cameron, V., de Haseth, P. L., and Uhlenbeck, O. C. (1983) Sequence-specific interaction of R17 coat protein with its ribonucleic acid binding site. *Biochemistry*, **22**, 2001.
28. Ryan, P. C. and Draper, D. E. (1989) Thermodynamics of protein–RNA recognition in a highly conserved region of the large subunit ribosomal RNA. *Biochemistry*, **28**, 9949.
29. Galas, D. J. and Schmidt, A. (1978) DNase footprinting: a simple method for the detection of protein–DNA binding specificity. *Nucleic Acids Res.*, **5**, 3157.
30. Ehresmann, C., Baudin, F., Mougel, M., Romby, P., Ebel, J.-P., and Ehresmann, B. (1987) Probing the structure of RNAs in solution. *Nucleic Acids Res.*, **15**, 9109.
31. Christiansen, J., Egebjerg, J., Larsen, N., and Garrett, R. A. (1990) Analysis of ribosomal RNA structure: experimental and theoretical considerations. In *Ribosomes and Protein Synthesis: A Practical Approach*. Spedding, G. (ed.). IRL Press, Oxford, p. 229.
32. Thurlow, D. L., Ehresmann, C., and Ehresmann, B. (1983) Nucleotides in 16S rRNA that are required in unmodified form for features recognized by ribosomal protein S8. *Nucleic Acids Res.*, **11**, 6787.
33. Ryan, P. C. and Draper, D. E. (1991) Recognition of the highly conserved GTPase center of 23 S ribosomal RNA by ribosomal protein L11 and the antibiotic thiostrepton. *J. Mol. Biol.*, **221**, 1257.
34. Bartel, D. P., Zapp, M. L., Green, M. R., and Szostak, J. W. (1991) HIV-1 Rev regulation involves recognition of non-Watson–Crick base pairs in viral RNA. *Cell*, **67**, 529.
35. Brimacombe, R., Maly, P., and Zwieb, C. (1983) The structure of ribosomal RNA and its organization relative to ribosomal protein. *Prog. Nucleic Acids Res. Mol. Biol.*, **28**, 1.
36. Hajnsdorf, E., Favre, A., and Expert-Bezançon, A. (1989) New RNA–protein cross-links in domains 1 and 2 of *E. coli* 30S ribosomal subunits obtained by means of an intrinsic photoaffinity probe. *Nucleic Acids Res.*, **17**, 1475.
37. Brimacombe, R., Greur, B., Mitchell, P., Osswald, M., Rinke-Appel, J., Schüler, D., and Stade, K. (1990) Three-dimensional structure and function of *Escherichia coli* 16S and 23S rRNA as studied by cross-linking techniques. In *The Ribosome: Structure, Function, and Evolution*. Hill, W., Dahlberg, A., Garrett, R. A., Moore, P. B., Schlessinger, D., and Warner, J. R. (eds). American Society for Microbiology, Washington, DC, p. 93.
38. Rould, M. A., Perona, J. J., Söll, D., and Steitz, T. A. (1989) Structure of *E. coli* glutaminyl-tRNA synthetase complexed with tRNA^{Gln} and ATP at 2.8 Å resolution. *Science*, **246**, 1135.
39. Cavarelli, J., Rees, B., Ruff, M., Thierry, J.-C., and Moras, D. (1993) Yeast tRNA^{Asp} recognition by its cognate class II aminoacyl-tRNA synthetase. *Nature*, **362**, 181.
40. Puglisi, J. D., Tan, R., Calnan, B. J., Frankel, A. D., and Williamson, J. R. (1992) Conformation of the TAR RNA–arginine complex by NMR spectroscopy. *Science*, **257**, 76.
41. Stern, S., Powers, T., Changchien, L.-M., and Noller, H. F. (1989) RNA–protein interactions in 30S ribosomal subunits: folding and function of 16S rRNA. *Science*, **244**, 783.

42. Egebjerg, J., Douthwaite, S. R., Liljas, A., and Garrett, R. A. (1990). Characterization of the binding sites of protein L11 and the L10.(L12)$_4$ pentameric complex in the GTPase domain of 23 S ribosomal RNA from *Escherichia coli*. *J. Mol. Biol.*, **213**, 275.
43. Gregory, R. J., Zeller, M. L., Thurlow, D. L., Gourse, R. L., Stark, M. J. R., Dahlberg, A. E., and Zimmermann, R. A. (1984) Interaction of ribosomal proteins S6, S8, S15, and S18 with the central domain of 16S ribosomal RNA from *Escherichia coli*. *J. Mol. Biol.*, **178**, 287.
44. Wower, I. and Brimacombe, R. (1983) The localization of multiple sites on 16S RNA which are cross-linked to proteins S7 and S8 in *Escherichia coli* 30S ribosomal subunits by treatment with 20-iminothiolane. *Nucleic Acids Res.*, **11**, 1419.
45. Mougel, M., Eyermann, F., Westhof, E., Romby, P., Expert-Bezançon, A., Ebel, J.-P., Ehresmann, B., and Ehresmann, C. (1987) Binding of *E. coli* ribosomal protein S8 to 16S rRNA. A model for the interaction and the tertiary structure of the RNA binding site. *J. Mol. Biol.*, **198**, 91.
46. Svensson, P., Changchien, L.-M., Craven, G. R., and Noller, H. F. (1988) Interaction of ribosomal proteins S6, S8, S15, and S18 with the central domain of 16S ribosomal RNA. *J. Mol. Biol.*, **200**, 301.
47. Mougel, M., Allmang, C., Eyermann, F., Cachia, C., Ehresmann, B., and Ehresmann, C. (1993) Minimal 16S rRNA binding site and role of conserved nucleotides in *E. coli* ribosomal protein S8 recognition. *Eur. J. Biochem.*, **215**, 787.
48. Cerretti, D. P., Mattheakis, L. C., Kearney, K. R., Vu, L., and Nomura, M. (1988) Translational regulation of the *spc* operon in *Escherichia coli*. Identification and structural analysis of the target site for S8 repressor protein. *J. Mol. Biol.*, **204**, 309.
49. Gregory, R. J., Cahill, P. B. F., Thurlow, D. L., and Zimmermann, R. A. (1988) Interaction of *Escherichia coli* ribosomal protein S8 with its binding sites in ribosomal RNA and messenger RNA. *J. Mol. Biol.*, **204**, 295.
50. Douthwaite, S., Garrett, R. A., Wagner, R., and Feunteun, J. (1979) A ribonuclease-resistant region of 5S RNA and its relation to the RNA binding sites of proteins L18 and L25. *Nucleic Acids Res.*, **6**, 2453.
51. Huber, P. W. and Wool, I. G. (1984) Nuclease protection analysis of ribonucleoprotein complexes: use of the cytotoxic ribonuclease α-sarcin to determine the binding sites for *Escherichia coli* ribosomal proteins L5, L18, and L25 on 5S rRNA. *Proc. Natl Acad. Sci. USA*, **81**, 322.
52. Gewirth, D. T. and Moore, P. B. (1988) Exploration of the L18 binding site on 5S RNA by deletion mutagenesis. *Nucleic Acids Res.*, **16**, 10 717.
53. Zhang, P. and Moore, P. (1989) An NMR study of the helix V–loop E region of the 5S RNA from *Escherichia coli*. *Biochemistry*, **28**, 4607.
54. Leontis, N. B. and Moore, P. B. (1986) Imino proton exchange in the 5S RNA of *Escherichia coli* and its complex with protein L25 at 490 MHz. *Biochemistry*, **25**, 5736.
55. Tang, R. S. and Draper, D. E. (1994) Bend and helical twist associated with a symmetric internal loop from 5S ribosomal RNA. *Biochemistry*, **33**, 10089.
56. Wimberly, B., Varani, G., and I. Tinoco, J. (1993) The conformation of loop E of eukaryotic 5S ribosomal RNA. *Biochemistry*, **32**, 1078.
57. Stern, S., Changchien, L.-M., Craven, G. R., and Noller, H. F. (1988) Interaction of proteins S16, S17, and S20 with 16S ribosomal RNA. *J. Mol. Biol.*, **200**, 291.
58. Cormack, R. S. and Mackie, G. A. (1991) Mapping ribosomal protein S20–16S rRNA interactions by mutagenesis. *J. Biol. Chem.*, **266**, 18 525.
59. Schmidt, F. J., Thompson, J., Lee, K., Dijk, J., and Cundliffe, E. (1981) The binding

site for ribosomal protein L11 within 23 S ribosomal RNA of *Escherichia coli*. *J. Biol. Chem.*, **256,** 12301.

60. El-Baradi, T. T. A. L., de Regt, V. H. C. F., Einerhand, S. W. C., Teixido, J., Planta, R. J., Ballesta, J. P. G., and Raué, H. A. (1987) Ribosomal proteins EL11 from *Escherichia coli* and L15 from *Saccharomyces cerevisiae* bind to the same site in both yeast 26 S and Mouse 28 S rRNA. *J. Mol. Biol.*, **195,** 909.

61. Stark, M. J. R., Cundliffe, E., Dijk, J., and Stöffler, G. (1980) Functional homology between *E. coli* ribosomal protein L11 and *B. megaterium* protein BM-L11. *Mol. Gen. Genet.*, **180,** 11.

62. Beauclerk, A. A. D., Hummel, H., Holmes, D. J., Böck, A., and Cundliffe, E. (1985) Studies of the GTPase domain of archaebacterial ribosomes. *Eur. J. Biochem.*, **151,** 245.

63. Musters, W., Gonçalves, P. M., Boon, K., Raué, H. A., van Heerikhuizen, H., and Planta, R. J. (1991) The conserved GTPase center and variable region V9 from *Saccharomyces cerevisiae* 26S rRNA can be replaced by their equivalents from other prokaryotes or eukaryotes without detectable loss of ribosomal function. *Proc. Natl Acad. Sci. USA*, **88,** 1469.

64. Thompson, J., Musters, W., Cundliffe, E., and Dahlberg, A. E. (1993) Replacement of the L11 binding region within *E. coli* 23S ribosomal RNA with its homologue from yeast: *in vivo* and *in vitro* analysis of hybrid ribosomes altered in the GTPase centre. *Eur. J. Biochem.*, **12,** 1499.

65. Ryan, P. C. and Draper, D. E. (1991) Detection of a key tertiary interaction in the highly conserved GTPase center of large subunit ribosomal RNA. *Proc. Natl Acad. Sci. USA*, **88,** 6308.

66. Lu, M. and Draper, D. E. (1993) Unpublished observations.

67. Rosendahl, G., and Douthwaite, S. (1993) Ribosomal proteins L11 and L10, (L12)$_4$ and the antibiotic thiostrepton interact with overlapping regions of the 23 S rRNA backbone in the ribosomal GTPase centre. *J. Mol. Biol.*, **234,** 1013.

68. Laing, L. G., Gluick, T. C., and Draper, D. E. (1993) Stabilization of RNA structure by Mg^{2+} ion: specific and non-specific effects. *J. Mol. Biol.*, **237,** 577.

69. Berg, J. M. (1990) Zinc finger domains: hypotheses and current knowledge. *Ann. Rev. Biophys. Biophys. Chem.*, **19,** 405.

70. Raué, H. A., Otaka, E., and Suzuki, K. (1989) Structural comparison of 26S rRNA-binding ribosomal protein L25 from two different yeast strains and the equivalent proteins from three eubacteria and two chloroplasts. *J. Mol. Evol.*, **28,** 418.

71. El-Baradi, T. T. A. L., Raué, H. A., de Regt, V. C. H. F., Verbee, E. C., and Planta, R. J. (1985) Yeast ribosomal protein L25 binds to an evolutionary conserved site on yeast 26S and *E. coli* 23S rRNA. *EMBO J.*, **4,** 2101.

72. El-Baradi, T. T. A. L., de Regt, V. H. C. F., Planta, R. J., Nierhaus, K. H., and Raué, H. A. (1987) Interaction of ribosomal proteins L25 from yeast and EL23 from *E. coli* with yeast 26S and mouse 28S rRNA. *Biochemie*, **69,** 939.

73. Gutell, R. R. and Woese, C. R. (1990) Higher order structural elements in ribosomal RNAs: pseudo-knots and the use of noncanonical pairs. *Proc. Natl Acad. Sci. USA*, **87,** 663.

74. Kooi, E. A., Rutgers, C. A., Mulder, A., Riet, J. V., Venema, J., and Raué, H. A. (1992) The phylogenetically conserved doublet tertiary interaction in domain III of the large subunit rRNA is crucial for ribosomal protein binding. *Proc. Natl Acad. Sci. USA*, **90,** 213.

75. Egebjerg, J., Christiansen, J., and Garrett, R. A. (1991) Attachment sites of primary

binding proteins L1, L2, and L23 on 23S ribosomal RNA of *Escherichia coli*. *J. Mol. Biol.*, **222**, 251.
76. Branlant, C., Widada, J. S., Krol, A., and Ebel, J.-P. (1977) RNA sequences in ribonucleoprotein fragments of the complex formed from ribosomal 23-S RNA and ribosomal protein L24 of *Escherichia coli*. *Eur. J. Biochem.*, **74**, 155.
77. Nowotny, V. and Nierhaus, K. H. (1982) Initiator proteins for the assembly of the 50S subunit from *Escherichia coli* ribosomes. *Proc. Natl Acad. Sci. USA*, **79**, 7238.
78. Nowotny, V. and Nierhaus, K. H. (1988) Assembly of the 30S subunit from *Escherichia coli* ribosomes occurs via two assembly domains which are initiated by S4 and S7. *Biochemistry*, **27**, 7051.
79. Ehresmann, C., Stiegler, P., Carbon, P., Ungewickell, E., and Garrett, R. A. (1980) The topography of the 5' end of 16-S RNA in the presence and absence of ribosomal proteins S4 and S20. *Eur. J. Biochem.*, **103**, 439.
80. Stern, S., Weiser, B., and Noller, H. F. (1988) Model for the three-dimensional folding of 16 S ribosomal RNA. *J. Mol. Biol.*, **204**, 447.
81. Mackie, G. A. and Zimmermann, R. A. (1978) RNA–protein interactions in the ribosome IV. Structure and properties of binding sites for proteins S4, S16/17 and S20 in the 16S RNA. *J. Mol. Biol.*, **121**, 17.
82. Sapag, A., Vartikar, J. V., and Draper, D. E. (1990) Dissection of the 16S rRNA binding site of ribosomal protein S4. *Biochim. Biophys. Acta*, **1050**, 34.
83. Sapag, A. and Draper, D. E. (1993). Unpublished results.
84. Tang, R. S. and Draper, D. E. (1990) Bulge loops used to measure the helical twist of RNA in solution. *Biochemistry*, **29**, 5232.
85. Wilson, K. S., Appelt, K., Badger, J., Tanaka, I., and White, S. W. (1986) Crystal structure of a prokaryotic ribosomal protein. *Proc. Natl Acad. Sci. USA*, **83**, 7251.
86. Leijonmarck, M., Ericksson, S., and Liljas, A. (1987) Structure of the C-terminal domain of the ribosomal protein L7/L12 from *Escherichia coli* at 1.7 Å. *J. Mol. Biol.*, **195**, 555.
87. Ramakrishnan, V. and Gerchman, S. E. (1991) Cloning, sequencing, and overexpression of genes for ribosomal proteins from *Bacillus stearothermophilus*. *J. Biol. Chem.*, **266**, 880.
88. Garber, M. B., Agalarov, S. C., Eliseikina, I. A., Fomenkova, N. P., Nikonov, S. V., Sedelnikova, S. E., Shikaeva, O. S., Vasiliev, D., Zhdanov, A. S., Liljas, A., and Svensson, L. A. (1992) Ribosomal proteins from *Thermus thermophilus* for structural investigations. *Biochemie*, **74**, 327.
89. Ramakrishnan, V. and White, S. W. (1992) The structure of ribosomal protein S5 reveals sites of interaction with 16S rRNA. *Nature*, **358**, 768.
90. Golden, B. L., Ramakrishnan, V., and White, S. W. (1993) Ribosomal protein L6: structural evidence of gene duplication from a primitive RNA-binding protein. *EMBO J.*, **12**, 4901.
91. Stern, S., Powers, T., Changchien, L.-M., and Noller, H. F. (1988) Interaction of ribosomal proteins S5, S6, S11, S18, and S21 with 16 S rRNA. *J. Mol. Biol.*, **201**, 683.
92. Leffers, H., Egebjerg, J., Andersen, A., Christensen, T., and Garrett, R. A. (1988) Domain VI of *Escherichia coli* 23 S ribosomal RNA. Structure, assembly and function. *J. Mol. Biol.*, **204**, 507.
93. Wower, I., Wower, J., Meinke, M., and Brimacombe, R. (1981) The use of 2-iminothiolane as an RNA–protein cross-linking agent in *Escherichia coli* ribosomes, and the localisation on 23S RNA of sites cross-linked to proteins L4, L6, L21, L23, L27, and L29. *Nucleic Acids Res.*, **9**, 4285.

94. Golden, B., Hoffman, D. W., Ramakrishnan, V., and White, S. W. (1993) Ribosomal Protein S17: characterization of the three-dimensional structure by ^{1}H and ^{15}N NMR. *Biochemistry*, **32**, 12812.
95. Mattaj, I. W. (1993) RNA recognition: a family matter? *Cell*, **73**, 837.
96. Chan, Y.-L., Suzuki, K., Olvera, J., and Wool, I. G. (1993) Zinc finger-like motifs in rat ribosomal proteins S27 and S29. *Nucleic Acids Res.*, **21**, 649.
97. Hoffman, D. W., Query, C. C., Golden, B. L., White, S. W., and Keene, J. D. (1991) RNA-binding domain of the A protein component of the U1 small nuclear ribonucleoprotein analyzed by NMR spectroscopy is structurally similar to ribosomal proteins. *Proc. Natl Acad. Sci. USA*, **88**, 2495.
98. Mougel, M., Philippe, C., Ebel, J.-P., Ehresmann, B., and Ehresmann, C. (1988) The *E. coli* 16S rRNA binding site of ribosomal protein S15: higher order structure in the absence and in the presence of the protein. *Nucleic Acids Res.*, **16**, 2825.
99. Vester, B. and Garrett, R. A. (1984) Structure of a protein L23–RNA complex located at the A-site domain of the ribosomal peptidyl transferase centre. *J. Mol. Biol.*, **179**, 431.
100. Stern, S., Wilson, R. C., and Noller, H. F. (1986). Localization of the binding site for protein S4 on 16S ribosomal RNA by chemical and enzymatic probing and primer extension. *J. Mol. Biol.*, **192**, 101.

5 | RNA–protein interactions in RNase P

VENKAT GOPALAN, SIMON J. TALBOT, and SIDNEY ALTMAN

1. Introduction

The endoribonuclease RNase P is a ubiquitous and essential ribonucleoprotein. It cleaves the 5′ terminal leader sequences of precursor tRNAs (ptRNAs) to generate mature tRNAs (*Figure 1*). Much of our current understanding of this enzyme is derived from studies of RNase P from *Escherichia coli* and *Bacillus subtilis* (1, 2). In *E. coli*, RNase P consists of an RNA subunit (M1 RNA, 377 nucleotides (nt)) and a protein subunit (C5 protein, 119 amino acid residues). M1 RNA and C5 protein are encoded by the *rnpB* and *rnpA* genes, respectively. Apart from ptRNAs, the precursors to 4.5S RNA (p4.5S RNA) (*Figure 1*) and 10Sa RNA (H. Inokuchi, personal communication) are also processed by RNase P from *E. coli in vitro* and *in vivo*. The observation that M1 RNA can catalyse the hydrolysis of ptRNAs *in vitro* even in the absence of C5 protein demonstrated that RNase P owed its catalytic potential to its RNA subunit and that the protein subunit is a co-factor (3). However, both M1 RNA and C5 protein are essential for the activity of RNase P *in vivo*. This latter observation emphasizes the importance of auxiliary role(s) played by the protein subunit *in vivo* (4, 5).

As might be expected of an RNA-binding protein, C5 protein is highly basic; indeed one-quarter of its composition consists of basic residues (6). The protein could therefore serve as a source of counter-ions which would neutralize the electrostatic repulsion between the polyanionic substrates and M1 RNA, the catalytic subunit of RNase P. However, it is apparent from several observations that M1 RNA is not only a more efficient, but also a more versatile enzyme in the presence of the protein subunit (1, 2). It is to be anticipated, then, that apart from a coulombic role, specific interactions must occur between the two subunits of RNase P. Consistent with this expectation is the finding that spermidine, a polycation, which can enhance the catalytic activity of M1 RNA (perhaps by alleviating the charge repulsion between the substrate and enzyme) is inadequate in altering the substrate specificity of M1 RNA or in influencing the choice of cleavage site in various substrates of M1 RNA (7).

In the ensuing sections we describe results that highlight the variety of effects elicited by C5 protein when present as part of the RNase P holoenzyme (that is, M1

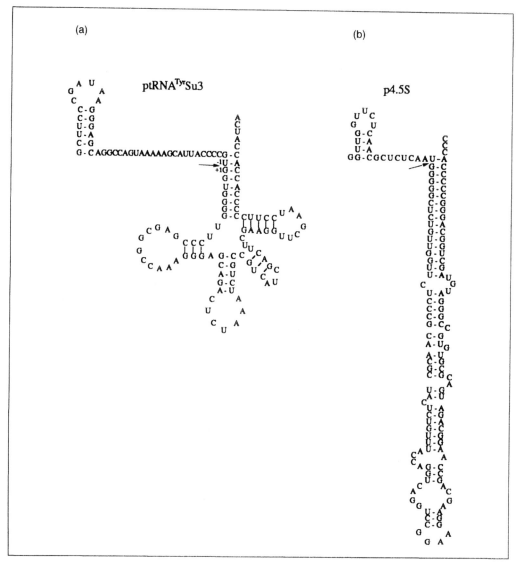

Fig. 1 Secondary structures of some substrates for *E. coli* RNase P are depicted. The arrowheads indicate the sites of cleavage by RNase P. (a) The precursor to tRNA^{Tyr}Su3 from *E. coli*. (b) The precursor to 4.5S RNA.

RNA plus C5 protein) and summarize our current understanding of the RNA–protein interactions that dictate the assembly of the holoenzyme.

2. Effects of C5 protein on catalysis by M1 RNA
2.1 Efficiency

The primary observations on *E. coli* RNase P catalysis *in vitro* included the finding that high ionic strength and high Mg^{2+} concentration obviate the need for the protein subunit during catalysis (3). However, it was soon demonstrated that the k_{cat} of the reaction catalysed by the holoenzyme was enhanced compared to that by M1 RNA alone. The various possibilities by which the protein could aid the catalytic reaction include

(1) providing functional groups for the chemical cleavage step or for chelation of metal ions that promote catalysis;
(2) accelerating the rate of cleavage by facilitating substrate binding or product release or by decreasing product inhibition (8).

Several investigations have now provided clues as to the role of C5 protein in RNase P catalysis.

A kinetic analysis of the reaction catalysed by M1 RNA in the absence and presence of C5 protein with the precursor to tRNA^{Tyr}Su3 as the substrate revealed that the protein decreased the K_m 2-fold and increased the k_{cat} at least 20-fold (Table 1). This result suggested that M1 RNA is the primary determinant in RNase P that controls substrate binding. However, several recent reports indicate that C5 protein can alter both K_m and k_{cat} in a substrate identity-dependent manner (Table 1).

Contrasting results to those obtained with ptRNA^{Tyr}Su3 were observed with two mutant derivatives of ptRNA^{Tyr}Su3 (9). These temperature-sensitive mutant substrates, ptRNA^{Tyr}Su3A2 and ptRNA^{Tyr}Su3A15, bear G2→A and G15→A changes, respectively. C5 protein was able to decrease the K_m values for these mutant substrates 6-fold compared to that with M1 RNA alone (Table 1) (9). When ptRNA^{His} served as the substrate, addition of C5 protein lowers the K_m at least 4-fold for the holoenzyme reaction compared to the reaction with M1 RNA alone

Table 1 Kinetic parameters of cleavage of various substrates by M1 RNA and RNase P

Substrate	Enzyme	K_m ($\times 10^9$) (M)	k_{cat} (min^{-1})	k_{cat}/K_m ($\times 10^{-6}$) (M^{-1} min^{-1})
ptRNA^{Tyr}Su3 [a]	M1 RNA	60	0.2	3.3
	RNase P	30	6.6	220
p4.5S [a]	M1 RNA	17 000	4.3	0.3
	RNase P	200	37	190
ptRNA^{Tyr}Su3 [b]	M1 RNA	45	0.5	10
	RNase P	13	6.6	510
ptRNA^{Tyr}Su3A2 [b]	M1 RNA	92	0.7	7.9
	RNase P	14	6.5	460
ptRNA^{Tyr}Su3A15 [b]	M1 RNA	390	0.6	1.6
	RNase P	68	5.0	74
ptRNA^{His} [c]	M1 RNA	150	6.4	43
	RNase P	39	35	900

[a] Adapted from Peck-Miller and Altman (11).
[b] Adapted from Kirsebom and Altman (9).
[c] Adapted from Kirsebom and Svärd (10).

(Table 1). The effect is even more dramatic (up to 20-fold) with certain mutant derivatives of ptRNA^{His} (10). Similarly, with p4.5S RNA as the substrate, the presence of C5 protein lowered the K_m at least 85-fold relative to the K_m of the reaction with M1 RNA alone (Table 1) (11). These data suggest that with some of the substrates, M1 RNA (by itself) can essentially achieve the most efficient conformation for recognition of substrates while for some others it requires C5 protein. Therefore, C5 protein appears to play a critical role in recognition/binding of some substrates by the RNase P holoenzyme.

At least two studies have shed light on how the protein subunit of eubacterial RNase P increases the k_{cat} of the reaction compared to the catalysis by the RNA subunit alone. In the case of B. subtilis RNase P, an increase in the concentrations of mono- and divalent cations (for example, NH_4^+ and Mg^{2+}) causes a decrease in the K_m, however, the presence of these cations has no significant effect on the k_{cat} (8). On the other hand, the holoenzyme exhibits a K_m value for ptRNA similar to and a k_{cat} value that is at least 20-fold higher than that of the RNA-alone reaction performed under optimal ionic conditions. Kinetic analyses revealed that in the RNA-alone reaction performed at high ionic strength and under multiple turnover conditions, the first round of substrate cleavage was much more rapid than subsequent rounds. In the holoenzyme reaction a burst was not observed and there was no difference in the rates of the first and subsequent rounds of cleavage. In the following scheme, these results could be interpreted as k_3 being the rate-limiting step for the reaction

$$E + S \underset{k_{-1}}{\overset{k_1}{\rightleftharpoons}} ES \xrightarrow{k_2} EP \xrightarrow{k_3} E + P$$

catalysed by the RNA subunit alone. In the presence of the protein subunit of RNase P, the rate of product release (k_3) must be comparable to the rate of chemical cleavage (k_2).

A recent study using RNase P from E. coli demonstrated that product release is the rate-limiting step in the cleavage of ptRNA^{Tyr}Su3 by M1 RNA in the absence of C5 protein (12). In the presence of C5 protein, the rate constants for the single and multiple turnover conditions were comparable; these results are identical to those obtained with the RNase P from B. subtilis. It was also found that C5 protein increases the K_i value for inhibition of RNase P by mature tRNA at least 18-fold relative to that observed with M1 RNA alone (12). Taken together, the results from these two studies reinforce the postulate that apart from the role of an electrostatic shield, the protein subunit of eubacterial RNase P stimulates k_{cat} by increasing k_3 (that is accelerating the dissociation of EP).

2.2 Versatility

Lumelsky and Altman (13) subjected the gene for M1 RNA to random chemical mutagenesis *in vitro* and analysed the M1 RNA mutants for catalytic activity in the absence or presence of C5 protein. This study revealed that in the absence of C5 protein the various mutants of M1 RNA displayed a wide range of catalytic activities, however, the presence of C5 protein was able to narrow the differences in the activities of the mutants (13). These results demonstrated that the presence of C5 protein can alleviate the deleterious effect of mutations in different parts of the M1 RNA molecule on the catalytic activity of M1 RNA, by altering either K_m or k_{cat} or both in a substrate identity-dependent manner.

Further evidence for direct participation of the C5 protein in the formation of an active conformation of M1 RNA was derived from a study in which the role of various 2'-hydroxyl groups in substrates was investigated using mixed RNA/DNA derivatives of a model substrate for RNase P from *E. coli* (14). The presence of 2'-hydroxyl groups at the -1, -2, and $+1$ positions (the cleavage site being between -1 and $+1$) as well as the first C in the 3'-terminal CCA sequence were determined to be important but not essential for catalysis. It is possible that these 2'-hydroxyl groups might serve as ligands in a tetra- or hexacoordinated Mg^{2+} complex which is presumed to generate the nucleophile for the catalytic reaction. However, the lack of 2'-hydroxyl groups in the substrate at these positions resulted in only a marginal decrease in cleavage efficiency when the holoenzyme was used instead of M1 RNA. This observation, in conjunction with the finding that the reaction with the *E. coli* holoenzyme requires only 10 mM Mg^{2+} as opposed to 100 mM Mg^{2+} needed for optimal activity in the reaction catalysed by M1 RNA alone, suggests that C5 protein could diminish the requirement for Mg^{2+}. It would be interesting to determine which amino acid residues in C5 protein reduce the requirements for Mg^{2+} and assist in maintenance of the tertiary structure of M1 RNA.

2.3 Does C5 protein influence the choice of cleavage site in substrates?

The location of the cleavage site in the precursor tRNA substrates by RNase P is dependent on several factors that include, but are not limited to, the identity of the nucleotides proximal to the cleavage site, the length of the acceptor stem, and the nucleotides involved in Mg^{2+} binding (10, 15–18).

Cleavage at the $+1$ position (see *Figure 1*) is called cleavage at the 'expected' site while cleavages at other sites are referred to as 'aberrant' cleavages. In some instances, C5 protein can correct the aberrant cleavages that are observed with the RNA subunit alone. For example, when $\Delta 92$ and $\Delta 92$, $\Delta 106$ M1 RNA were assayed with a ptRNA^{Tyr}Su3 substrate that lacked the 3' CCA sequence, cleavage was observed at sites four, five, and six nucleotides upstream from the normal site of cleavage (7). However, holoenzymes bearing these mutant derivatives of M1 RNA predominantly cleaved at the expected site ($+1$ site). If $\Delta 92$ M1 RNA is purified on a denaturing gel and renatured, the aberrant cleavages are no longer observed (C. Guerrier-Takada, personal communication). Apparently, renaturation of the RNA subunit results in a conformation that yields the expected cleavage.

In another study, various changes were introduced into a eukaryotic pre-tRNA^{Ser} from *Schizosaccharomyces pombe* and the effect on the cleavage site was analysed with M1 RNA or the *E. coli* or *S. pombe* holoenzyme as the catalyst (16). Remarkably, with all the substrates tested, the cleavage specificity at the +1 site of the RNase P holoenzyme from *S. pombe* was unaltered. With the *E. coli* holoenzyme, however, the aberrant cleavages observed when M1 RNA alone functioned as the catalyst were corrected in most but not all the substrates. Nevertheless, it is interesting however, that with every substrate examined with the holoenzyme the amount of the product generated by cleavage at the +1 site was greater than that observed with M1 RNA alone. The data from this study also suggest that, apart from the normal site (a purine–pyrimidine base pair between nucleotides +1 and +72), the holoenzyme can tolerate a pyrimidine–purine base pair on the 3′ side of the scissile bond with a few of the substrates.

The observation that C5 protein can influence the K_m for the substrates of RNase P in a substrate-dependent manner even though it does not physically interact with them (see below) is consistent with the following interpretation. In the absence of C5 protein, it is likely that M1 RNA can adopt a range of conformations including those that may lead to non-productive complexes with some of the substrates. M1 RNA, when present in a complex with C5 protein, adopts a more constrained structure (19). The resulting decrease in conformational flexibility could eliminate the conformations that lead to non-productive binding. The conformation adopted by M1 RNA in the presence of C5 protein may also be more efficient in denaturation of the acceptor stem of the substrate, an apparent requirement for cleavage to occur.

3. Gel-retardation analysis of RNA–protein interactions in the RNase P holoenzyme

Many important cellular events including pre-mRNA splicing, mRNA translation, pre-tRNA processing, and RNA localization are mediated by the specific interaction of proteins with RNA (20). Whereas DNA-binding proteins primarily recognize sequence, RNA-binding proteins recognize both sequence and structure in their target sites. To understand these phenomena, many groups have reproduced the binding reaction with purified components in order to observe the effect of changing solution conditions on the equilibrium dissociation constant (K_d). Applying this approach to DNA–protein binding systems (21, 22) and RNA–protein interactions (23–26), inferences have been made about the mechanism and molecular basis for these interactions.

We have described an RNA-binding assay that is useful in the analysis of RNA–protein interactions in the RNase P holoenzyme from *E. coli* (27, 28). Several deletion derivatives of M1 RNA were examined for their ability to bind C5 protein and participate in the cleavage of ptRNAs. The equilibrium dissociation constant for the C5 protein–M1 RNA interaction was determined under a variety of solution conditions. The kinetics of the binding reaction and the unit stoichiometry of M1 RNA and the C5 protein in the complex were also determined from the gel-retardation assay. The results were interpreted in terms of contacts made between the protein and nucleic acid and a working model for the interaction was proposed.

C5 protein exhibits a non-specific affinity ($K_d \geq 10$ nM) for RNAs other than M1 RNA, but a high specific affinity for M1 RNA ($K_d \leq 0.4$ nM) and closely related homologues (for example, P RNA from B. subtilis, $K_d = 1.2$ nM) as determined with the gel-retardation assay (27). Using a filter-binding assay, Vioque et al. (29) reported similar results for the interaction of M1 RNA and C5 protein. It is perhaps not surprising that C5 and P proteins have a high specific affinity for their heterologous RNA subunits, since it has been demonstrated previously that catalytically active hybrids can be formed from an RNA subunit and a protein subunit derived from either enzyme (3). Under conditions where C5 protein forms a specific complex with M1 RNA, a complex with H1 RNA (the RNA component of human RNase P) could not be detected even though H1 RNA can be folded into a secondary structure similar to that of M1 RNA (27, 30). Since the human RNase P holoenzyme is known to have a higher proportion of its mass provided by the protein moiety (as much as 50%, compared to the 10% for prokaryotic RNase P) (31), it is likely that the nature of and mechanism for complex formation is different in the prokaryotic and eukaryotic enzymes.

3.1 Ionic strength dependence and ion requirement for formation of RNase P holoenzyme

The affinity of C5 protein for M1 RNA determined with the gel-retardation assay increases 500-fold with increasing ionic strength in the range 0.1–1.0 M ($\partial \log K_a / \partial \log [NH_4^+] = 2.66 \pm 0.13$) (28). This is the opposite of the ionic strength dependence observed with other RNA–protein interactions such as the R17 coat protein translational operator interaction (23), tat-derived peptides binding to the TAR region of HIV-1 RNA (26), the S4–16S ribosomal RNA complex (24), and E. coli ribosomal protein L11 binding to 23S rRNA (25). In these examples the affinity of the protein for its RNA ligand decreases with increasing ionic strength, reflecting the contribution of ionic contacts to the RNA–protein interaction. Increasing the ionic strength leads to greater occupancy of the anion-binding sites on the protein. If the sites occupied by the anions and RNA are the same, the competition between these ligands will result in an increase in the K_d (that is, decreased affinity) for the RNA–protein interaction.

The salt dependence of the C5 protein–M1 RNA interaction suggests that ionic contacts may not contribute significantly to the formation of the specific complex, but rather, hydrophobic interactions may be the driving force for the complex formation. The change in entropy (ΔS) for complex formation is +6.4 cal mol^{-1} deg^{-1} at 37 °C (Table 2). This entropy change could result from configurational changes in M1 RNA or C5 protein and/or the freeing of water molecules upon binding of C5 protein to M1 RNA. Hydrophobic interactions are also associated

Table 2 Summary of kinetic and thermodynamic parameters for formation of the RNase P holoenzyme complex[a]

A.	Equilibrium binding constant, K_d	$\leq 4 \times 10^{-10}$ M
	Dissociation rate constant, k_{off}	$6.0 \pm 0.1 \times 10^{-3}$ min^{-1}
	Association rate constant, k_{on}	1.5×10^7 M^{-1} min^{-1}
	$t_{1/2}$	115 ± 10 min
B.	ΔG	-13.3 kcal mol^{-1} (-55.6 kJ mol^{-1})
	ΔH	-11.3 kcal mol^{-1} (-44.7 kJ mol^{-1})
	ΔS	$+6.4$ cal mol^{-1} deg^{-1} ($+26.8$ J mol^{-1} deg^{-1})

[a] Adapted from Talbot and Altman (28).

with a relatively small ΔH (compared to ΔG) and a positive ΔS. Since the values for the parameters for C5 protein binding to M1 RNA fit these characteristics, it is possible that hydrophobic interactions form the basis for the thermodynamic stability of the holoenzyme complex.

The effect of different monovalent ions on the formation of RNase P holoenzyme is qualitatively different from that observed with the R17 coat protein–translational operator interaction (23) or from that of the *lac* repressor binding to its operator (21). In both these examples the identity of the anion has a large effect on complex formation, presumably due to the differential interaction of the anions with the protein. In contrast, specific monovalent and divalent cations are required to promote formation of the RNase P holoenzyme. The identity of the anion shows only a slight effect on K_d (up to 3-fold), whereas different monovalent cations have a significant effect on K_d (14-fold). The identity of the divalent cation in the binding buffer has an even greater effect on the K_d (50-fold) (*Table 3*).

The RNA subunit of eubacterial RNase P has a strict requirement for Mg^{2+} for optimal activity. Cations play an important role in allowing RNAs to adopt specific structures either by general charge neutralization or by interacting with specific binding pockets formed within the RNA tertiary structure. Interestingly, the effects elicited by cations in the reaction catalysed by M1 RNA parallel those observed in the M1 RNA–C5 protein-binding assay. In the presence of Mg^{2+}, Ca^{2+}, or Mn^{2+} it appears that M1 RNA can adopt the specific three-dimensional (3-D) structure required for catalysis and C5 protein recognition. This is consistent with the report of Kazakov and Altman (32) which concluded, on the basis of metal ion-induced cleavage of M1 RNA, that specific binding sites for Mg^{2+}, Ca^{2+}, and Mn^{2+} must exist in M1 RNA. The failure to observe wild-type affinity of C5 protein for M1 RNA with other cations (for example, Co^{2+}, $K_d = 2.1$ nM; Ni^{2+} and Cu^{2+}, $K_d > 40$ nM) may reflect their inability to occupy tight binding sites for metal ions in M1 RNA (28). It is noteworthy that these cations are all ineffective in substituting for Mg^{2+} in the reaction catalysed by M1 RNA (33).

Table 3 Effects of various salts of monovalent anions and mono- and divalent cations on the K_d for formation of the RNase P holoenzyme[a]

	Salt	K_d ($\times 10^9$) (M)
Anions	NH_4NO_3	0.4
	NH_4OAc	0.4
	NH_4Cl	0.6
	NH_4Br	0.6
	NH_4I	1.2
Monovalent cations	NH_4OAc	0.4
	LiOAc	5.4
	NaOAc	2.7
	KOAc	0.6
Divalent cations	$MgCl_2$	0.4
	$MnCl_2$	0.4
	$CaCl_2$	0.4
	$CdCl_2$	0.6
	$CoCl_2$	2.1

[a] Adapted from Talbot and Altman (28).

3.2 Kinetic and thermodynamic parameters affecting the C5 protein–M1 RNA interaction

We have investigated several properties of the interaction between C5 protein and M1 RNA in the formation of the RNase P holoenzyme (27, 28). The types of experiment and methods of data analysis used have been described previously to study both DNA–protein interactions (21, 34, 35) and RNA–protein interactions (23, 24, 36). Many of these are examples of regulatory proteins which operate by binding specifically to a relatively small site in the context of a large polynucleotide. Since C5 protein requires the presence of a complex 3-D M1 RNA structure in order to bind it specifically and form the RNase P holoenzyme (27, 29), formation of the RNase P holoenzyme might be expected to follow very different kinetics from the transient nature of regulatory protein–operator complexes.

The dissociation rate for the C5 protein–M1 RNA interaction is relatively slow. At 37 °C, the half-life of the complex is ~110 min (28). This value is 100-fold higher than the dissociation rates observed for other RNA–protein complexes such as the HIV-1 tat–TAR (36) or the R17 coat protein–translational operator interactions (23). The fast on–off rates of these interactions probably reflects their function as transient translational or transcriptional regulators *in vivo*. A precise determination of the forward rate is not possible using the gel-retardation assay, but the k_{on} value of 1.5×10^7 M^{-1} min^{-1} can be calculated from the experimentally determined equilibrium constant (K_d = 0.4 nM) and the dissociation rate constant (k_{off} = 6 × 10^{-3} min^{-1}) for formation of the RNase P holoenzyme.

The value obtained for the association rate constant is somewhat lower than that expected (10^8–10^9 M^{-1} min^{-1}) for a diffusion-controlled reaction (37) between molecules the size of C5 protein and M1 RNA. Therefore, it is possible that rather than a simple bimolecular equilibrium, one or more first-order intermediate steps occur in the association pathway. An 'induced-fit' binding mechanism could be envisioned, where a conformational change in M1 RNA is required before C5 protein can bind specifically. This conformational change could be induced by an initial non-specific interaction of basic residues on C5 protein with the phospho-diester backbone, which then leads to the formation of the specific complex with high affinity. This mechanism of binding is analogous to the 'bind and slide' mechanism of DNA binding by the *lac* repressor (38, 39). In the case of the *lac* repressor the non-specific binding mode is thought to be an important mediator of the specific binding mode (38, 39).

The values of the thermodynamic parameters describing the C5 protein–M1 RNA interaction are summarized in *Table 2*. The interaction is characterized by a large favourable ΔG and ΔH and a favourable ΔS at 37°C. A favourable ΔH generally results from the establishment of relatively weak contacts promoted by van der Waals interactions and hydrogen bonds, while ionic bonds and hydrophobic forces contribute favourably to ΔS, resulting from the release of bound ions and water molecules (40). As stated earlier, examination of the ionic strength dependence of the interactions reveals that the affinity of C5 protein for M1 RNA increases with increasing ionic strength. Taken together with the thermodynamic parameters, these data suggest that hydrophobic and stacking interactions play a more significant role in complex formation than do ionic contacts, which would be expected to be disrupted upon increasing ionic strength. The favourable ΔS could also arise from conformational changes induced in M1 RNA upon binding of the C5 protein causing the release of ions and water molecules bound to the RNA and/or the protein.

4. Binding site for C5 protein on M1 RNA

The recognition site for C5 protein on M1 RNA is complex (27). In contrast with simpler systems in which a protein recognizes a short stem–loop structure (for example, R17 coat protein (23), HIV tat protein (41), and T4 gp32 (42)), the formation of RNase P holoenzyme is analogous to ribosomal protein–RNA interactions. In these latter cases, the RNA-binding proteins generally bind non-contiguous regions of their target RNA molecules, thereby complicating the analysis (see Chapter 4). Several approaches were used to identify regions in M1 RNA that are critical for binding to C5 protein. The results of studies using enzymatic and chemical probes to examine the accessibility of bases in the M1 RNA–C5 protein complex are reviewed below.

4.1 Footprint analyses

Two main regions that are protected from RNase T1 digestion span positions 82–96 and 170–200. Also, two G residues at 270 and 271 are protected (*Figure 2*). The protection was observed under conditions (high salt and limiting C5 protein) that ensure specific binding of C5 protein to M1 RNA (27).

We have also employed dimethyl sulphate (DMS), a chemical reagent that methylates the N1 of adenine, the N3 of cytosine, and the N7 of purines, to identify the residues involved in M1 RNA–C5 protein interactions. Modified residues were identified by primer extension. By analysing the modification patterns in the absence or presence of C5 protein, four distinct regions of M1 RNA interacting with C5 protein were identified: residues 41–66, 79–99, 168–198 and 266–271/282–287 (*Figure 2*) (27). These results are qualitatively consistent with the earlier study using RNase T1; however, it was observed that RNase T1 protection of residues in the stem–loop 19–61 was observed only under conditions of low ionic strength or high concentrations of C5 protein, implying that this region may be available for non-specific binding. Residues 266–271/282–287 were only weakly protected from DMS modification in the presence of C5 protein.

The differences in reactivity of nucleotides in M1 RNA to chemical and enzymatic probes in the presence of C5 protein have been used to map the footprint of the protein on M1 RNA. However, the possibility that some of these changes may be caused by conformational alterations in M1 RNA induced by C5 protein must also be considered. It is therefore important to confirm the footprint through use of cross-linking procedures.

4.2 Chemical mutagenesis

The chemical mutagenesis of M1 RNA yielded some mutants (for example, C→U93, A→G200, and C→U277) which displayed only a modest increase in their activity when C5 protein was included in the assay (13). These mutations in M1 RNA coincide with sites of interaction with C5 protein (see *Figure 2*). Some of these mutant holoenzymes (C→U93 or A→G200 plus C5 protein) also displayed temperature sensitivity; they exhibited lower catalytic activities at 42 °C relative to 37 °C. In fact a thermosensitive mutation (termed *rnpA49*) in the *rnpA* gene was one of the first mutations identified as being detrimental for RNase P activity (4). An Arg→His46 mutation in the protein subunit adversely affects holoenzyme assembly at higher temperatures and presumably leads to the thermosensitive phenotype of the A49 mutant (43, 44). One of the thermosensitive mutants of the *rnpB* gene, ts709, has two G→A substitutions, at positions 89 and 365 (45). The G→A substitution at position 89 in M1 RNA leads to a defect in the association with the protein subunit at 40 °C (45). Therefore, it is conceivable that mutations in either the RNA (for example, G→A89) or the protein subunit (for example, Arg→His46) could make the holoenzyme assembly thermosensitive due to weaker inter-

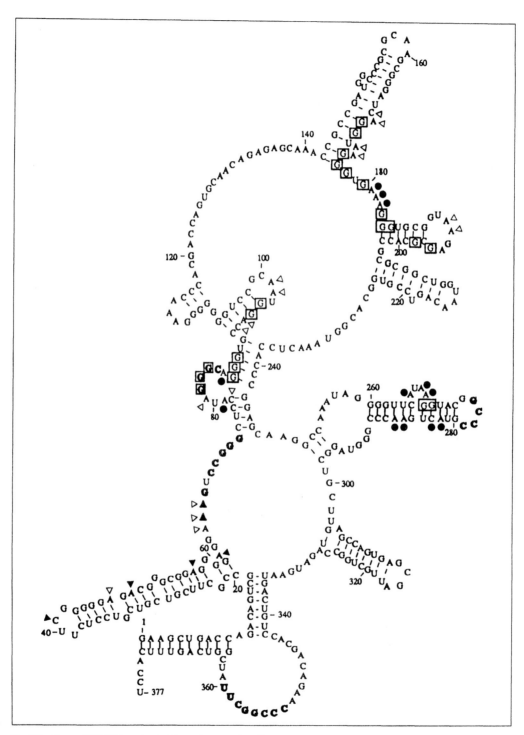

Fig. 2 Regions of M1 RNA protected from cleavage by RNase T1 (29) or from modification in the presence of C5 protein (27). Boxed regions show decreased sensitivity to RNase T1 and residues with symbols next to them indicate decreased sensitivity to DMS: ▲, strongly protected; Δ, moderately protected; ●, weakly protected. (Reprinted from reference 26 with permission.)

actions in these mutant ribonucleoprotein complexes compared to the wild-type holoenzyme.

4.3 Gel-retardation analyses

We have assessed the effect of deletions of regions in M1 RNA, determined to be important for binding to C5 protein, on the K_d values for the interaction with C5 protein. Deletion of C5 protein footprint regions Δ79–86, Δ93–104, Δ185–201, and Δ266–269 had different effects on the dissociation constant. The deletion of nucleotides Δ185–201 had the most drastic effect, a 100-fold increase, on the K_d value. This region in M1 RNA may therefore be the most important determinant for binding to the C5 protein. Deletion of the other C5 protein footprint regions resulted in modest increases of 4–6-fold relative to the wild-type K_d value of 0.4 nM (27).

In a second approach involving gel-shift assays, antisense oligodeoxynucleotides directed against certain regions in M1 RNA were used as inhibitors of holoenzyme assembly primarily to assess if these regions were critical for binding to the C5 protein (27). Subsequent to hybridization of a specific oligonucleotide to a low concentration of radiolabelled M1 RNA, the ability of the resulting M1 RNA–oligo complex to bind C5 protein was investigated. The oligonucleotides were used at a 5-fold higher concentration than their K_d values for M1 RNA to ensure ~100% formation of an M1 RNA-oligonucleotide complex. A gel-shift assay, developed in our laboratory, to detect enzyme–substrate complexes between M1 RNA and its ptRNA substrates, enabled the calculation of K_d values for the interaction of these oligonucleotides and M1 RNA. As expected, oligonucleotides complementary to regions in M1 RNA protected in footprinting experiments inhibited the formation of RNase P holoenzyme; however, their inhibitory potential, as reflected in the change in K_d values for the interaction of M1 RNA and C5 protein, was vastly different (27). For instance, the oligonucleotide complementary to nucleotides from 82–100 and 152–179 elicited a 4- and >100-fold change, respectively, in the K_d values for the interaction of M1 RNA and C5 protein.

While the increase in K_d values is a reflection of the inability of C5 protein to have access to its usual binding site, the data must be interpreted with caution since it is also possible that binding of the oligonucleotide to M1 RNA drastically alters the structure of M1 RNA and accentuates the increase in K_d values. Control experiments established that oligonucleotides complementary to other regions of M1 RNA had no significant effect on K_d values for the formation of an M1 RNA–C5 complex.

5. Demarcation of the role of different regions in M1 RNA

The availability of a gel-shift assay has enabled us to demarcate domains in M1 RNA that are important for catalysis from those that are important for binding to

C5 protein. These data have to be examined in conjunction with data obtained from other approaches such as cross-linking studies and structure–function analyses performed with M1 RNA alone.

Some of the M1 RNA mutants (for example Δ1–54, Δ1–163, Δ62–108, Δ94–204, and Δ169–377) which bind C5 protein with a $K_d > 40$ nM failed to display any catalytic activity in the absence or presence of C5 protein (27). This suggests that gross alterations in the structure of M1 RNA, in these mutants, induced by the respective deletion has led to deletion of the active site and the binding site for the protein.

Residues 83–86 have been proposed to base pair through a pseudoknot with nucleotides 275–279 (46). Deletion of nucleotides from either 79–86 or 273–281 resulted in mutants with only a 4-fold lower affinity for C5 protein, relative to the wild type. However, the effects of these mutations on catalysis are different. The M1 RNA mutant Δ79–86 was catalytically inactive both in the absence or presence of C5 protein regardless of whether the substrate tested was ptRNA^{Tyr}Su3 or p4.5S RNA (27). The finding that a UV-induced cross-link occurs between residue C92 in M1 RNA and residue -3 in the ptRNA substrate (7) suggests that deletion of nucleotides 79–86 may have eliminated or significantly altered a region important for RNase P catalysis.

The deletion mutant Δ273–281 displays wild-type activity with ptRNA^{Tyr}Su3, both in the absence and presence of C5 protein; however, it displays wild-type activity with p4.5S as the substrate, but only in the absence of C5 protein (47). The latter finding implies that when p4.5S is the substrate the increase in k_{cat} and the decrease in K_m achieved by C5 protein, relative to the reaction catalysed by M1 RNA alone, might stem from interactions of C5 protein with this region (residues 273–281) in M1 RNA. Furthermore, interactions with ptRNA substrates are severely affected by various deletions of nucleotides between 80 and 205 while cleavage of p4.5S is only partially affected. It is reasonable then to conclude that different nucleotides in M1 RNA, alone or as part of the holoenzyme, are responsible for binding to ptRNA^{Tyr}Su3 and p4.5S (27, 47).

6. Model for the formation of the RNase P holoenzyme

The results of experiments using these various approaches enable us to infer the sites in M1 RNA that are responsible for specific recognition by C5 protein. Footprinting studies have demonstrated that the contacts made by C5 protein on M1 RNA

(1) are spatially dispersed in the currently accepted secondary structure model;

(2) span contiguous single-stranded and double-stranded regions, where the bases are likely to be more accessible to amino acid contacts.

In particular, it is increasingly evident that stem–loop structures and single-stranded bulges are important sites for specific RNA–protein interactions (for example, R17 coat protein (23), ribosomal proteins S8 and S15 (48), and HIV tat protein (36)). Computer-assisted molecular modelling has been used to elaborate a tertiary structure for M1 RNA (19). This model positions the sites of interaction of C5 protein in M1 RNA to be in proximity despite their being non-contiguous in the primary and secondary structures. Furthermore, this model predicts that the substrate-binding site is oriented such that the substrate is not able to make any contacts with C5 protein. This prediction is borne out by binding data which show that the affinity of C5 protein for M1 RNA is unaltered in the presence of the substrate. Using various oligonucleotides complementary to M1 RNA in a gel-shift assay it has been possible to establish which regions in M1 RNA are most accessible to binding; it appears that regions thus identified coincide with those protected by C5 protein in footprinting experiments. Hence, these nucleotides must be positioned near the surface of the molecule.

As indicated by Vioque *et al.* (29) and S. J. Talbot and S. Altman (27) there is a considerable degree of similarity between two regions in M1 RNA that are part of the C5 protein footprint. The nucleotide sequences 81–101 and 174–194 are compared below:

```
  81–101: 5' AGGGC . . . AGGGUGCCAGGUAACG 3'
             ||||        |||||||  ||||| |
 174–194: 5' AGGGUGAAAGGGUGC . . GGUAA . G 3'
```

The residues that are identical are in bold letters. The structures of these two regions in the secondary structure model are different. Since studies on stoichiometry of the M1 RNA–C5 protein complex established that there is only one molecule of C5 protein bound to a molecule of M1 RNA (27), the possibility of two molecules of C5 protein binding separately to the two sites appears remote. The relevance of the similarity is not obvious and the individual contribution of these two regions to the binding affinity of C5 protein for M1 RNA remains to be established.

The consensus-binding motifs identified in other RNA-binding proteins (for example, heterogeneous nuclear ribonucleoproteins (hnRNPs)) (see Chapter 6; 49, 50) are absent in C5 protein. Arginine-rich sequences present in a subset of RNA-binding proteins have been demonstrated to mediate specific recognition of their RNA ligands. The alignment of sequences for seven protein subunits of RNase P from different prokaryotes reveals only a 7% identity (*Figure 3*). Despite this poor overall similarity, a core of highly conserved basic residues, as well as a number of conserved and semi-conserved hydrophobic residues, is apparent from the alignment. The presence of conserved aromatic and hydrophobic residues lends credence to the postulate that hydrophobic interactions play a critical role in holoenzyme formation. Indeed preliminary evidence from site-directed mutagenesis of the C5 protein suggests that a Phe → Ala18, Trp → Ala109 alteration leads to loss of RNase P function *in vivo*, however, the Phe → Trp18, Trp → Phe109 mutant is functional *in vivo* (V. Gopalan and S. Altman, unpublished observations).

```
M. luteus       M----LPRDR RVRTPAEFRH LGRTG-TRAG RRTVVV---S VATDPDQTRS    42
S. bikiniensis  M----LPTEN RLRRREDFAT AVRRG-RRAG RPLLVVHRLS GATDPH---A    42
E. coli         MVKLAFPREL PLLTPSCFTF VFQQP-QRAG TPQITILGRL NSLGHP----    45
P. mirabilis    MVKLAFPREL RLLTPKHFNF VFQQP-QRAS SPEVTILGRQ NELGHP----    45
P. putida       MSQ-DFSREK RLLTPRHFKA VFDSPTGKVP GKNLLILARE NGLDHP----    45
B. aphidicola   MLNYFFKKKS KLLKSTNFQY VFSNPCNKNT F-HINILGRS NLLGHP----    45
B. subtilis     MSH--LKKRN RLKKNEDFQK VFKHG-TSVA NRQF-VL--- YTLDQPENDE    43
Consensus       M..-.FPRE. RLLTP.JF.. VF..P-.RA. .....IL.R. N.LDHP----    50

M. luteus       TSPSAPRPRA GFVVSKAVGN AVTRNRVKRR IRAVV---AE QMRLPPLRDL    89
S. bikiniensis  PGESAPPTRA GFVVSKAVGG AVVRNQVKRR IRHLV---CD RL--SALPP-    86
E. coli         -RIGLT---- --VAKKNVRR AHERNRIKRL TRESF----- RLRQHELPAM    83
P. mirabilis    -RIGLT---- --IAKKNVKR AHERNRIKRL AREYF----- RLHQHQLPAM    83
P. putida       -RLGLV---- --IGKKSVKL AVQRNRIKRL MRDSF----- RLNQQLLAGL    83
B. aphidicola   -RLGLS---- --ISHKNIKH AYHRNKIKRL IRETF----- RLLQHRLISM    83
B. subtilis     LRVGLS---- --VSHK-IGN AVMRNRIKRL IRQAFLEEKE RLKEK-----    81
Consensus       -R.GL.---- --V.KK.V.. AV.RNRIKRL IR..F----- RL.Q..L...   100

M. luteus       PVLVQVR-AL PAAAEADYA- ----LLRRET -VGALGKALK PHLPAASEHA   132
S. bikiniensis  GSLVVVR-AL PGAGDADHA- ----QLARD- ----LDAALQ -RLLGGGTR-   123
E. coli         DFVVAKKGV ADLDNRALSE ALEKLWRRHC R--------- ----L-ARGS   119
P. mirabilis    DFVVLVRKGV AELDNHQLTE VLGKLWRRHC R--------- ----L-AQKS   119
P. putida       DIVIVARKGL GEIENPELHQ HFGKLWKRLA RSRPTPAVTA NSAGVDSQDA   133
B. aphidicola   DFVVIAKKNI VYLNNKKIVN ILEYIWSNYQ ---------- ----------   114
B. subtilis     DYIIIARKPA SQLTYEETKK SLQHLFRKSS LY-------- ----KKSSSK   119
Consensus       D.VV.ARK.. ..L.N..... .L..LWRR.. R--------- ----......   150
```

Fig. 3 Alignments of amino acid sequences of the homologues of C5 protein of RNase P from prokaryotic sources. The alignments were generated by Geneworks software (Intelligenetics, Inc., Palo Alto, CA). Conserved residues are boxed. (Reprinted from reference 28 with permission.)

A working model describing the interaction could involve initial binding via ionic interaction between basic amino acid side chains and the phosphodiester backbone, followed by formation of specific contacts between hydrophobic residues and the bases. This latter type of interaction would confer the high specificity observed for C5 protein binding and any disruption of M1 RNA tertiary structure would significantly affect recognition. The model proposed has a precedent in the interaction of U1 snRNP (small nuclear ribonucleoprotein) A with U1 RNA (51). It has been shown in this case that central to the RNA–protein interaction is a cluster of aromatic residues which is part of the RNP1 and RNP2 sequences of the so-called RNA recognition motif (RRM) or RNP motif which is conserved between the U1 A and U2 B" proteins (see Chapters 6 and 7; 51, 52). In addition to these aromatic residues, which are believed to stabilize the complex by van der Waals interactions, there are also a number of basic amino acid residues which are critical for the interaction (52). Although C5 protein does not have any sequence homology with the RRM identified in these snRNPs, the mechanism of RNA binding would appear to have significant similarities.

7. RNase P: evolutionary considerations

Phylogenetic comparisons have enabled the formulation of a secondary structure for the RNA subunit of eubacterial RNase P (53, 54). The absence of conserved nucleotide motifs coupled with the prevalence of covariant changes in hydrogen bonded regions validates the postulate that the emphasis during evolution has been on preservation of overall structure. The key to the catalytic ability of the RNase P from various sources must therefore lie in the tertiary structure of the RNA and/or protein components of RNase P rather than their primary structure. A model for the RNA component of RNase P, derived from this analysis, is reasonably consistent with data obtained using enzymatic and chemical probes of the structure of M1 RNA.

While structure–function analyses have been extensively performed with the RNA component of RNase P from prokaryotes, our understanding of the role of the protein subunit in RNase P catalysis is limited. In the following section we discuss the similarities and differences in RNase P from bacteria, Archaea, and Eukarya and also speculate on the possible roles of the respective protein subunits in the different enzymes.

7.1 Role of protein subunit in RNase P function

In contrast to the RNA subunits of RNase P from eubacteria, those of RNase P from archaebacteria and eukaryotes fail to exhibit catalytic activity, under the conditions examined so far, in the absence of protein *in vitro* (54). The higher protein : nucleic acid mass ratio of the latter holoenzymes, relative to their eubacterial counterparts, suggests that some of the functions originally present in the RNA component could have been relegated to the protein component during evolution. It would then become obligatory for both components of archaebacterial and eukaryotic RNase P to be present for catalysis to occur. However, since the Archaea and Eukarya diverged so early during evolution, the protein subunit requirement for RNase P catalysis to occur in archaebacteria and eukaryotes might be due to different reasons.

The archaebacterial enzymes display characteristics that are a mosaic of prokaryotic and eukaryotic features. This could be adventitious and is presumably related to the unusual habitats (that is, high salt or temperature) of archaebacteria which necessitates that their macromolecules adopt diverse characteristics. RNase P from *Sulfolobus solfataricus* exhibits a protein : nucleic acid mass ratio similar to that of eukaryotes whereas the RNase P from *Haloferax volcanii* resembles that of the *E. coli* holoenzyme (55–57). With the various archaebacterial enzymes, it is possible that the harsh intracellular conditions make it obligatory for the protein subunit to be present as a scaffold that aids in the maintenance of the active conformation of their RNA subunits.

In eukaryotes, RNase P activity has been identified in nuclei, mitochondria, and chloroplasts (58–61). The data on RNase P from chloroplasts and mitochondria seem to belie the premise that these enzymes would resemble their bacterial predecessors. However, it must be noted that the mitochondrial RNase P from yeast *Saccharomyces cerevisiae* is assembled from a mitochondrial-encoded RNA subunit and a nuclear-encoded protein subunit and that the human mitochondrial RNase P is coded for entirely by the nuclear genome. The chloroplast enzyme from spinach has been reported to be free of RNA, based on the resistance of the activity to micrococcal nuclease and on the buoyant density of the enzyme in caesium sulphate gradients (58). Further characterization of the enzyme may help in establishing the claim that a protein is solely responsible for the tRNA processing activity of RNase P in chloroplasts. The mitochondrial RNase P from *S. cerevisiae* remains the best studied eukaryotic enzyme (59, 60). Foremost among the differences between this organellar enzyme and the eubacterial enzyme is the inability of the RNA component to display activity in the absence of the protein. The protein:nucleic acid mass ratio of the yeast mitochondrial enzyme is higher than that of the eubacterial holoenzyme. This is evidenced by the lower buoyant density (as measured in caesium sulphate gradients) of the yeast mitochondrial enzyme (1.28 g cm^{-3}) compared to that of *E. coli* RNase P (1.55 g cm^{-3}).

It has recently been reported that a 105 kDa protein is required for mitochondrial RNase P activity in *S. cerevisiae* and a 100 kDa protein is associated with nuclear RNase P activity in *S. pombe* (60, 61). It is evident from these two reports that each of these proteins co-fractionate with RNase P activity. While the experiments in these studies demonstrate that these proteins may be present in a complex associated with RNase P activity, it is unclear whether these proteins, either alone or in conjunction with other proteins, interact with the respective RNA subunits to form the catalytically active holoenzyme complex. Rigorous proof that these proteins are indeed the protein subunits of RNase P will depend on the ability to functionally reconstitute these proteins with *in vitro*-transcribed RNA components of the corresponding enzymes.

Gene disruption experiments in *S. cerevisiae* revealed that interrupting the gene RPM2, which encodes the 105 kDa putative RNase P protein subunit, was not lethal and results in respiratory-deficient cells (60). An accumulation of tRNA$^{f\text{-met}}$ precursor was interpreted to imply that the RPM2 gene is necessary for RNase P activity. Since this result could also imply that the 105 kDa protein could be involved in biosynthesis of either subunit of RNase P or the holoenzyme, it leaves unresolved the question of how the 105 kDa protein contributes to RNase P function *in vivo*.

With regard to the role of the protein in eukaryotes, as in the case of yeast mitochondrial RNase P, it has been postulated that the A+U-rich nature of the RNA may require a large protein to maintain the RNA in an active conformation. Of course, the protein could be directly involved in substrate binding or in catalysis (54, 60, 62).

If the 105 kDa protein is a subunit of yeast mitochondrial RNase P then the lack of apparent similarity between it and the bacterial RNase P subunits in primary sequence and in physical size (105 versus 14 kDa) raises the possibility that an as yet unrecognized function has been accommodated by the acquisition of additional domains in the eukaryotic protein (60). Such a hypothesis does not preclude a domain of the 105 kDa mitochondrial protein subunit from forming a catalytic site with its cognate RNA component, similar to that observed with the *E. coli* holoenzyme.

7.2 Conservation of surface epitopes in various RNase P holoenzymes

The structural features that dictate the functional attributes of macromolecules are often conserved during evolution across diverse species. Accordingly, it was hypothesized that certain surface epitopes in the protein subunits of RNase P holoenzymes from different sources may be related. Immunological studies have indeed confirmed this speculation. Rabbit antibodies raised against C5 protein, apart from binding to C5, recognize the protein component of RNase P from *B. subtilis* in immunoblots and solid-phase immunoassays. Interestingly, immunoprecipitation experiments demonstrated that these antibodies can deplete the enzymatic activity from preparations of RNase P from not only *E. coli* but also human cells (63).

In an independent study, it was determined that sera from certain patients with systemic lupus erythematosus (SLE) and related rheumatic disorders contained antibodies which were effective in removing the RNase P activity from a HeLa cell extract (64). It is noteworthy that the RNA component of human RNase P was also selectively immunoprecipitated. This finding has a precedent in the ability of antibodies from sera of certain patients with autoimmune disorders precipitating other ribonucleoproteins such as Sm snRNPs. These human RNase P antibodies can also inactivate the purified holoenzyme from *E. coli* and a partially purified archaebacterial RNase P isolated from *H. volcanii* (65, 66)

Since the RNA subunits of RNase P from various sources exhibit homologies in their secondary structures and the protein subunits of RNase P from several prokaryotes exhibit conserved amino acid residues, it is possible that similar nucleoprotein interactions in the different RNase P holoenzymes enable the formation of a catalytic centre whose tertiary structure is comparable. In this regard, it is noteworthy that

(1) the RNA subunit from *Salmonella typhimurium* can form a functional RNase P complex with C5 protein (67);

(2) the protein subunits of the RNase P holoenzymes from *E. coli* and *B. subtilis* can be reconstituted with heterologous RNAs to form functional holoenzymes (3);

(3) M1 RNA exhibits catalytic activity when combined with the protein subunit of the RNase P from the archaebacterium *H. volcanii* (65).

The finding that an immunologically recognized determinant is shared among the RNase P holoenzymes from the three major phylogenetic domains, namely

Bacteria, Archaea, and Eukarya, attests to the ancient nature of RNase P. The drift in the primary sequences of the RNase P subunits, taken together with data regarding their shared antigenic determinants, provides a classical illustration of how selective pressures have preserved structural domains essential for biological function across the three major phylogenetic divisions, even though other domains in the same macromolecule have diverged and possibly acquired new functions.

7.3 Relationship of C5 protein to other RNA-binding proteins

The structures of several proteins that are components of ribonucleoprotein complexes have now been elucidated. It is pertinent for this discussion to consider a list that is comprised of three prokaryotic ribosomal proteins S5, L12 CTF (C terminal fragment), and L30 (68–70) and the RNA recognition motif (RRM) present in the human U1 snRNP and the RNA-binding domain in hnRNP C (heterogeneous nuclear RNA-binding protein C) (49, 51, 71). It has been observed that all of them have similar tertiary structures involving helices packed against one side of an antiparallel β-pleated sheet. Of course, important differences (which might be responsible for their unique functions) do exist despite preservation of overall structure. Computer algorithms predict a secondary structure consisting of three β-strands followed by a helix for the N-terminal domain of C5 protein. This is similar to the tertiary structure of the N-terminal domain reported for the ribosomal protein S5. Despite the poor similarity between the primary structures of these two proteins, if the 3-D structure of C5 protein resembles the ribosomal protein S5 it would be an interesting parallel to the similarity observed between the putative ptRNA-binding site in M1 RNA and the 'E' site in *E. coli* 23S rRNA (7). It would also prove informative as to how these RNA-binding proteins with little sequence similarity have a common structure that is well suited for the purpose of binding to RNA.

Acknowledgements

We thank our colleagues for valuable discussions and comments on this manuscript. Research in the laboratory of S.A. is supported by a grant from the US National Institutes of Health (GM19422). S.J.T. was supported by a Bristol-Myers Squibb postdoctoral fellowship and V.G. is supported by a postdoctoral fellowship from the Patrick and Catherine Weldon Donaghue Medical Research Foundation.

References

1. Altman, S., Kirsebom, L., and Talbot, S. (1993) Recent studies of ribonuclease P. *FASEB J.*, **7**, 7.
2. Pace, N. R. and Smith, D. (1990) Ribonuclease P: function and variation. *J. Biol. Chem.*, **265**, 3587.
3. Guerrier-Takada, C., Gardiner, K., Marsh, T., Pace, N., and Altman, S. (1983) The RNA moiety of ribonuclease P is the catalytic subunit of the enzyme. *Cell*, **35**, 849.

4. Schedl, P. and Primakoff, P. (1973) Mutants of *Escherichia coli* thermosensitive for the synthesis of transfer RNA. *Proc. Natl Acad. Sci. USA*, **70**, 2091.
5. Sakano, H., Yamada, S., Ikemura, T., Shimura, Y., and Ozaki, H. (1974) Temperature-sensitive mutants of *Escherichia coli* for tRNA biosynthesis. *Nucleic Acids Res.*, **1**, 355.
6. Hansen, F., Hansen, E., and Atlung, T. (1985) Physical mapping and nucleotide sequence of the *rnpA* gene that encodes the protein component of ribonuclease P in *Escherichia coli*. *Gene*, **38**, 85.
7. Guerrier-Takada, C., Lumelsky, N., and Altman, S. (1989) Specific interactions in RNA enzyme–substrate complexes. *Science*, **286**, 1578.
8. Reich, C., Olsen, G. J., Pace, B., and Pace, N. R. (1988) Role of the protein moiety of RNase P, a ribonucleoprotein enzyme. *Science*, **239**, 179.
9. Kirsebom, L. A. and Altman, S. (1989) Reaction *in vitro* of some mutants of RNase P with wild-type and temperature-sensitive substrates. *J. Mol. Biol.*, **207**, 837.
10. Kirsebom, L. A. and Svärd, S. G. (1992) The kinetics and specificity of cleavage by RNase P is mainly dependent on the structure of the amino acid acceptor stem. *Nucleic Acids Res.*, **20**, 425.
11. Peck-Miller, K. and Altman, S. (1991) Kinetics of the processing of the precursor to 4.5 S RNA, a naturally occurring substrate for RNase P from *Escherichia coli*. *J. Mol. Biol.*, **221**, 1.
12. Tallsjö, A. and Kirsebom, L. A. (1993) Product release is a rate-limiting step during cleavage by the catalytic RNA subunit of *Escherichia coli* RNase P. *Nucleic Acids Res.*, **21**, 51.
13. Lumelsky, N. and Altman, S. (1988) Selection and characterization of randomly produced mutants in the gene coding for M1 RNA. *J. Mol. Biol.*, **202**, 443.
14. Perreault, J.-P. and Altman, S. (1992) Important 2'-hydroxyl groups in model substrates for M1 RNA, the catalytic subunit of RNase P from *Escherichia coli*. *J. Mol. Biol.*, **226**, 399.
15. Burkard, U., Willis, I., and Söll, D. (1988) Processing of histidine transfer RNA precursors. *J. Biol. Chem.*, **263**, 652.
16. Krupp, G., Kahle, D., Vogt, T., and Char, S. (1991) Sequence changes in both flanking sequences of a pre-tRNA influence the cleavage specificity of RNase P. *J. Mol. Biol.*, **217**, 637.
17. Svärd, S. G. and Kirsebom, L. A. (1992) Several regions of a tRNA precursor determine the *Escherichia coli* RNase P cleavage site. *J. Mol. Biol.*, **227**, 1019.
18. Svärd, S. G. and Kirsebom, L. A. (1993) Determinants of *Escherichia coli* RNase P cleavage site selection: a detailed *in vitro* and *in vivo* analysis. *Nucleic Acids Res.*, **21**, 427.
19. Westhof, E. and Altman, S. (1994) Three-dimensional working model of M1 RNA, the catalytic RNA subunit of ribonuclease P from *Escherichia coli*. *Proc. Natl Acad. Sci. USA*, **91**, 5133.
20. Frankel, A. D., Mattaj, I. W., and Rio, D. C. (1991) RNA–protein interactions. *Cell*, **67**, 1041.
21. Riggs, A. D., Suzuki, H., and Bourgeois, S. (1970) *lac* repressor–operator interactions. I. Equilibrium studies. *J. Mol. Biol.*, **48**, 67.
22. Lohman, T. M., de Haseth, P. L., and Record, M. T. (1980) Pentalysine–deoxyribonucleic acid interactions: a model for the general effects of ion concentration on the interactions of proteins with nucleic acids. *Biochemistry*, **19**, 3522.
23. Carey, J. and Uhlenbeck, O. C. (1983) Kinetic and thermodynamic characterization of the R17 coat protein–ribonucleic acid interaction. *Biochemistry*, **22**, 2610.
24. Vartikar, J. V. and Draper, D. E. (1989) S4–16 S ribosomal RNA complex: binding

constant measurements and specific recognition of a 460 nucleotide region. *J. Mol. Biol.*, **209**, 221.
25. Ryan, P. C. and Draper, D. E. (1989) Thermodynamics of protein–RNA recognition in a highly conserved region of the large subunit ribosomal RNA. *Biochemistry*, **28**, 9949.
26. Weeks, K. M. and Crothers, D. M. (1992) RNA binding assays for Tat-derived peptides: implications for specificity. *Biochemistry*, **31**, 10 281.
27. Talbot, S. J. and Altman, S. (1993) Gel retardation analysis of the interaction between C5 protein and M1 RNA in the formation of the ribonuclease P holoenzyme from *Escherichia coli*. *Biochemistry*, **33**, 1399.
28. Talbot, S. J. and Altman, S. (1993) Kinetic and thermodynamic analysis of RNA–protein interactions in the RNase P holoenzyme from *Escherichia coli*. *Biochemistry*, **33**, 1406.
29. Vioque, A., Arnez, J., and Altman, S. (1988) Protein–RNA interactions in the RNase P holoenzyme from *Escherichia coli*. *J. Mol. Biol.*, **202**, 835.
30. Forster, A. C. and Altman, S. (1990) Similar cage-shaped structures for the RNA components of all ribonuclease P and ribonuclease MRP enzymes. *Cell*, **62**, 407.
31. Bartkiewicz, M., Gold, H., and Altman, S. (1989) Identification and characterization of an RNA molecule that copurifies with RNase P activity from HeLa cells. *Genes Devel.*, **3**, 488.
32. Kazakov, S. and Altman, S. (1991) Site-specific cleavage by metal ion cofactors and inhibitors of M1 RNA, the catalytic subunit of RNase P. *Proc. Natl Acad. Sci. USA*, **88**, 9193.
33. Guerrier-Takada, C., Haydock, K., Allen, L., and Altman, S. (1986) Metal ion requirements and other aspects of the reaction catalyzed by M1 RNA, the RNA subunit of ribonuclease P from *Escherichia coli*. *Biochemistry*, **25**, 1509.
34. Fried, M. and Crothers, D. M. (1981) Equilibria and kinetics of *lac* repressor–operator interactions by polyacrylamide gel electrophoresis. *Nucleic Acids Res.*, **9**, 6505.
35. Carey, J. (1988) Gel retardation at low pH resolves *trp* repressor–DNA complexes for quantitative study. *Proc. Natl Acad. Sci. USA*, **85**, 975.
36. Weeks, K. M. and Crothers, D. M. (1991) RNA recognition by Tat-derived peptides: interactions in the major groove. *Cell*, **66**, 577.
37. Alberty, R. A. and Hammes, G. G. (1958) Application of the theory of diffusion-controlled reactions to enzyme kinetics. *J. Phys. Chem.*, **62**, 154.
38. Berg, O. G., Winter, R. B., and von Hippel, P. H. (1981) Diffusion-driven mechanisms of protein translocation on nucleic acids. 1. Models and theory. *Biochemistry*, **20**, 6929.
39. Winter, R. B. and von Hippel, P. H. (1981) Diffusion-driven mechanisms of protein translocation on nucleic acids. 2. The *Escherichia coli* repressor–operator interaction: equilibrium measurements. *Biochemistry*, **20**, 6948.
40. Beaudette, N. V. and Langerman, N. (1980) Thermodynamics of nucleotide binding to proteins. *CRC Crit. Rev. Biochem.*, **9**, 145.
41. Weeks, K. M., Ampe, C., Schultz, S. C., Steitz, T. A., and Crothers, D. M. (1990) Fragments of the HIV-1 Tat protein specifically bind TAR RNA. *Science*, **249**, 1281.
42. von Hippel, P. H., Kavalozykowski, S. C., Lonberg, N., Newport, J. W., Paul, L. S., Stormo, G. D., and Gold, L. (1982) Autoregulation of gene expression: quantitative evaluation of the expression and function of the bacteriophage T4 gene 32 (single-stranded DNA binding) protein system. *J. Mol. Biol.*, **162**, 795.
43. Kirsebom, L. A., Baer, M. F., and Altman, S. (1988) Differential effects of mutations in the RNA and protein moieties of RNase P on the efficiency of suppression by various tRNA suppressors. *J. Mol. Biol.*, **204**, 879.
44. Baer, M. F., Wesolowski, D., and Altman, S. (1989) Characterization *in vitro* of the

defect in a temperature-sensitive mutant of the protein subunit of RNase P from *Escherichia coli*. *J. Bacteriol.*, **171**, 6862.
45. Shiraishi, H. and Shimura, Y. (1986) Mutations affecting two distinct functions of the RNA component of RNase P. *EMBO J.*, **5**, 3673.
46. Haas, E. S., Morse, D. P., Brown, J. W., Schmidt, F. J., and Pace, N. R. (1992) Long-range interactions in ribonuclease P RNA. *Science*, **254**, 853.
47. Guerrier-Takada, C. and Altman, S. (1992) Reconstitution of enzymatic activity from fragments of M1 RNA. *Proc. Natl Acad. Sci. USA*, **89**, 1266.
48. Gregory, L. and Zimmermann, T. (1986) Site-directed mutagenesis of the binding site for ribosomal protein S8 within 16 S ribosomal RNA from *Escherichia coli*. *Nucleic Acids Res.*, **14**, 5761.
49. Dreyfuss, G., Matunis, M. J., Piñol-Roma, S., and Burd, C. G. (1993) hnRNP proteins and the biogenesis of mRNA. *Ann. Rev. Biochem.*, **62**, 289.
50. Lazinski, D., Grzadielska, E., and Das, A. (1989) Sequence-specific recognition of RNA hairpins by bacteriophage antiterminators requires a conserved arginine-rich motif. *Cell*, **59**, 207.
51. Nagai, K., Oubridge, C. J., Jessen, T-H., Li, J., and Evans, P. R. (1990) Crystal structure of the RNA-binding domain of the U1 small nuclear ribonucleoprotein A. *Nature*, **348**, 515.
52. Jessen, T.-H., Oubridge, C. J., Teo, C. H., Pritchard, C., and Nagai, K. (1991) Identification of molecular contacts between the U1 A small nuclear ribonucleoprotein and U1 RNA. *EMBO J.*, **10**, 3447.
53. Brown, J. W. and Pace, N. R. (1992) Ribonuclease P RNA and protein subunits from bacteria. *Nucleic Acids Res.*, **20**, 1451.
54. Darr, S. C., Brown, J. W., and Pace, N. R. (1992) The varieties of ribonuclease P. *Trends Biochem. Sci.*, **17**, 178.
55. Darr, S. C., Pace, B., and Pace, N. R. (1990) Characterization of ribonuclease P from the archaebacterium *Sulfolobus solfataricus*. *J. Biol. Chem.*, **265**, 12927.
56. LaGrandeur, T. E., Darr, S. C., Haas, E. S., and Pace, N. R. (1993) Characterization of the RNase P RNA of *Sulfolobus acidocaldarius*. *J. Bacteriol.*, **175**, 5043.
57. Nieuwlandt, D. T., Haas, E. S., and Daniels, C. J. (1991) The RNA component of RNase P from the archaebacterium *Haloferax volcanii*. *J. Biol. Chem.*, **266**, 5689.
58. Wang, M. J., Davis, N. W., and Gegenheimer, P. (1988) Novel mechanisms for maturation of chloroplast tRNA precursors. *EMBO J.*, **7**, 1567.
59. Morales, M. J., Dang, Y. L., Lou, Y. C., Sulo, P., and Martin, N. (1992) A 105-kDa protein is required for yeast mitochondrial RNase P activity. *Proc. Natl Acad. Sci. USA*, **89**, 9875.
60. Dang, Y. L. and Martin, N. (1993) Yeast mitochondrial RNase P: sequence of the RPM2 gene and demonstration that its product is a protein subunit of the enzyme. *J. Biol. Chem.*, **268**, 19791.
61. Zimmerly, S., Drainas, D., Sylvers, L., and Söll, D. (1993) Identification of a 100 kDa protein associated with nuclear ribonuclease P activity in *Schizosaccharomyces pombe*. *Eur. J. Biochem.*, **217**, 501.
62. Nichols, M., Söll, D., and Willis, I. (1988) Yeast RNase P: catalytic activity and substrate binding are separate functions. *Proc. Natl Acad. Sci. USA*, **85**, 1379.
63. Mamula, M. J., Baer, M., Craft, J., and Altman, S. (1989) An immunological determinant of RNase P protein is conserved between *Escherichia coli* and humans. *Proc. Natl Acad. Sci. USA*, **86**, 8717.

64. Gold, H. A., Craft, J., Hardin, J. A., Bartkiewicz, M., and Altman, S. (1988) Antibodies in human serum that precipitate ribonuclease P. *Proc. Natl Acad. Sci. USA*, **85**, 5483.
65. Lawrence, N., Wesolowski, D., Gold, H., Bartkiewicz, M., Guerrier-Takada, C., McClain, W. H., and Altman, S. (1987) Characterization of RNase P from various organisms. *Cold Spring Harbor Symp. Quant. Biol.*, **52**, 325.
66. Gold, H. A. (1988) *Studies of RNase P from Hela cells*. PhD Thesis, Yale University.
67. Baer, M. and Altman, S. (1985) A catalytic RNA and its gene from *Salmonella typhimurium*. *Science*, **228**, 999.
68. Ramakrishnan, V. and White, S. W. (1992) The structure of ribosomal protein S5 reveals sites of interaction with 16S rRNA. *Nature*, **358**, 768.
69. Wilson, K. S., Appelt, K., Badger, J., Tanaka, I., and White, S. W. (1986) Crystal structure of a prokaryotic ribosomal protein. *Proc. Natl Acad. Sci. USA*, **83**, 7251.
70. Leijonmarck, M. and Liljas, A. (1987) Structure of the C-terminal domain of the ribosomal protein L7/L12 from *Escherichia coli* at 1.7 Å. *J. Mol. Biol.*, **195**, 555.
71. Hoffman, D. W., Query, C. C., Golden, B. L., White, S. W., and Keene, J. D. (1991) RNA-binding domain of the A protein component of the U1 small nuclear ribonucleoprotein analyzed by NMR spectroscopy is structurally similar to ribosomal proteins. *Proc. Natl Acad. Sci. USA*, **88**, 2495.

6 | Structure and function of hnRNP proteins

MEGERDITCH KILEDJIAN, CHRISTOPHER G. BURD, MATTHIAS GÖRLACH, DOUGLAS S. PORTMAN, and GIDEON DREYFUSS

1. Introduction

The primary transcripts of eukaryotic RNA polymerase II, termed hnRNAs or pre-mRNAs, undergo a complex and highly-regulated series of events in the nucleus as they mature into functional, cytoplasmic mRNAs (1, 2). Prominent among these events are the excision of introns, generation of the 5'- and 3'-terminal structures and export of the processed mRNA into the cytoplasm. Upon the emergence of the nascent transcript from the transcription complex and throughout its nuclear lifetime, pre-mRNA exists as a complex with a group of nuclear pre-mRNA-binding proteins and it is in this form that the RNA undergoes nuclear RNA-processing reactions. The RNA-binding proteins which associate with nuclear pre-mRNAs but are not stable components of other nuclear structures (for example, snRNPs (small nuclear ribonucleoprotein particles)) are known as hnRNP proteins (3). The structure formed upon the association of these proteins with a pre-mRNA molecule is known as an hnRNP complex (reviewed in Dreyfuss *et al.* (4, 5)). hnRNP complexes, therefore, occupy a central position in the pathway of gene expression and the study of their structure and function is of great interest.

Early studies of hnRNP proteins were performed with hnRNP complexes and proteins isolated from nuclear extracts using sedimentation methods. Though this technique has inherent shortcomings (for example, exposure to RNases, proteases, and centrifugal drag), its use successfully identified the so-called 'core hnRNP proteins', those of the A, B, and C groups (see *Figure 1* and *Table 1*), which are among the major nuclear pre-mRNA-binding proteins (for example Wilk *et al.* (6)). More recent approaches have identified proteins which are associated with nuclear poly(A)$^+$ RNA *in vivo* by using UV light to covalently cross-link the protein to the RNA of these complexes in living cells and subsequently examining the proteins which have become cross-linked to the RNA. While this technique also has shortcomings (for example, heterogeneity in the abilities of RNA-binding proteins to cross-link to RNA and that unpolyadenlyated RNA cannot be examined), it

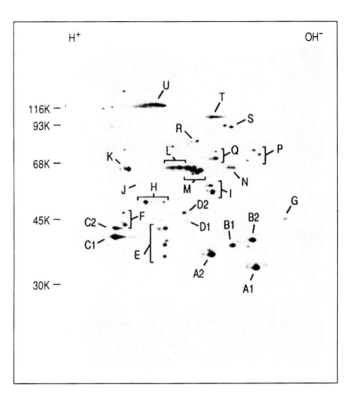

Fig. 1 Protein composition of hnRNP complexes immunopurified with a monoclonal antibody, 4F4, to hnRNP C1/C2. The hnRNP complexes were immunopurified from the nucleoplasm of [^{35}S]methionine-labelled HeLa cells (9, 21). The proteins were separated by non-equilibrium pH gradient gel electrophoresis (NEPHGE) in the first dimension and by SDS–PAGE in the second dimension and visualized by fluorography. Figure courtesy of S. Piñol-Roma.

allowed the additional identification of hnRNP proteins of molecular weight 53, 68, and 120 kDa (7, 8). The third and most sensitive technique for analysing proteins bound to pre-mRNA was made possible by generating monoclonal antibodies against UV-cross-linked RNA-binding proteins. By immunoprecipitating from nucleoplasm under conditions which preserve the integrity of the pre-mRNA, these reagents can in a specific and sensitive manner isolate RNA–protein complexes containing an assortment of ~20 major nuclear proteins (9). These proteins, denoted hnRNP A1 through to hnRNP U, are shown in *Figure 1* and listed in *Table 1*. These proteins constitute the major pre-mRNA-binding or hnRNP proteins in HeLa cells.

The study of the structure and function of this large complement of proteins has proven interesting and revealing. hnRNP proteins are largely modular in nature, containing regions which function as RNA-binding domains and others as auxiliary domains (see below; 4, 10, 11). Most of the hnRNP proteins bind RNA directly (rather than associate with the complex solely through protein–protein interaction) and, moreover, many of them exhibit sequence specificity in this binding (12–15). These findings have led to the suggestion that the composition of hnRNP complexes is not fixed but rather is transcript specific, an idea which has recently been supported experimentally, both *in vitro* and *in vivo* (16–18).

With several notable exceptions, the hnRNP proteins exhibit a uniform and

Table 1 The major hnRNP proteins

Protein	Molecular weight (kDa)[a]	pI[b]	Structural motifs	Comments
A1	34	9.0–9.1	2 × RNP-CS, RGG box	Contains DMA[c], may be phosphorylated
A2/B1	36/38	8.4–8.8	2 × RNP-CS, RGG box	Contains DMA[c], B1 identical to A2 except for an 11 amino acid insert
B2	39	9.0	–	
C1/C2	41/43	5.9	RNP-CS–AspGlu	Phosphorylated, avid binding to poly(U), nuclear localization signal, C2 identical to C1 except for a 13 amino acid insert
D	44–48	7.7–7.8	–	
E	36–43	7.3	–	Avid binding to poly(G)
G	43	9.5	RNP-CS, RGG box	
F/H (142)	53/56	6.1–7.1	3 × RNP-CS	Avid binding to poly(G)
I	59	8.5	4 × RNP-CS[d]	Identical to PTB
K/J	62/68	6.1–6.7	3 × KH, RGG	Avid binding to poly(G)
L	68	7.4–7.7	4 × RNP-CS[d]	
M	68	7.8–8.2	4 × RNP-CS[d]	
N	70	8.7–8.9	–	
P	72	9.0	–	Avid binding to poly(A)
Q	76–77	8.3	–	
R	82	8.0	–	
S	105	8.8	–	
T	113	8.4	–	
U	120	6.6–7.2	RGG box	Phosphorylated, nuclear localization signal

For references not listed, see Dreyfuss et al. (4) and references therein.
[a] Molecular weight estimated from SDS–polyacrylamide gel electrophoresis.
[b] pI estimated from isoelectrofocusing gel electrophoresis.
[c] N^G,N^G-Dimethylarginine.
[d] Non-canonical RNP consensus sequence.

diffuse nucleoplasmic distribution, consistent with their roles as pre-mRNA-binding proteins (4). However, recent studies have revealed a surprising property, namely, that some of these proteins continuously shuttle between the nucleus and the cytoplasm, perhaps as part of an RNA–protein complex. This has raised the interesting possibility that these proteins may be involved in the transport of mature mRNA out of the nucleus (19).

Studies in non-human vertebrate cells have shown that there is a high degree of conservation, with respect both to the assortment of hnRNP proteins as well as their individual sequences, in these organisms (20–28). However, examination of hnRNP complexes from more distant organisms, particularly *Drosophila melanogaster*, has revealed some interesting differences. *Drosophila* lacks the diversity of vertebrate hnRNP proteins, containing instead a smaller set of nuclear pre-mRNA-binding proteins, most of which are similar to the human A/B family (that is, 2×RBD–Gly; see below; 29–32). Notably, an important role in embryonic development has recently been demonstrated for a particular *Drosophila* hnRNP protein, hrp40.1, which has been shown to be essential for normal dorsal–ventral axis development of the *Drosophila* oocyte (33–36). Studies in the fungi *Saccharomyces cerevisiae* and *Schizosaccharomyces pombe* have recently identified promising candidates for yeast hnRNP proteins, such as Pub1p, Nab2p, and Hrp60 (35–37; J. P. O'Conner, M. J. Matunis, and G. Dreyfuss, submitted), however, further studies will be required in these organisms to illuminate their full complement of pre-mRNA-binding proteins and their functions.

2. Structural motifs in hnRNP proteins

The cloning and sequencing of cDNAs for most of the hnRNP proteins has revealed a modular structure and allowed for the identification of several different motifs. Human hnRNP proteins characterized thus far contain one or more RNA-binding modules and at least one other region whose function is not fully known and which is termed the 'auxiliary domain'. RNA-binding motifs that have been identified to date are the RNP consensus sequence, the RGG box, and the KH domain. Identification of these motifs as RNA-binding domains has provided a predictive measure for this activity in proteins of unknown function that contain such domains. The auxiliary domains may mediate protein–protein interactions and specify intracellular localization. The following section summarizes our knowledge of the different RNA-binding domains as well as some of the auxiliary domains found in hnRNP proteins.

2.1 The RNP consensus sequence (RNP-CS) RNA-binding domain (RBD)

The most widely found and best characterized RNA-binding motif to date is the evolutionarily conserved RNP consensus sequence RNA-binding domain (RNP-CS RBD; 10) of 90–100 amino acids and is also referred to as the RNA recognition motif (RRM) (38, 39), RNP 80 (40), or RNP motif (4, 11). This domain or multiple non-identical copies thereof is found in many pre-mRNA-binding proteins (reviewed in Dreyfuss *et al.* (4, 5)) and a host of other RNA-binding proteins in animals, plants, fungi and cyanobacteria. Proteins with this domain localize to the nucleus, the nucleolus, the cytoplasm, and to cytoplasmic organelles and they bind to a wide variety of RNAs, including pre-mRNA, pre-rRNA, snRNA, and mRNA (Dreyfuss *et al.* (4) and references within). The number, the diverse localization, and the surprising multitude of functions of these proteins makes the RNP-CS proteins an interesting and the largest RNA-binding protein family.

The RNP-CS RBD, though only moderately conserved overall, is readily identified by its highly conserved hallmark RNP 1 octamer and RNP 2 hexamer sequences with their characteristic array of aromatic amino acids (Phe, Tyr) and a number of interspersed, well-conserved, and mostly hydrophobic amino acids. The RNP 1 and RNP 2 consensus sequences are separated by ~ 30 amino acids and the least conserved residues within the RNP-CS are those just preceding RNP 1 (see *Figure 2*;

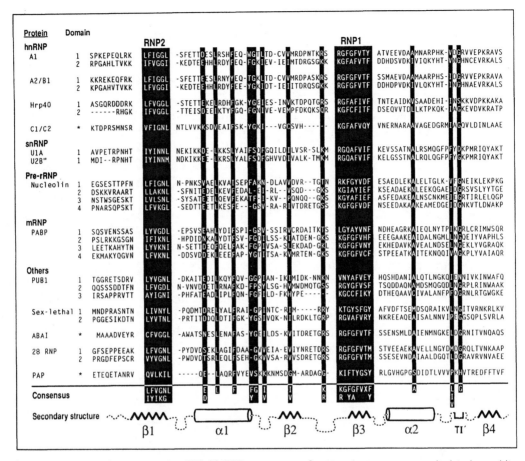

Fig. 2 Alignment of representative RNP-CS RBD sequences. Conserved sequences are depicted as white residues on a black background. The RNP1 and RNP2 consensus sequences are indicated on the top of the alignment. For proteins with more than one RNP-CS RBD the individual RBDs are numbered and * indicates proteins with only one RNP-CS RBD. The conserved positions are depicted at the bottom of the figure along with the approximate position of the conserved secondary structural elements (β1–β4) β-strands, (α1–α2) α-helices, (TI′) type I′ tight turn, and the stippled lines represent loops and extended N- and C-terminal regions. For a detailed analysis of the 3-D structure of the RNP-CS RBD refer to Nagai et al. (43), Hoffman et al. (44), and Wittekind et al. (45). Sequences are taken from human hnRNP A1 (A1; 119), human hnRNP A2/B1 (A2/B1; 68), D. melanogaster Hrp40 (Hrp; 30, 40), human hnRNP C1 and C2 (C1/C2; 68, 120), snRNP U1A protein (U1A; 121), snRNP U2B″ protein (U2B″; 121), hamster nucleolin (122), S. cerevisiae poly(A)-binding protein (PABP; 46, 47), S. cerevisiae poly(U)-binding protein (PUB1; 35, 36), D. melanogaster sex-lethal gene product (sex-lethal; 123, 124), maize absissic acid-induced protein (ABAI; 125), chloroplast 28 kDa RNP protein (28 RNP; 126), and S. cerevisiae poly(A) polymerase (PAP; 127). For a comprehensive listing and alignment refer to Bandziulis et al. (10), Kenan et al. (39), and Birney et al. (143).

10, 38–40). Several hnRNP proteins, such as hnRNP I (41), L (18), and M (42) contain multiple copies of non-canonical RNP-CSs. The overall structure of these domains is likely to be very similar to the structure of the canonical RNP-CS RBD even though the RNP 1 and RNP 2 consensus sequences are less well conserved (18, 39, 41).

The three-dimensional (3-D) structures of two different RNP-CS RBDs have recently been determined by X-ray crystallography and by multidimensional nuclear magnetic resonance (NMR). The N-terminal RBD of the U1 snRNP A protein (U1A; see Chapter 7; 43, 44) and the single hnRNP C1/C2 RBD (45) fold into a βαββαβ structure, forming an antiparallel four-stranded β-sheet which is packed against two perpendicularly oriented α-helices which are positioned on one side of the β-sheet (*Figures 2* and *3*). The RNP1 and RNP2 consensus sequences are juxtaposed on the two central β-strands (β3 and β1) of the folded domain. The β-sheet exhibits a pronounced right-handed twist and a number of other structural irregularities, notably β-bulges in β1 and β4, which are likely to be common features of all RNP-CS RBDs (45). Structural refinement studies revealed a new secondary structural element between α2 and β4 of the hnRNP C RBD, a short two-stranded antiparallel β-sheet with a type I′ tight turn which is likely to be a general structural feature of many RNP-CS RBDs (M. Wittekind, M. Görlach, G. Dreyfuss, and L. Mueller, in preparation). The side chains of the C-terminal aromatic residue of RNP 1 and of other highly conserved mostly aromatic residues contributed by α1 and α2 appear to play a crucial role in positioning the two α-helices of the RBD perpendicular relative to each other. This feature, therefore, is probably one of the major structural elements causing the particular fold of the RNP-CS RBD (Wittekind *et al.* (45) and references within).

A major question with respect to the RNA-binding activity of the RNP-CS RBD centres around the identification of the amino acids which constitute the RNA-

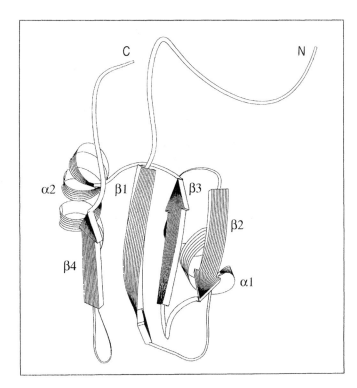

Fig. 3 A schematic representation of the 3-D structure of the hnRNP C RNP-CS RBD is shown with its four-stranded antiparallel β-sheet (β1–β4) packed against the two α-helices (α1 and α2). For details see Wittekind *et al.* (45).

binding site. Binding studies showed that the RBD is a bona fide RNA-binding domain (9, 10, 12, 13, 38, 46–57) which binds RNA with affinities ranging from 10^{-11} to 10^{-7} M (14, 15, 54, 57, 58; see also section below). Delineation studies revealed that the actual physical limits of an RBD may extend beyond the 90 amino acid region of the domain homology (14, 15, 38, 52, 54). Early UV-cross-linking experiments, mutational analysis, and peptide-binding studies indicated that amino acids located on the four-stranded β-sheet are important for RNA binding. In particular, the RNP 1 and RNP 2 consensus sequences including their solvent-exposed aromatic amino acid residues were found to play a central role in binding RNA (43, 50, 52, 57, 59, 60). A recent NMR study of the hnRNP C protein RBD yielded the first direct physical evidence about the location and dimensions of the RNA-binding surface of an RBD. A large number of candidate residues for the interaction with the RNA ligand reside in the four-stranded β-sheet and notably in the contiguous N- and C-terminal regions (61). These results suggest that the RNA engages with the RBD on an open 'platform' rather than in a binding 'crevice', thereby leaving the bound RNA exposed and apparently accessible for interactions with other RNAs or RNA-binding proteins (see Görlach et al. (61) for a discussion). This interpretation is consistent with the finding that the hnRNP C RNP-CS RBD promotes the annealing of complementary RNAs which may be a result of increased accessibility of the RNA strands upon protein binding (see section 4; 62). It is interesting to note that nucleic acid-binding domains of other, unrelated proteins also contain β-sheet structures, for example transcription factor IID (TFIID), bacteriophage MS2 coat protein, and *Bacillus subtilis* cold shock protein (63–66) suggesting that nucleic acid–protein interaction through β-sheet structures may be a general theme.

Though many of the RNA–protein interactions of the RNP-CS RBD are governed by amino acids within the conserved antiparallel four-stranded β-sheet, the specificity for their RNA ligands requires also residues outside of that conserved structural element. For example, the specificities of U1A and the U2 snRNP B" protein (U2B") largely reside in the loop connecting β2 and β3 (see Chapter 7; 53, 67). In the case of the hnRNP C1/C2, which does not contain this loop, RNA-binding specificity resides within the ten amino acids immediately C-terminal to the 94 amino acid minimal RBD (15). Deleting these ten residues removes the binding specificity but does not abolish the general binding activity of that RBD for RNA. The interesting conclusion from these findings is that the elements within an RBD which carry general RNA-binding activity (that is, the antiparallel four-stranded β-sheet) and those that confer specificity to the RBD (that is, the loop connecting β2 and β3 or sequences immediately C-terminal to the RBD) are distinct. In addition, several hnRNP proteins (hnRNP A2/B1 and hnRNP C1/C2) differ from each other only by small peptide inserts (68) adjacent to the RBD, suggesting that those amino acids may modulate sequence-specific RNA binding. The RNA ligands for RNP-CS RBD proteins are structurally diverse, for example stem–loop RNA for snRNP U1A or ssRNA for hnRNP C (12, 15, 52, 61). The modular architecture of the RNP-CS proteins provides a rationale for how different members of this class of proteins can bind such diverse RNA ligands specifically, despite the fact that they contain such highly conserved elements. A somewhat extreme case for the requirements of sequences outside the RBD for specific RNA-binding are the specificity requirements of U2B", which by itself binds RNA non-specifically but needs the contact of another protein, the U2 snRNP A' protein (U2A'), in order to specifically recognize its cognate RNA ligand (see Chapter 7; 53, 67).

In proteins with multiple RNP-CS RBDs, specificity appears to be contributed by analogous sequences outside of and connecting the RBDs. For example, in the poly(A)-binding protein (PABP) which contains four RBDs, the first two RBDs together exhibit a strong preference for binding poly(A) (54, 55), thereby reflecting the specific binding activity of the entire protein (69). The sequences outside of and connecting the first two RBDs of the PABP are absolutely conserved from yeast to man (55) and it is possible that they contribute to the poly(A)-binding specificity to the PABP. In hnRNP A1 the connecting sequences between its two RBDs also appear to play a major role in determining the RNA-binding specificity of the contiguous two domains of that protein (14).

2.2 The RGG box

The RGG box, initially identified as the RNA-binding domain of hnRNP U, is a conserved motif which has been subsequently found in numerous RNA-binding proteins (70). It is a region of ~20–25 amino acids containing several arginine–glycine–glycine (RGG) tripeptide repeats interspersed with other, often aromatic amino acids (*Figure 4*). Fusion of RGG box-containing regions from several proteins to non-nucleic acid-binding proteins imparts RNA-binding activity to the hybrid proteins (37, 70). Furthermore, the hnRNP U RGG box alone can bind RNA homopolymers and single-stranded DNA (M. Kiledjian and G. Dreyfuss, unpublished observations), demonstrating that the RGG box is an autonomous RNA-binding domain. RGG boxes of different proteins possess unique nucleic acid-binding characteristics. For example, the hnRNP U RGG box retains differential ribohomopolymer-binding properties (70) while the RGG box of nucleolin appears to bind RNA indiscriminately (71, 72). In nucleolin, the RGG box is essential for high-affinity RNA binding while the specificity of binding resides in amino acids outside of this region (that is, the RNP-CSs; 72). In contrast, the RGG box is the main (if not only) RNA-binding component of hnRNP U and it confers both specific and high-affinity RNA binding (70).

Little is known about the structure of the RGG box and its interaction with RNA. Ghisolfi *et al.* (71) have shown that the nucleolin RGG box can unstack RNA bases and unfold RNA secondary structures. Spectroscopic analysis and subsequent molecular modelling predicts that the consecutive RGGF repeats in the nucleolin RGG box form a 'β-spiral' structure rather than an unstructured random coil (71, 73, 74). A more detailed picture of the RGG box structure awaits NMR and X-ray crystallographic analysis.

Protein																								
hnRNP U		M	R G G	N F		R G G				A P G N	R G G	Y N	R R G	N
SSB-1		G	R G G	F		R G G		F	R G G Y	R G G	F R G R G	N
Fibrillarin		G	R G G	F G D	R G G			R G G		R G G	F G G G R	G
Nucleolin		G	R G G	F G G	R G G		G R G G		R G G	F G G R G	R
hnRNP A1		G	R G G	N F S G	R G G			F G G S	R G G G	G Y G G	T
hnRNP G		L	R G G		R G G			S G G T	R G P P	S R G G	H
EWS	1	G	R G G	F D	R G G	M S	R G G		R G G		G R G G	M
EWS	2	G	R G G	P G G M	R G G			R G G	L M D		R G G	P
EWS	3	G	P G G	M F	R G G			R G G	D R G G	F	R G G	R
TLS	1	G	R G G		G	R G G		R G G	M G G S D	R G G	F
TLS	2	N	R G G	G N G	R G G	R G	R G G		P M G		R G G	Y
TLS	3	D	R G G	Y R G	R G G	D	R G G	F R G G			R G G	G
FMR-1		L	R R G	D G R R	R G G	G	G R G	Q G G R		G R G G	G
RNA helicase		D	F G H	S		G R G G		R G G	D R G G	D D	R R G	G
VASA		F	R G G	E G G F	R G G		Q G G S	R G G		Q G G	S
GAR-1	1	F	R G G	N		R G G		R G G	F R G G	F	R G G	R
GAR-1	2	G	R G G	S		R G G		F R G G	R G G S S	R G G	G
GAR-1	3	F	R G G	S		R G G	S F	R G G	S R G G	S	R G G	F
NLP3/NOP3		N	R G G	F R G	R G G		F R G G	F R G G	F		R G G	F
NAB2		Q	R G G	G A V G	K N R			R G G	R G G	N	R G G	R
HSV-1 LRP1		P	R G S	R G	R G G	R G	R G G		R G G		G R R G	R
HSV-2 ORF-1		R	R G G			R G G		Q V C G	R G G			R R G	G
PRV ORF-3		G	G R G			R G G			R G G	R G G	R G R G G	G
EBNA 1		G	R G G	S	G	G R G		R G G S	G G R		G R G G	S
Consensus		G	R G G	N F / S		G R G G		R G G	R G G	F	G R R G G / G

Fig. 4 Alignment of representative RGG box sequences. Conserved residues are depicted in white on a black background. For proteins with more than one RGG box the individual RGG boxes are numbered. A consensus sequence derived from conserved positions of the RGG boxes are depicted at the bottom of the figure. The sequences are taken from hnRNP U (70), *S. cerevisiae* single-stranded DNA-binding protein (SSB; 128), human fibrillarin (129), hnRNP A1 (119), human hnRNP G (130), human Ewing's sarcoma-related gene product (EWS; 82), human myxoid liposarcoma-related gene product (TLS; 84), human FMR-1 protein (80), *D. melanogaster* RNA-helicase (131), *D. melanogaster* vasa gene product (132), *S. cerevisiae* GAR 1 protein (133), *S. cerevisiae* NPL3/NOP3 protein (134, 135), *S. cerevisiae* hnRNP NAB2 protein (37), herpes simplex virus latency related protein 1 (HSV-1 LRP1; 136), HSV-2 open reading frame protein 1 (HSV-2 ORF-1; 137), pseudorabies virus ORF 3 protein (138) and Epstein–Barr virus nuclear antigen 1 (EBNA-1; 81). For a comprehensive listing and alignment refer to Kiledjian and Dreyfuss (69).

It is striking that the strong positive charge of the RGG box is almost exclusively carried by arginine, implying a conserved and unique function for this residue. By analogy to the tat RNA-binding protein of HIV in which an arginine makes unique contacts with RNA that cannot be made by other amino acid side chains (75, 76), the arginines in the RGG box may make similar contacts with RNA. It is interesting to note that several of the RNA-binding proteins listed in *Figure 4* contain the modified amino acid N^G,N^G-dimethylarginine (77). In fact, RGG box-containing peptides derived from nucleolin, fibrillarin, and hnRNP U are substrates for this modification (78; Q. Liu and G. Dreyfuss, in preparation). While the significance of arginine methylation is not known, it is possible that it could serve to modulate the RNA-binding activity of RGG box-containing proteins. Another potential mechanism of regulating the RGG box RNA-binding activity is phosphorylation. The binding of the hnRNP A1 C-terminal fragment, which contains the RGG box, is abrogated by phosphorylation of a serine just N-terminal to the RGG box (79).

It has recently become apparent that RNA-binding proteins, in particular RGG box-containing proteins, are involved in various human abnormalities and malignancies. For example, a protein implicated in fragile X syndrome (FMR-1) (80) contains an RGG box as well as two KH domains (see *Figures 4* and *5*; see below). The Epstein–Barr virus nuclear antigen-1 (EBNA-1; 81) which is necessary for the pathogenicity of Epstein–Barr virus also contains an RGG box, suggesting that this protein is involved in RNA metabolism in addition to transcription and DNA replication. The recently identified EWS gene is involved in Ewing's sarcomas and malignant melanomas of soft parts (82, 83) and a closely related gene, TLS, is involved in myxoid liposarcomas (84). Both of these genes encode proteins with three RGG boxes as well as an RNP-CS. The roles of the RGG box proteins in these disorders remain unclear.

Protein																										
hnRNP K	1	R	I	L	Q	S	K	N	A	G	A	V	I	G	K	G	G	K	N	I	K	A	L	R	T	D	Y N A S	V	S	V	P
	2	R	L	L	I	H	Q	S	L	A	G	G	I	I	G	V	K	G	A	K	I	K	E	L	R	E	N T Q T T	I	K	L	F
	3	Q	V	T	I	P	K D	L	A	G	S	I	I	G	K	G	G	Q	R	I	K	Q	I	R	H	E S G A S	I	K	I	D E
FMR 1	1	Q	F	I	V	R	E	D	L	M	G	L	A	I	G	T	H	G	A	N	I	Q	Q	A	R	K	V P G V T A	I	D	L	D E
	2	V	I	Q	V	P	R	N	L	V	G	K	V	I	G	K	N	G	K	L	I	Q	E	I	V	D	K S G V V R	V	R	I	E A
HhaS3		Q	I	V	L	K	A	E	K	P	G	M	V	I	G	K	G	G	K	N	I	R	K	I	T	T	Q L E E R	F	D	L	E D
EcoS3		R	V	T	I	H	T	A	R	P	G	I	V	I	G	K	K	G	E	D	V	E	K	L	R	K	V V A D I	A	G	V	P A
PRNT α		T	I	K	I	N	P	D	K	I	K	D	V	I	G	K	G	G	S	V	I	R	A	L	T	E	E T G T T	I	E	I	E D
MER-1		E	I	K	L	N	K	T	Q	I	T	F	L	I	G	A	K	G	T	R	I	E	S	L	R	E	K S G A S	I	K	I	I P
GAPAp62		K	Q	Y	P	K	F	N	F	V	G	K	I	L	G	P	Q	G	N	T	I	K	R	L	Q	E	E T G A K	I	S	V	L G
GRP33		D	Q	F	P	K	Y	N	F	L	G	K	L	L	G	P	G	G	S	T	M	K	Q	L	Q	D	E T M T K	I	S	I	G R
HX	1	T	V	N	I	P	A	N	S	V	A	R	L	I	G	N	K	G	S	N	L	Q	Q	I	R	E	K F A C Q	I	D	I	P N
HX	2	E	L	I	V	P	V	K	F	H	G	S	L	I	G	P	H	G	T	Y	R	N	R	L	Q	E	K Y N V F	I	N	F	P R
HX	3	V	I	N	V	P	A	E	H	V	P	R	I	I	G	K	N	G	D	N	I	N	D	I	R	A	E Y G V E	M	D	F	L Q
HX	4	T	I	D	I	P	A	E	R	K	G	A	L	I	G	P	G	G	I	V	R	Q	L	E	S	E	F N I N	L	F	V	P N
Consensus			I L V	I L V							G		L V	I	G	K	K	G			I L V			I L V	R E E Q D K			I L V	I L V	

Fig. 5 Alignment of representative KH domain sequences. Conserved residues are depicted in white on a black background. For proteins with more than one KH domain the individual domains are numbered. The conserved positions of the KH domain are depicted at the bottom of the figure. Sequences were taken from human hnRNP K (85), ribosomal S3 proteins from *Halobacterium halobium* (HhaS3; 86) and from *Escherichia coli* (EcoS3; 87), *E. coli* polyribonucleotide–orthophosphate nucleotidyltransferase α-subunit (PRNTα; 139), *S. cerevisiae* MER-1 (88), human GTPase-activating protein-associated tyrosine phosphoprotein p62 (GAPAp62; 91), *Artemia salina* glycine-rich putative hnRNP GRP33 (GRP33; 140), and *S. cerevisiae* HX protein (HX; 141). For a comprehensive listing and alignment refer to Gibson *et al.* (89).

2.3 The KH domain

The K homology (KH) domain, initially identified in hnRNP K (27, 85), is an evolutionarily conserved domain present in numerous other RNA-associated proteins including the ribosomal S3 proteins (86, 87) and the yeast alternative splicing factor, Mer 1p (88; *Figure 5*). The KH domain is ~50 amino acids long, contains a conserved octapeptide, Ile–Gly–X_2–Gly–X_2–Ile (where X is any amino acid) and is often repeated within a protein. Computer predictions of the KH domain secondary structure suggest a three-stranded β-sheet exposed on one surface and two α-helices on the opposite surface (89, 90). This structure may provide an RNA-binding surface similar to the RNP-CS RBD. It is currently unknown whether the KH domain directly binds RNA and is itself an autonomous RNA-binding domain or whether it influences RNA binding only indirectly. However, there are currently several indications suggesting that the KH domain is involved in RNA binding. A KH domain-containing fragment of the GAP-associated tyrosine phosphoprotein p62 can bind RNA *in vitro* (91). Mutational analysis of the conserved octapeptide in hnRNP K and FMR-1 indicates that the KH domain is essential for the RNA-binding activities of these two proteins (92).

2.4 Auxiliary domains in hnRNP proteins

The hnRNP proteins and RNA-binding proteins in general have a modular structure, containing one or more RNA-binding domains and additional regions termed auxiliary domains. Potential functions for auxiliary domains include protein–protein interactions, intracellular localization, and modulation of RNA-binding activity.

A common denominator for the hnRNP proteins identified thus far is their predominant nuclear localization. Immunofluorescence studies have shown that most of the hnRNP proteins are nucleoplasmic and are excluded from the nucleoli and cytoplasm. Several of the proteins, including hnRNP U and C, contain an SV40 T antigen-type or nucleoplasmin-type bipartite nuclear localization signal (93, 94). Other hnRNP proteins, including hnRNP A1, do not have a classical NLS (nuclear localization signal), suggesting they utilize other signals for nuclear localization. Interestingly, a novel type of nuclear targeting sequence has recently been found in the glycine-rich auxiliary domian of hnRNP A1 (H. Siomi, W. M. Micael, and G. Dreyfuss, submitted).

A glycine-rich region is found in many pre-mRNA-binding proteins and appears to be the most recurrent hnRNP auxiliary domain identified thus far. There is evidence that this region is involved in protein–protein interactions (26, 95). The glycine-rich regions of the hnRNP A1 and U are also involved in RNA binding and this binding appears to be mediated through the RGG box. An additional property of the glycine-rich region of hnRNP A1 is its ability to promote complementary nucleic acid annealing *in vitro* (79, 96, 97). This activity may be a result of both the RNA-binding and protein–protein interaction functions of the glycine-rich domain.

Many pre-mRNA-binding proteins contain a cluster of serine (S) and arginine (R) dipeptide repeats (SR motif or RS motif; 98–100). All the SR motif-containing proteins identified to date are involved in pre-mRNA splicing. The function of the SR motif is not known, however, a current hypothesis suggests that this region is important for mediating protein–protein interactions during splicing (100, 101). The importance of the SR motif in splicing has been demonstrated in two different SR proteins, U2 snRNP auxiliary factor (U2AF) and alternative splicing factor/splicing factor 2 (ASF/SF2). Mutations of this motif disrupt the ability of these proteins to function in constitutive splicing yet do not interfere with their ability to bind RNA (99, 102, 103).

Auxiliary domains of several hnRNP proteins bear an intriguing resemblance in amino acid composition to that of transcription factors. Both hnRNP C and U contain regions rich in acidic residues, reminiscent of several transcription activation domains (104). A glutamine-rich region similar to the activation domain of the transcription factor SP1 (105) is also found in hnRNP U, while hnRNP K and L have proline-rich regions that resemble that of the CCAAT-binding transcription factor CTF.

3. RNA-binding specificities of hnRNP proteins

The function of hnRNP proteins in mRNA biogenesis is certain to be reflected in their RNA-binding properties. A wide variety of experiments have demonstrated that hnRNP proteins are arranged in a non-uniform manner on the hnRNA (16, 18, 30), implying that there are important differences in the RNA-binding specificities of hnRNP proteins. *In vitro* RNA-binding experiments with individual hnRNP proteins are consistent with these observations (12, 13). For example, many hnRNP proteins exhibit differential binding to various ribonucleotide homopolymers (12) and binding studies with segments of pre-mRNAs have directly extended these conclusions to natural RNA sequences (13, 16). These studies have been extended further by selection/amplification experiments that identified high-affinity binding sites for hnRNP A1 and C1 (14, 15). The consensus hnRNP A1 high-affinity binding site, UAGGGA/U, is similar to the consensus sequences of vertebrate 5' and 3' splice sites (14) and the hnRNP C1 high-affinity binding site, UUUUUU (15) conforms to the polypyrimidine stretch found at the 3' end of most vertebrate introns. These experiments, as well as experiments described below, directly confirm that these hnRNP proteins bind RNA with sequence specificity and suggests a role for these proteins in pre-mRNA processing.

The affinities of several hnRNP proteins for different RNA sequences have been measured and they have provided considerable insight into the arrangement of hnRNP proteins on pre-mRNA. Both hnRNP A1 and hnRNP C1 bind high-affinity sites with $\sim 10^{-9}$–10^{-8} M affinity (equilibrium dissociation constant, K_d) (14, 15).

Binding to RNAs containing no high-affinity sites is with relatively low affinity ($K_d \simeq 10^{-7}$–10^{-6} M) and apparently without strong sequence preferences (49, 106, 107). RNAs containing sequences similar, but not identical to high-affinity sites are bound with an intermediate affinity. The important conclusion to be drawn from this experiment is that individual hnRNP proteins bind different RNA sequences with a range of affinities. Potential binding sites that conform closely to high-affinity sites will be bound with greatest affinity, with a commensurate loss in affinity for sites as they diverge from the high-affinity binding site sequence.

The identification of high-affinity binding sites for hnRNP proteins makes it possible to predict where on the pre-mRNA each protein is bound *in vivo*. There are two important considerations in making these predictions. First, it is necessary to know the specificity and affinities of other RNA-binding factors whose binding sites may overlap with the hnRNP protein in question and, second, one must know the intracellular concentration of each relevant factor. The RNA-binding specificity of hnRNP A1 overlaps with 5' splice site-binding factors (for example, U1 snRNP) and hnRNP C1-binding sites overlap with those of other polypyrimidine tract-binding proteins (for example, U2AF and the polypyrimidine tract-binding protein (hnRNP I/PTB)). *Table 2* lists the abundance (determined by quantitative Western blotting) of hnRNP A1 and C1 in human HeLa cells, as well as the previously determined abundance of several other nuclear components. We estimate the nuclear concentration of both hnRNP A1 and C1 to be in the order of 10^{-5} M in HeLa cells. By comparison to other major nuclear constituents such as histones, this value underscores the fact that hnRNP complexes are major nuclear structures. Although we do not know the concentration of high-affinity RNA-binding sites, it is likely that hnRNP A1 and C1 and probably most other abundant hnRNP proteins are in excess of the number of high-affinity and intermediate-affinity binding sites. With such large amounts of hnRNP proteins present, it is

Table 2 Abundances of hnRNP proteins and other major cellular components

Component	Molecules per HeLa cell
hnRNP A1	70×10^6
hnRNP C1	90×10^6
hnRNA (~6–10 Kb)	0.3–0.5×10^6
U1 snRNA	1×10^6
Core histones	70×10^6
Ribosomes	4×10^6

Abundances of hnRNP A1 and C1 were calculated by quantitative Western blotting using bacterially-expressed, purified proteins as standards.

likely that they will also be bound to low-affinity sites. A consequence of the large amount of hnRNP proteins and their overall RNA-binding properties, would be to form a protein–RNA fibril in which most, if not all, of the pre-mRNA is coated with hnRNP proteins (4).

4. Functions of hnRNP proteins

The structures, RNA-binding properties, abundance, and intracellular trafficking of hnRNP proteins suggest that they have important functions in mRNA metabolism. So far, however, only a few specific roles of individual hnRNP proteins have been elucidated. This situation is changing with the rapid emergence of new tools and experimental systems to study RNA metabolism and current evidence suggests that some hnRNP proteins directly participate in pre-mRNA processing and transport.

As some hnRNP proteins exhibit high-affinity sequence-specific RNA binding, they could directly influence the binding of other *trans*-acting factors to these sites. In this context, hnRNP proteins can be thought of as serving analogous functions to those of transcription factors, in that sequence-specific binding by hnRNP proteins may influence the formation of specialized RNA–protein complexes. At least two modes of action can be envisioned; hnRNP proteins could recruit other factors to the bound sites via protein–protein interactions or they could hinder the binding of such factors. In *in vitro* pre-mRNA-splicing assays, for example the choice between multiple 5′ splice sites can be decided by the competing actions of hnRNP A1, which activates distal splice sites and the essential splicing factor SF2/ASF, which activates proximal splice sites (108). Binding of snRNPs to pre-mRNAs can be modulated *in vitro* by other RNA-binding factors, including hnRNP A1 and in yeast by Mer1p (109–111).

As a consequence of the RNA-binding properties and the tremendous abundance of hnRNP proteins *in vivo*, pre-mRNAs are bound with large amounts of hnRNP proteins (4). What are the consequences to the pre-mRNA of such binding and how is it possible for other less abundant pre-mRNA-processing factors to recognize and bind appropriate sites on the RNA? Insight into this issue has come from a recent study (61) in which HeLa cell nucleoplasm was fractionated and assayed for RNA-annealing activity–in this case, the hybridization of a pre-mRNA with an antisense RNA probe complementary to 60 nucleotides (nt) encompassing the 3′ splice site. This strategy identified at least nine proteins which had RNA-annealing activity. Surprisingly, eight of these activities were found to co-purify with known hnRNP proteins. In addition to hnRNP A1, which had previously been shown to have such activity (96, 97, 112) hnRNP C1/C2, hnRNP P and hnRNP U also have RNA-annealing activity *in vitro* (62). The finding that many hnRNP proteins have the ability to promote RNA–RNA interactions suggests that a function of hnRNP proteins *in vivo* may be to modulate such interactions and also demonstrates that it is possible for *trans*-acting factors to recognize pre-mRNA sequences or structures even when these sequences are bound by hnRNP proteins.

Moreover, the finding that the RNP-CS domain alone of the hnRNP C1/C2 proteins can promote complementary RNA interactions (62) suggests that this domain has the ability to modulate pre-mRNA conformation and accessibility by binding to RNA. Thought of in this way, hnRNP proteins can be considered 'RNA chaperones' (62) that stabilize favourable, presumably productive RNA conformations and thereby profoundly influence pre-mRNA processing.

Until recently, discussions of hnRNP protein functions have considered strictly nuclear events. The possibility of cytoplasmic functions, however, was recently raised by the observation that several hnRNP proteins shuttle between the nucleus and cytoplasm (19; S. Piñol-Roma and G. Dreyfuss, in preparation). In the cytoplasm, the shuttling proteins (for example, hnRNP A1) are bound to polyadenylated RNA, inviting the possibility that they could influence gene expression in both compartments of the cell (19). A precedent for cytoplasmic functions of hnRNP proteins is the proposition that hnRNP I (a shuttling hnRNP protein) regulates the translation of picornavirus mRNAs in extracts of infected cells by mediating internal ribosome entry (113). Since mRNA is transported through nuclear pores as a ribonucleoprotein complex (114), the shuttling of hnRNP proteins is consistent with a role as carriers of mRNA to the cytoplasm (19).

As a final note, several hnRNP proteins have appreciable affinity for specific DNA sequences and could therefore play a role in transcription and DNA metabolism (115–117). The hnRNP A, B and D groups are the major proteins in nuclear extract that bind to oligonucleotides containing the vertebrate telomere repeat sequence (116, 118). Surprisingly, the high-affinity hnRNP A1-binding site sequence matches exactly the human telomere repeat sequence (14). Oligonucleotides derived from a regulated c-*myc* promoter are bound by hnRNP K and inhibit transcription from this promoter *in vitro* (115). Further experiments are necessary to investigate the significance of these *in vitro* interactions.

References

1. Darnell, J. E., Jr (1982) Variety in the level of gene control in eukaryotic cells. *Nature*, **297**, 365.
2. Nevins, J. R. (1983) The pathway of eukaryotic mRNA formation. *Ann. Rev. Biochem.*, **52**, 441.
3. Dreyfuss, G., Swanson, M. S., and Piñol-Roma, S. (1988) Heterogeneous nuclear ribonucleoprotein particles and the pathway of mRNA formation. *Trends Biochem. Sci.*, **13**, 86.
4. Dreyfuss, G., Matunis, M. J., Piñol-Roma, S., and Burd, C. G. (1993) hnRNP proteins and the biogenesis of mRNA. *Ann. Rev. Biochem.*, **62**, 289.
5. Görlach, M., Burd, C. G., Portman, D. S., and Dreyfuss, G. (1993) The hnRNP proteins. *Mol. Biol. Rep.*, **18**, 73.
6. Wilk, H. E., Herr, H., Friedrich, D., Kiltz, H. H., and Schäfer, K. P. (1985) The core proteins of 35S hnRNP complexes: characterization of nine different species. *Eur. J. Biochem.*, **46**, 71.
7. Dreyfuss, G., Adam, S. A., and Choi, Y. D. (1984) Physical change in cytoplasmic

messenger ribonucleoproteins in cells treated with inhibitors of mRNA transcription. *Mol. Cell. Biol.*, **4**, 415.
8. Dreyfuss, G., Choi, Y. D., and Adam, S. A. (1984) Characterization of heterogeneous nuclear RNA–protein complexes *in vivo* with monoclonal antibodies. *Mol. Cell. Biol.*, **4**, 1104.
9. Piñol-Roma, S., Choi, Y. D., Matunis, M. J., and Dreyfuss, G. (1988) Immunopurification of heterogeneous nuclear ribonucleoprotein particles reveals an assortment of RNA-binding proteins. *Genes Devel.*, **2**, 215.
10. Bandziulis, R. J., Swanson, M. S., and Dreyfuss, G. (1989) RNA-binding proteins as developmental regulators. *Genes Devel.*, **3**, 431.
11. Mattaj, I. W. (1993) RNA recognition: a family matter? *Cell*, **73**, 837.
12. Swanson, M. S. and Dreyfuss, G. (1988) Classification and purification of proteins of heterogeneous nuclear ribonucleoprotein particles by RNA-binding specificities. *Mol. Cell. Biol.*, **8**, 2237.
13. Swanson, M. S. and Dreyfuss, G. (1988) RNA binding specificity of hnRNP proteins: a subset bind to the 3' end of introns. *EMBO J.*, **11**, 3519.
14. Burd, C. G. and Dreyfuss, G. (1993) RNA-binding specificity of hnRNP A1: significance of hnRNP A1 high-affinity binding sites in pre-mRNA splicing. *EMBO J.*, **13**, 1197.
15. Görlach, M., Burds, C., and Dreyfuss, G. (1994) The determinants of RNA-binding specificity of the heterogeneous nuclear ribonucleoprotein C proteins. *J. Biol. Chem.*, **269**, 23074.
16. Bennett, M., Piñol-Roma, S., Staknis, D., Dreyfuss, G., and Reed, R. (1992) Differential binding of heterogeneous nuclear ribonucleoproteins to mRNA precursors prior to spliceosome assembly *in vitro*. *Mol. Cell. Biol.*, **12**, 3165.
17. Matunis, E. L., Matunis, M. J., and Dreyfuss, G. (1993) Association of individual hnRNP proteins and snRNPs with nascent transcripts. *J. Cell Biol.*, **121**, 219.
18. Piñol-Roma, S., Swanson, M. S., Gall, J. G., and Dreyfuss, G. (1989) A novel heterogeneous nuclear RNP protein with a unique distribution on nascent transcripts. *J. Cell Biol.*, **109**, 2575.
19. Piñol-Roma, S. and Dreyfuss, G. (1992) Shuttling of pre-mRNA binding proteins between the nucleus and the cytoplasm. *Nature*, **355**, 730.
20. Beyer, A. L., Christensen, M. E., Walker, B. W., and LeStourgeon, W. M. (1977) Identification and characterization of the packaging proteins of core 40S hnRNP particles. *Cell*, **11**, 127.
21. Karn, J., Vidali, G., Boffa, L. C., and Allfrey, V. G. (1977) Characterization of the non-histone nuclear proteins associated with rapidly labeled heterogeneous nuclear RNA. *J. Biol. Chem.*, **252**, 7307.
22. Choi, Y. D. and Dreyfuss, G. (1984) Isolation of the heterogeneous nuclear RNA–ribonucleoprotein complex (hnRNP): a unique supramolecular assembly. *Proc. Natl Acad. Sci. USA*, **81**, 7471.
23. Leser, G. P., Escara-Wilke, J., and Martin, T. E. (1984) Monoclonal antibodies to heterogeneous nuclear RNA–protein complexes. *J. Biol. Chem.*, **259**, 827.
24. Brunel, C. and Lelay, M. N. (1979) Two-dimensional analysis of proteins associated with heterogenous nuclear RNA in various animal cell lines. *Eur. J. Biochem.*, **99**, 273.
25. Kay, B. K., Sawhney, R. K., and Wilson, S. H. (1990) Potential for two isoforms of the A1 ribonucleoprotein in *Xenopus laevis*. *Proc. Natl Acad. Sci. USA*, **87**, 1367.
26. Cobianchi, F., SenGupta, D. N., Zmudzka, B. Z., and Wilson, S. H. (1986) Structure of rodent helix-destabilizing protein revealed by cDNA cloning. *J. Biol. Chem.*, **261**, 3536.
27. Siomi, H., Matunis, M. J., Michael, W. M., and Dreyfuss, G. (1993) The pre-mRNA

binding K protein contains a novel evolutionarily conserved motif. *Nucleic Acids Res.*, **21**, 1193.
28. Preugschat, F. and Wold, B. (1988) Isolation and characterization of a *Xenopus laevis* C protein cDNA: structure and expression of a heterogeneous nuclear ribonucleoprotein core protein. *Proc. Natl Acad. Sci. USA*, **85**, 9669.
29. Matunis, M. J., Matunis, E. L., and Dreyfuss, G. (1992) Isolation of hnRNP complexes from *Drosophila melanogaster*. *J. Cell Biol.*, **116**, 245.
30. Matunis, E. L., Matunis, M. J., and Dreyfuss, G. (1992) Characterization of the major hnRNP proteins from *Drosophila melanogaster*. *J. Cell Biol.*, **116**, 257.
31. Haynes, S. R., Johnson, D., Raychaudhuri, G., and Beyer, A. L. (1991) The *Drosophila* Hrb87F gene encodes a new member of the A and B hnRNP protein group. *Nucleic Acids Res.*, **19**, 25.
32. Haynes, S. R., Raychaudhuri, G., and Beyer, A. L. (1990) The *Drosophila* Hrb98DE locus encodes four protein isoforms homologous to the A1 protein of mammalian heterogeneous nuclear ribonucleoprotein complexes. *Mol. Cell. Biol.*, **10**, 316.
33. Kelley, R. L. (1993) Initial organization of the *Drosophila* dorsoventral axis depends on an RNA-binding protein encoded by the *squid* gene. *Genes Devel.*, **7**, 948.
34. Matunis, E. L., Kelley, R., and Dreyfuss, G. (1993) Essential role for an hnRNP protein in oogenesis: hrp40 is absent from the germline in the dorsoventral mutant *squid*. *Proc. Natl Acad. Sci. USA*, in press.
35. Matunis, M. J., Matunis, E. L., and Dreyfuss, G. (1993) PUB1: a major yeast poly(A)$^+$ RNA-binding protein. *Mol. Cell. Biol.*, **13**, 6114.
36. Anderson, J. T., Paddy, M. R., and Swanson, M. S. (1993) PUB1 is a major nuclear and cytoplasmic polyadenylated RNA-binding protein in *Saccharomyces cerevisiae*. *Mol. Cell. Biol.*, **13**, 6102.
37. Anderson J. T., Wilson, S. M., Datar, K. V., and Swanson, M. S. (1993) NAB2: a yeast nuclear polyadenylated RNA-binding protein essential for cell viability. *Mol. Cell. Biol.*, **13**, 2730.
38. Query, C. C., Bentley, R. C., and Keene, J. D. (1989) A common RNA recognition motif identified within a defined U1 RNA binding domain of the 70K U1 protein. *Cell*, **57**, 89.
39. Kenan, D. J., Query, C. C., and Keene, J. D. (1991) RNA recognition: towards identifying determinants of specificity. *Trends Biochem. Sci.*, **16**, 214.
40. Mattaj, I. W. (1989) A binding consensus: RNA–protein interactions in splicing, snRNPs, and sex. *Cell*, **57**, 1.
41. Ghetti, A., Piñol-Roma, S., Michael, W. M., Morandi, C., and Dreyfuss, G. (1992) hnRNPI, the polypyrimidine tract-binding protein: distinct nuclear localization and association with hnRNAs. *Nucleic Acids. Res.*, **20**, 3671.
42. Datar, K. V., Dreyfuss, G., and Swanson, M. S. (1993) The human hnRNP M proteins: identification of a methionine/arginine-rich repeat motif in ribonucleoproteins. *Nucleic Acids Res.*, **21**, 439.
43. Nagai, K., Oubridge, C., Jessen, T. H., Li, J., and Evans, P. R. (1990) Crystal structure of the RNA-binding domain of the U1 small nuclear ribonucleoprotein A. *Nature*, **346**, 515.
44. Hoffman, D. W., Query, C. C., Golden, B. L., White, S. W., and Keene, J. D. (1991) RNA-binding domain of the A protein component of the U1 small nuclear ribonucleoprotein analyzed by NMR spectroscopy is structurally similar to ribosomal proteins. *Proc. Natl Acad. Sci. USA*, **88**, 2495.

45. Wittekind, M., Görlach, M., Friedrichs, M., Dreyfuss, G., and Mueller, L. (1992) ^{1}H, ^{13}C, and ^{15}N NMR assignments and global folding pattern of the RNA binding domain of the human hnRNP C proteins. *Biochemistry*, **31**, 6254.
46. Adam, S. A., Nakagawa, T. Y., Swanson, M. S., Woodruff, T., and Dreyfuss, G. (1986) mRNA polyadenylate-binding protein: gene isolation and sequencing and identification of a ribonucleoprotein consensus sequence. *Mol. Cell. Biol.*, **6**, 2932.
47. Sachs, A. B., Bond, M. W., and Kornberg, R. D. (1986) A single gene from yeast for both nuclear and cytoplasmic polyadenylate-binding proteins: domain structure and expression. *Cell*, **45**, 827.
48. Bugler, B., Bourbon, H., Lapeyre, B., Wallace, M. O., Chang, J.-H., Amalric, F., and Olson, M. O. J. (1987) RNA binding fragments from nucleolin contain the ribonucleoprotein consensus sequence. *J. Biol. Chem.*, **262**, 10922.
49. Cobianchi, F., Karpel, R. L., Williams, K. R., Notario, V., and Wilson, S. H. (1988) Mammalian heterogeneous nuclear ribonucleoprotein complex protein A1. *J. Biol. Chem.*, **263**, 1063.
50. Merrill, B. M., Stone, K. L., Cobianchi, F., Wilson, S. H., and Williams, K. R. (1988) Phenylalanines that are conserved among several RNA-binding proteins form part of a nucleic acid-binding pocket in the A1 heterogeneous nuclear ribonucleoprotein. *J. Biol. Chem.*, **263**, 3307.
51. Wilusz, J., Feig, D. I., and Shenk, T. (1988) The C proteins of heterogeneous nuclear ribonucleoprotein complexes interact with RNA sequences downstream of polyadenylation cleavage sites. *Mol. Cell. Biol.*, **8**, 4477.
52. Scherly, D., Boelens, W., van Venrooij, W. J., Dathan, N. A., Hamm, J., and Mattaj, I. W. (1989) Identification of the RNA binding segment of human U1A protein and definition of its binding site on U1 snRNA. *EMBO J.*, **8**, 4163.
53. Scherly, D., Boelens, W., Dathan, N. A., van Venrooij, W. J., and Mattaj, I. W. (1990) Major determinants of the specificity of interaction between small nuclear ribonucleoproteins U1A and U2B" and their cognate RNAs. *Nature*, **345**, 502.
54. Lutz-Freyermuth, C., Query, C. C., and Keene, J. D. (1990) Quantitative determination that one of two potential RNA-binding domains of the A protein component of the U1 small nuclear ribonucleoprotein complex binds with high affinity to stem–loop II of U1 RNA. *Proc. Natl Acad. Sci. USA*, **87**, 6393.
55. Nietfeld, W., Mentzel, H., and Pieler, T. (1990) The *Xenopus laevis* poly(A) binding protein is composed of multiple functionally independent RNA binding domains. *EMBO J.*, **9**, 3699.
56. Burd, C. G., Matunis, E. L., and Dreyfuss, G. (1991) The multiple RNA-binding domains of the mRNA poly(A)-binding protein have different RNA-binding activities. *Mol. Cell. Biol.*, **7**, 3419.
57. Jessen, T.-H., Oubridge, C., Teo, C. H., Pritchard, C., and Nagai, K. (1991) Identification of molecular contacts between the U1 A small nuclear ribonucleoprotein and U1 RNA. *EMBO J.*, **10**, 3447.
58. Hall, K. B. and Stump, W. T. (1992) Interaction of N-terminal domain of U1A protein with an RNA stem/loop. *Nucleic Acids Res.*, **20**, 4283.
59. Surowy, C. S., van Santen, V. L., Scheib-Wixted, S. M., and Spritz, R. A. (1989) Direct, sequence-specific binding of the human U1-70K ribonucleoprotein antigen protein to loop I of U1 small nuclear RNA. *Mol. Cell. Biol.*, **9**, 4179.
60. Schwemmle, M., Görlach, M., Bader, M., Sarre, T. F., and Hilse, K. (1989) Binding of

mRNA by an oligopeptide containing an evolutionarily conserved sequence from RNA binding proteins. *FEBS Lett.*, **251**, 117.
61. Görlach, M., Wittekind, M., Beckman, R. A., Mueller, L., and Dreyfuss, G. (1992) Interaction of the RNA-binding domain of the hnRNP C proteins with RNA. *EMBO J.*, **11**, 3289.
62. Portman, D. S. and Dreyfuss, G. (1994) RNA annealing activities in HeLa nuclei. *EMBO J.*, **13**, 213.
63. Nikolov, D. B., Hu, S.-H., Lin, J., Gasch, A., Hoffmann, A., Horikoshi, M., Chua, N.-H., Roeder, R. G., and Burley, S. K. (1992) Crystal structure of TFIID TATA-box binding protein. *Nature*, **360**, 40.
64. Peabody, D. S. (1992) The RNA binding site of bacteriophage MS2 coat protein. *EMBO J.*, **12**, 595.
65. Schindelin, H., Marahiel, M. A., and Heinemann, U. (1993) Universal nucleic acid-binding domain revealed by crystal structure of the *B. subtilis* major cold-shock protein. *Nature*, **364**, 164.
66. Schnuchel, A., Wiltscheck, R., Czisch, M., Herrler, M., Willimsky, G., Graumann, P., Marahiel, M. A., and Holak, T. A. (1993) Structure in solution of the major cold-shock protein from *Bacillus subtilis*. *Nature*, **364**, 169.
67. Scherly, D., Dathan, N. A., Boelens, W., van Venrooij, W. J., and Mattaj, I. W. (1990) The U2B" RNP motif as a site of protein–protein interaction. *EMBO J.*, **9**, 3675.
68. Burd, C. G., Swanson, M. S., Görlach, M., and Dreyfuss, G. (1989) Primary structures of the heterogeneous nuclear ribonucleoprotein A2, B1, and C2 proteins: a diversity of RNA binding proteins is generated by small peptide inserts. *Proc. Natl Acad. Sci. USA*, **86**, 9788.
69. Görlach, M., Burd, C. G., and Dreyfuss, G. (1994) The mRNA poly(A)-binding protein: localization, abundance and RNA-binding specificity. *Exp. Cell Res.*, **211**, 400.
70. Kiledjian, M. and Dreyfuss, G. (1992) Primary structure and binding activity of the hnRNP U protein: binding RNA through RGG box. *EMBO J.*, **11**, 2655.
71. Ghisolfi, L., Joseph, G., Amalric, F., and Erard, M. (1992) The glycine-rich domain of nucleolin has an unusual supersecondary structure responsible for its RNA-helix-destabilizing properties. *J. Biol. Chem.*, **267**, 2955.
72. Ghisolfi, L., Kharrat, A., Joseph, G., Amalric, F., and Erard, M. (1992) Concerted activities of the RNA recognition and the glycine-rich C-terminal domains of nucleolin are required for efficient complex formation with pre-ribosomal RNA. *Eur. J. Biochem.*, **209**, 1.
73. Urry, D. W. (1982) Characterization of soluble peptides of leastin by physical techniques. *Methods Enzymol.*, **82**, 673.
74. Matsushima, N., Creutz, C. E., and Kretsinger, R. H. (1990) Polyproline, β-turn helices. Novel secondary structures proposed for the tandem repeats within rhodopsin, synaptophysin, synexin, gliadin, RNA polymerase II, hordein, and gluten. *Proteins*, **7**, 125.
75. Calnan, B. J., Tidior, B., Biancalana, S., Hudson, D., and Frankel, A. D. (1991) Arginine-mediated RNA recognition: the arginine fork. *Science*, **252**, 1167.
76. Puglisi, J. D., Tan, R., Calnan, B. J., Frankel, A. D., and Williamson, J. R. (1992) Confirmation of the TAR RNA–arginine complex by NMR spectroscopy. *Science*, **257**, 76.
77. Paik, W. K. and Kim, S. (1989) Amino acid sequence of arginine methylation. In *Protein Methylation*. Lischew, M. A. (ed.). CRC Press, Boca Raton, FL, p. 98.

78. Najbauer, J., Johnson, B. A., Young, A. L., and Aswad, D. W. (1993) Peptides with sequences similar to glycine, arginine-rich motifs in proteins interacting with RNA are efficiently recognized by methyltransferase(s) modifying arginine in numerous proteins. *J. Biol. Chem.*, **268**, 10501.
79. Cobianchi, F., Calvio, C., Stoppini, M., Buvoli, M., and Riva, S. (1993) Phosphorylation of human hnRNP protein A1 abrogates *in vitro* strand annealing activity. *Nucleic Acids Res.*, **21**, 949.
80. Verkerk, A. J. M. H., Pieretti, M., Sutcliffe, J. S., Fu, Y.-H., Kuhl, D. P. A., Pizzuti, A., Reiner, O., Richards, S., Victoria, M. F., Zhang, F., Eussen, B. E., van Ommen, G.-J. B., Blonden, L. A. J., Riggins, G. J., Chastain, J. L., Kunst, C. B., Galijarrd, H., Caskey, C. T., Nelson, D. L., Oostra, B. A., and Warren, S. T. (1991) Identification of a gene (FMR-1) containing a CGG repeat coincident with a breakpoint cluster region exhibiting length variation in fragile X syndrome. *Cell*, **65**, 905.
81. Baer, B. W., Bankier, A. T., Biggin, M. D., Deininger, P. L., Farrell, P. J., Gibson, T. J., Hatfull, G., Hudson, G. S., Satchwell, S. C., Séguin, C., Tuffnell, P. S., and Barrell, B. G. (1984) DNA sequence and expression of the B95-8 Epstein-Barr virus genome. *Nature*, **310**, 207.
82. Delattre, O., Zucman, J., Plougastel, B., Desmaze, C., Melot, T., Peter, M., Kovar, H., Jaoubert, I., de Jong, P., Rouleau, G., Aurias, A., and Thomas, G. (1993) Gene fusion with an ETS DNA-binding domain caused by chromosome translocation in human tumours. *Nature*, **359**, 162.
83. Zucman, J., Delattre, O., Desmaze, C., Epstein, A. L., Stenman, G., Speleman, F., Fletchers, C. D. M., Aurias, A., and Thomas, G. (1993) EWS and ATF-1 gene fusion induced by t(12;22) translocation in malignant melanoma of soft parts. *Nature Genet.*, **4**, 341.
84. Crozat, A., Aman, P., Mandahl, N., and Ron, D. (1993) Fusion of CHOP to a novel RNA-binding protein in human myxoid liposarcoma. *Nature*, **363**, 640.
85. Matunis, M. J., Matunis, E. L., and Dreyfuss, G. (1992) Characterization and primary structure of the poly(C)-binding heterogeneous nuclear ribonucleoprotein complex K protein. *Mol. Cell. Biol.*, **12**, 164.
86. Spiridonova, V. A., Akhomanova, S. A., Kagramanova, V. K., Köpke, A. K., and Mankin, A. S. (1989) Ribosomal protein gene cluster of *Halobacterium halobium*: nucleotide sequence of the genes coding for S3 and L29 equivalent ribosomal proteins. *Can. J. Microbiol.*, **35**, 153.
87. Brauer, D. and Röming, R. (1979) The primary structure of protein S3 from the small ribosomal subunit of *Escherichia coli*. *FEBS Lett.*, **106**, 352.
88. Engebrecht, J. and Roeder, G. S. (1990) MER1, a yeast gene required for chromosome pairing and genetic recombination, is induced in meiosis. *Mol. Cell. Biol.*, **10**, 2379.
89. Gibson, T. J., Thompson, J. D., and Heringa, J. (1993) The KH domain occurs in a diverse set of RNA-binding proteins that include the antiterminator NusA and is probably involved in binding nucleic acid. *FEBS Lett.*, **324**, 361.
90. Gibson, T. J., Rice, P. M., Thompson, J. D., and Heringa, J. (1993) KH domains within the FMR1 sequence suggest that fragile X syndrome stems from a defect in RNA metabolism. *Trends Cell Biol.*, **18**, 331.
91. Wong, G., Müller, O., Clark, R., Conroy, L., Moran, M. F., Polakis, P., and McCormick, F. (1992) Molecular cloning and nucleic acid binding properties of the GAP-associated tyrosine phosphoprotein p62. *Cell*, **69**, 551.
92. Siomi, H., Choi, M., Siomi, M. C., Nussbaum, R. L., and Dreyfuss, G. (1994) Essential

role for KH domains in RNA binding: impaired RNA binding by a mutantion in the KH domain of FMR1 that causes fragile X syndrome. *Cell*, **77**, 33.
93. Dingwall, C. and Laskey, R. A. (1991) Nuclear targeting sequences — a consensus? *Trends Biochem. Sci.*, **16**, 478.
94. Garcia-Bustos, J., Heitman, J., and Hall, M. N. (1991) Nuclear protein localization. *Biochem. Biophys. Acta*, **1071**, 83.
95. LeStourgeon, W. M., Barnett, S. F., and Northington, S. J. (1990) Tetramers of the core proteins of 40S nuclear ribonucleoprotein particles assemble to package nascent transcripts into a repeating array of regular particles. In *The Eukaryotic Nucleus: Structure and Function.* Strauss, P. and Wilson, S. (eds). The Telford Press, Caldwell, NJ, Vol. 2, p. 477.
96. Kumar, A. and Wilson, S. H. (1990) Studies of the strand-annealing activity of mammalian hnRNP complex protein A1. *Biochemistry*, **29**, 10717.
97. Munroe, S. H. and Dong, X. (1992) Heterogeneous nuclear ribonucleoprotein A1 catalyzes RNA–RNA annealing. *Proc. Natl Acad. Sci. USA*, **89**, 895.
98. Zahler, A. M., Lane, W. S., Stolk, J. A., and Roth, M. B. (1992) SR proteins: a conserved family of pre-mRNA splicing factors. *Genes Devel.*, **6**, 837.
99. Zamore, P. D., Patton, J. G., and Green, M. R. (1992) Cloning and domain structure of the mammalian splicing factor U2AF. *Nature*, **355**, 609.
100. Lamm, G. M. and Lamond, A. I. (1993) Non-snRNP protein splicing factors. *Biochim. Biophys. Acta*, **1173**, 247.
101. Fu, X.-D. (1993) Specific commitment of different pre-mRNAs to splicing by single SR proteins. *Nature*, **365**, 82.
102. Cáceres, J. F. and Krainer, A. R. (1993) Functional analysis of pre-mRNA splicing factor SF2/ASF structural domains. *EMBO J.*, **12**, 4715.
103. Zuo, P. and Manley, J. L. (1993) Functional domains of the human splicing factor ASF/SF2. *EMBO J.*, **12**, 4727.
104. Mitchell, P. J. and Tjian, R. (1989) Transcriptional regulation in mammalian cells by sequence-specific DNA binding proteins. *Science*, **245**, 371.
105. Courey, A. J., Holtzman, D. A., Jackson S. P., and Tjian, R. (1989) Synergistic activation by the glutamine-rich domains of human transcription factor Sp1. *Cell*, **59**, 827.
106. Nadler, S. G., Merrill, B. M., Roberts, W. J., Keating, K. M., Lisbin, M. J., Barnett, S. F., and Williams, K. R. (1991) Interactions of the A1 heterogeneous nuclear ribonucleoprotein and its proteolytic derivative, UP1, with RNA and DNA: evidence for multiple RNA binding domains and salt-dependent binding mode transitions. *Biochemistry*, **30**, 2968.
107. Kumar, A., Casas-Finet, J. R., Luneau, C. J., Karpel, R. L., Merrill, B. M., Williams, K. R., and Wilson, S. H. (1990) Mammalian heterogeneous nuclear ribonucleoprotein A1: nucleic acid binding properties of the COOH-terminal domain. *J. Biol. Chem.*, **265**, 17094.
108. Mayeda, A. and Krainer, A. R. (1992) Regulation of alternative pre-mRNA splicing by hnRNP A1 and splicing factor SF2. *Cell*, **68**, 365.
109. Eperon, I. C., Ireland, D. C., Smith, R. A., Mayeda, A., and Krainer, A. R. (1993) Pathways for selection of 5' splice sites by U1 snRNPs and SF2/ASF. *EMBO J.*, **12**, 3607.
110. Nandabalan, K., Price, L., and Roeder, G. S. (1993) Mutations in U1 snRNA bypass the requirement for a cell type-specific RNA splicing factor. *Cell*, **73**, 407.
111. Buvoli, M., Cobianchi, F., and Riva, S. (1992) Interaction of hnRNP A1 with snRNPs

and pre-mRNAs: evidence for a possible role of A1 RNA annealing activity in the first steps of spliceosome assembly. *Nucleic Acids Res.*, **20**, 5017.

112. Pontius, B. W. and Berg, P. (1990) Renaturation of complementary DNA strands mediated by purified mammalian heterogeneous nuclear ribonucleoprotein A1 protein: implications for a mechanism for rapid molecular assembly. *Proc. Natl Acad. Sci. USA*, **87**, 840.

113. Hellen, C. U. T., Witherell, G. W., Schmid, M., Shin, S. H., Pestova, T. T., Gil, A., and Wimmer, E. (1993) A cytoplasmic 57-kDa protein that is required for the translation of picornavirus RNA by internal ribosome entry is identical to the nuclear pyrimidine tract-binding protein. *Proc. Natl Acad. Sci. USA*, **90**, 7642.

114. Mehlin, H., Daneholt, B., and Skoglund, U. (1992) Translocation of a specific pre-messenger ribonucleoprotein particle through the nuclear pore studied with electron microscope tomography. *Cell*, **69**, 605.

115. Takimoto, M., Tomonaga, T., Matunis, M., Avigan, M., Krutzsch H., Dreyfuss, G., and Levens, D. (1993) Specific binding of heterogeneous ribonucleoprotein particle protein K to the human c-*myc* promoter, *in vitro*. *J. Biol. Chem.*, **268**, 18 249.

116. Ishikawa, F., Matunis, M. J., Dreyfuss, G., and Cech, T. R. (1993) Nuclear proteins that bind the pre-mRNA 3' splice site sequence r(UUAG/G) and the human telomeric DNA sequence d(TTAGGG)$_n$. *Mol. Cell. Biol.*, **13**, 4301.

117. Brunel, F., Alzari, P. M., Ferrara, P., and Zakin, M. M. (1991) Cloning and sequencing of PYBP, a pyrimidine-rich specific single strand DNA-binding protein. *Nucleic Acids Res.*, **19**, 5237.

118. McKay, S. J. and Cooke, H. (1992) hnRNP A2/B1 binds specifically to single stranded vertebrate telomeric repeat TTAGGG$_n$. *Nucleic Acids Res.*, **20**, 6461.

119. Buvoli, M., Biamonti, G., Tsoulfas, P., Bassi, M. T., Ghetti, A., Riva, S., and Morandi, C. (1988) cDNA cloning of human hnRNP protein A1 reveals the existence of multiple mRNA isoforms. *Nucleic Acids Res.*, **16**, 3751.

120. Swanson, M. S., Nakagawa, T. Y., LeVan, K., and Dreyfuss, G. (1987) Primary structure of human nuclear ribonucleoprotein particle C proteins: conservation of sequence and domain structures in heterogeneous nuclear RNA, mRNA, and pre-rRNA-binding proteins. *Mol. Cell. Biol.*, **7**, 1731.

121. Sillekens, P. T. G., Habets, W. J., Beijer R. P., and van Venrooij, W. J. (1987) cDNA cloning of the human U1 snRNA-associated A protein: extensive homology between U1 and U2 snRNP-specific proteins. *EMBO J.*, **6**, 3841.

122. Lapeyre, B., Bourbon, H., and Amalric, F. (1987) Nucleolin, the major nucleolar protein of growing eukaryotic cells: an unusual protein structure revealed by the nucleotide sequence. *Proc. Natl Acad. Sci. USA*, **84**, 1472.

123. Bell, L. R., Maine, E. M., Schedl, P., and Cline, T. W. (1988) *Sex-lethal*, a *Drosophila* sex determination switch gene, exhibits sex-specific RNA splicing and sequence similarity to RNA binding proteins. *Cell*, **55**, 1037.

124. Inoue, K., Hoshijima, K., Sakamoto, H., and Shimura, Y. (1990) Binding of the *Drosophila sex-lethal* gene product to the alternative splice site of *transformer* primary transcript. *Nature*, **344**, 461.

125. Gómez, J., Sánchez-Martínez, D., Stiefel, V., Rigau, J., Puigdoménech, P., and Pagés, M. (1988) A gene induced by the plant hormone abscisic acid in response to water stress encodes a glycine-rich protein. *Nature*, **334**, 262.

126. Schuster, G. and Gruissem, W. (1991) Chloroplast mRNA 3' end processing requires a nuclear-encoded RNA-binding protein. *EMBO J.*, **10**, 1493.

127. Lingner, J., Kellermann, J., and Keller, W. (1991) Cloning and expression of the essential gene for poly(A) polymerase from *S. cerevisiae*. *Nature*, **354**, 496.
128. Jong, A. Y., Clark, M. W., Gilbert, M., Oehm, A., and Campbell, H. L. (1987) *Saccharomyces cerevisiae* SSB1 protein and its relationship to nucleolar RNA-binding proteins. *Mol. Cell. Biol.*, **7**, 2947.
129. Aris, J. P. and Blobel, G. (1991) cDNA cloning and sequencing of human fibrillarin, a conserved nucleolar protein recognized by autoimmune antisera. *Proc. Natl Acad. Sci. USA*, **88**, 931.
130. Soulard, M., Della Valle, V., Siomi, M. C., Piñol-Roma, S., Codogno, P., Bauvy, C., Bellini, M., Lacroix, J.-C., Monod, G., Dreyfuss, G., and Larsen, C.-J. (1993) hnRNP G: sequence and characterization of glycosylated RNA-binding protein. *Nucleic Acids Res.*, **21**, 4210.
131. Dorer, D. R., Christensen, A. C., and Johnson, D. H. (1990) A novel RNA helicase gene tightly linked to the *triplo-lethal* locus of *Drosophila*. *Nucleic Acids Res.*, **18**, 5489.
132. Hay, B., Jan, L. Y., and Jan, Y. N. (1988) A protein component of *Drosophila* polar granules is encoded by *vasa* and has extensive sequence similarity to ATP-dependent helicases. *Cell*, **55**, 577.
133. Girard, J.-P., Lehtonen, H., Caizergues-Ferrer, M., Amalric, F., Tollervey, D., and Lapeyre, B. (1992) GAR1 is an essential small nucleolar RNP protein required for pre-rRNA processing in yeast. *EMBO J.*, **11**, 673.
134. Bossie, M. A., DeHoratius, C., Barcelo, G., and Silver, P. (1992) A mutant nuclear protein with similarity to RNA binding proteins interferes with nuclear import in yeast. *Mol. Cell. Biol.*, **3**, 875.
135. Russell, I. D. and Tollervey, D. (1992) NOP3 is an essential yeast protein which is required for pre-rRNA processing. *J. Cell Biol.*, **119**, 737.
136. Wechsler, S. L., Nesburn, A. B., Zwaagstra, J., and Ghiasi, H. (1989) Sequence of the latency-related gene of herpes simplex virus type 1. *Virology*, **168**, 168.
137. Krause, P. R., Ostrove, J. M., and Straus, S. E. (1991) The nucleotide sequence, 5' end promoter domain, and kinetics of expression of the gene encoding the herpes simplex virus type 2 latency-associated transcript. *J. Virology*, **65**, 5619.
138. Vlcek, C., Kozmik, Z., Paces, V., Schirm, S., and Schwyzer, M. (1990) Pseudorabies virus immediate-early gene overlaps with an oppositely oriented open reading frame: characterization of their promoter and enhancer regions. *Virology*, **179**, 365.
139. Régnier, P., Grunberg-Manago, M., and Portier, C. (1987) Nucleotide sequence of the *pnp* gene of *Escherichia coli* encoding polynucleotide phosphorylase: homology of the primary structure of the protein with the RNA-binding domain of ribosomal protein S1. *J. Biol. Chem.*, **262**, 63.
140. Cruz-Alvarez, M. and Pellicer, A. (1987) Cloning of full-length complementary DNA for an *Artemia salina* glycine protein. *J. Biol. Chem.*, **262**, 13377.
141. Delahodde, A., Becam, A. M., Perea, L., and Jacq, C. (1986) A yeast protein HX has homologies with the histone H2AF expressed in chicken embryo. *Nucleic Acids Res.*, **14**, 9213.
142. Matunis, M. J., Xing, J., and Dreyfuss, G. (1993) The hnRNP F protein: unique primary structure, nucleic acid-binding properties, and subcellular localization. *Nuc. Acids Res.*, **22**, 1059.
143. Birney, E., Kumar, S., and Krainer, A. R. (1993) Analysis of the RNA-recognition motif and RS and RGG domains: Conservation in metazoan pre-mRNA splicing factors. *Nucleic Acids Res.*, **21**, 5803.

7 | RNA–protein interactions in the splicing snRNPs

KIYOSHI NAGAI and IAIN W. MATTAJ

1. Introduction

In 1977, several groups made the striking discovery that some protein-coding eukaryotic genes are discontinuous, being interrupted by non-coding intervening sequences now known as introns (1–3). The entire length of these genes, including the introns, is transcribed into pre-mRNA. The introns are subsequently excised within the nucleus to produce continuous protein-coding mRNA sequences prior to transport to the cytoplasm. This process is called pre-mRNA splicing and its molecular mechanism has been subject to intensive research. The establishment of *in vitro* splicing-assay systems using either nuclear extracts of HeLa cells (a transformed human epithelial cell line) or yeast cell extracts has allowed detailed investigation of the catalytic mechanism as well as biochemical characterization of cellular components involved in this process (4–8). Yeast genetics has also facilitated identification of protein and RNA-splicing factors and characterization of interactions between them (6–9).

The excision of introns proceeds by two successive *trans*-esterification (phosphodiester bond exchange) reactions (*Figure 1*). In the first step the 2'-OH group of a specific adenosine within the intron, the branch-point adenosine, attacks the phosphodiester bond at the 5' splice site such that the 5' exon is cleaved off and a new phosphodiester bond is formed between the 5' end of the intron and the 2'-OH group of the adenosine. This results in the formation of a branched, lariat-shaped intron intermediate. In the second step, the 3'-OH group of the 5' exon attacks the phosphodiester bond at the 3' splice site and the 5' and 3' exons are ligated via formation of a standard 3'–5' phosphodiester linkage. The lariat form of the intron is liberated and normally rapidly degraded. Nuclear pre-mRNA splicing requires the participation of several RNA–protein complexes called small nuclear ribonucleoprotein particles or snRNPs (10, 11). Five small nuclear RNA species, U1, U2, U4, U5, and U6 snRNAs, are integral to the spliceosomal snRNPs. snRNPs are named after their RNA components, for example U1 snRNP and U5 snRNP contain U1 and U5 snRNAs, respectively.

In pre-mRNAs, the nucleotide sequences around the 5' splice site, branch-point, and the 3' splice site (*Figure 1*) are conserved, although to different extents in pre-

INTRODUCTION | 151

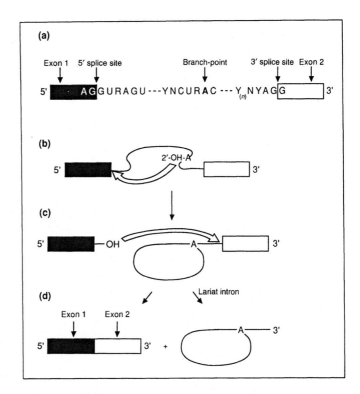

Fig. 1 Excision of introns from nuclear pre-mRNA by two successive transesterification reactions. (a) The consensus sequences at the 5' and 3' splice sites and the branch point. In the first step of splicing (b) the 2'-OH group of a specific adenosine (branch-point) within the intron attacks the phosphodiester bond at the 5' splice site (exon–intron junction). In the second step (c) the 3'-OH group of the 5' exon attacks the phosphodiester bond at the 3' splice site which results in ligation of two exons and liberation of the lariat intron. (d) The nucleotide sequences of the 5' and 3' splice sites and the branch-point are conserved and transiently interact with sequences within snRNAs during splice-site selection and the splicing reaction.

mRNAs from different species. The consensus sequences around the 5' and 3' splice sites and the branch-points of metazoan pre-mRNA are **AG**GURAGU, $Y_{(n)}$N**YAGG**, and YNCURA*C, respectively, where characters in bold type are exon sequences, A* shows the branch-point adenosine, and Y, R, and N denote pyrimidine, purine, and any nucleotide, respectively.

Genetic and biochemical experiments have revealed transient interactions between pre-mRNA and snRNAs and between pairs of snRNAs during the splicing cycle. U1 snRNP and U2 snRNP form base-pairing interactions with the 5' splice site and the branch-point, respectively. Subsequent binding of the U4/U6–U5 triple snRNP is necessary for the splicing reaction to occur. Of the late-joining snRNAs, U5 and U6 also interact directly with the pre-mRNA via base pairing (reviewed in Moore *et al.* (8) and Newman (12)). Extensive base pairing between U4 and U6 snRNPs is unwound and parts of the U2 and U6 snRNAs become paired before pre-mRNA undergoes the splicing reaction. Spliceosome assembly is a dynamic process which undoubtedly involves an ordered series of multiple protein–protein, RNA–protein, and RNA–RNA interactions (5–8, 12). Many of these involve snRNP components. This chapter reviews our current knowledge of the architecture of snRNPs, concentrating on the characterized examples of RNA–protein interactions within the particles. Where relevant, we will discuss briefly how these interactions relate to snRNP recognition of conserved sequences within mRNA and to pre-mRNA splicing in general.

2. RNA components of spliceosomal snRNPs

Five distinct small nuclear RNA species, U1, U2, U4, U5, and U6 snRNAs, are components of the spliceosomal snRNPs. U1, U2, U4, and U5 snRNAs are transcribed by RNA polymerase II and, as a consequence, the 5' end of these RNAs co-transcriptionally acquires an N7-methyl guanosine (m7G) cap. The m7G cap facilitates the export of the RNAs from the nucleus (13) presumably via recognition by a nuclear cap-binding protein (14). In the cytoplasm, the snRNAs become complexed with a subset of the snRNP proteins (15). The m7G cap of the snRNAs is then further methylated at position 2 of the guanine base to generate the characteristic 2,2,7-trimethyl-guanosine (m_3G) cap, a structure that is specific to the U snRNAs. Depending on cell type, this modification participates in the transport of snRNPs to the nucleus by either a direct, signalling mechanism or indirectly by preventing retention of the pre-snRNP in the cytoplasm via interaction with cytoplasmic cap-binding proteins (reviewed in Izaurralde and Mattaj (16)). In contrast, the U6 snRNA gene, like 5S ribosomal RNA and tRNA genes, is transcribed by RNA polymerase III. Thus, the primary U6 transcript has a triphosphate 5' end, that, after transcription, is methylated at the γ-phosphate position (17). This modification increases the stability of U6 snRNA in the nucleus (18), but has no other known function. The 3' end of U6 snRNA is often formed by an unusual 2'–3' cyclic phosphate linkage (19).

The snRNAs fold into secondary structures via intramolecular and, in the case of U4/U6, intermolecular base pairing between complementary sequences (20). *Figures 2–5* show the predicted secondary structures of the HeLa snRNAs. U1, U2, U4, U5, and U6 snRNAs from a wide variety of species have been sequenced and most, but not all, of the secondary structure elements in the mammalian RNAs (*Figures 2–5*) have been found to be conserved, even though the lengths of the individual U snRNAs and much of their primary nucleotide sequences diverge (20). This provides strong support for the correctness of the predicted secondary structures. Like tRNAs, snRNAs contain many modified nucleotides (not included in *Figures 2–5*) (21) but the functional significance of these modifications is not yet known.

3. Protein components of snRNPs

The most common source of snRNPs for biochemical characterization is the nuclear extract of HeLa cells. SnRNPs are purified from the nuclear extract by immuno-affinity chromatography using an antibody against the m_3G cap. U6 snRNA is co-purified with U4 snRNA as a large fraction of these two RNAs are found together in a single snRNP. SnRNP fractions can be further purified by glycerol density gradient centrifugation and mono-Q ion exchange chromatography (22). The protein components associated with each snRNA vary depending on the purification procedure. In particular, the temperature at which the purification is carried out, as well as the ionic strength and magnesium ion concentration of the buffer solutions

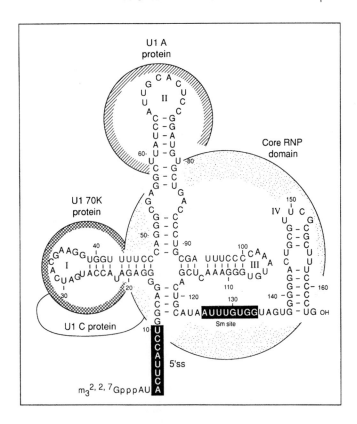

Fig. 2 Schematic representation of U1 snRNA and its interaction with various protein components. U1 snRNA folds into a secondary structure with four stem–loop structures (I, II, III, and IV). The core proteins, B, B', D1, D2, D3, E, F, G, and 69K assemble around the Sm site to form the core RNP domain. U1 70K and U1A protein bind to stem–loops I and II, respectively and form two protuberances as seen by electron microscopy (see *Figure 6*). U1C protein assembly into the particle requires U1 70K protein. Note that the positions of the proteins here and in the following three figures are represented very schematically. The precise RNA contacts have, in most cases, not been established. The nucleotides near the 5' end of U1 snRNA (5'SS) base pair with the conserved sequence at the 5' splice site of pre-mRNAs.

employed, can change the protein compositions of some U snRNPs dramatically. This is because some protein components are more loosely associated than others and tend to be lost in high-salt buffers or at elevated temperatures. *Table 1* summarizes the protein components associated with each snRNP (4, 22, 23).

A short discussion of snRNP protein nomenclature is necessary. The snRNP protein subunits first discovered by means of immunoprecipitation with the sera of patients suffering from particular autoimmune disorders were named alphabetically according to their mobility on SDS–polyacrylamide gels (24). Later analyses have necessitated additions to and changes in this scheme. The D-band, for example has now been resolved into three distinct proteins on a highly cross-linked polyacrylamide gel (4) and these bands are named D1, D2, and D3. As research has identified more and more snRNP proteins and yielded more information on each protein the nomenclature has been overtaken by events and is now a source of confusion to the uninitiated. For example, the U1A protein is related to the U2B" protein but not to the U2A' protein, while the U2B" protein bears no similarity to the B or B' proteins. As can be seen from *Table 1*, however, there is one logical major division of the snRNP proteins into two classes. The first class is that of the common (or core or Sm) proteins that are present in each U snRNP. These proteins are often called the Sm proteins as a subset is recognized by antibodies of the anti-Sm

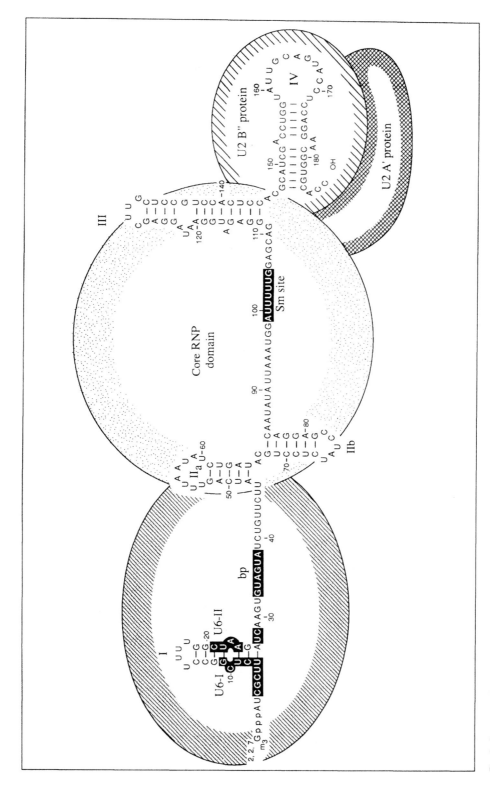

Fig. 3 Proposed folding of U2 snRNA with stem–loop structures, I, IIa, IIb, III, and IV. The core proteins assemble around the Sm site and U2B" and U2A' proteins bind to stem–loop IV. U6-I and U6-II interact with sequences within U6 snRNA (U2-I and U2II, respectively) in the spliceosome and bp represents the sequence that pairs with the branch-point sequence within introns.

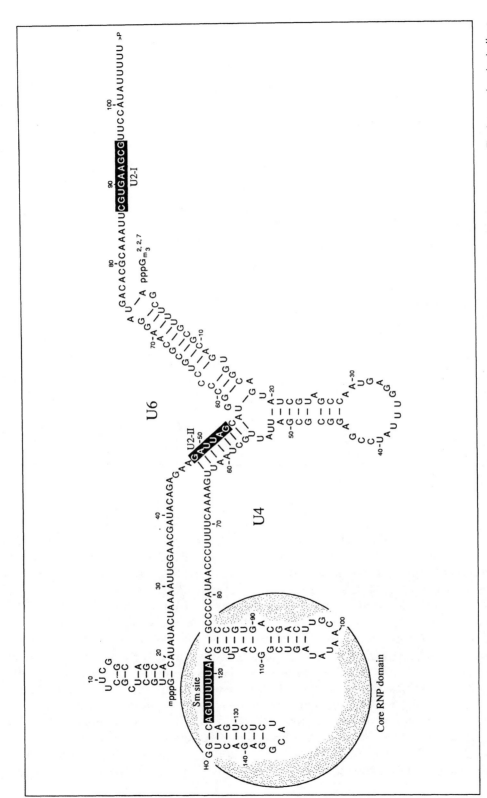

Fig. 4 U4 and U6 snRNAs form extensive intermolecular helices. The Sm site present in U4 snRNA is the binding site for the core proteins. The intermolecular helices between U4 and U6 snRNA are unwound and short sequences within U6 snRNA (U2-I and U2-II) form base pairs with U2 snRNA, in the process of splicing.

156 | RNA–PROTEIN INTERACTIONS IN THE SPLICING snRNPs

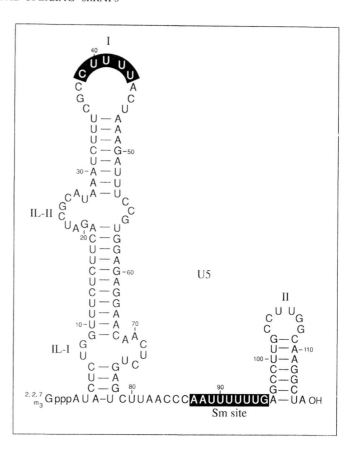

Fig. 5 Secondary structure of U5 snRNA. The Sm site is the binding site for the core proteins but removal of internal loop I (IL-I) has a substantial effect on the binding of the core proteins. The conserved nucleotides within hairpin loop I (boxed) are known to play an important role in 5' and 3' splice-site selection.

serotype from patients with autoimmune disorders such as systemic lupus erythematosus (24) and their binding site on the RNA is often referred to as the Sm-binding site (see below). The second class are the proteins unique to an individual snRNP. In this chapter, we will name the common proteins B, B', D1, etc., while the snRNP-specific proteins will be called U1A, U2B", etc. This, together with reference to *Table 1*, should help the non-expert reader.

Several general comments about the data summarized in *Table 1*, are appropriate before we continue with a more detailed discussion of RNA–protein interactions. The first concerns the common proteins. Most of the set of common-protein components forms a globular multiprotein complex around the Sm site present in U1, U2, U4, and U5 snRNAs, as will be discussed below. At least part of this complex assembles in the cytoplasm in the absence of snRNA (4, 15, 25), revealing the existence of extensive interactions between the core proteins. The second point relates to the various forms of snRNPs listed in *Table 1*. The functional form of the HeLa cell U1 snRNP contains only three identified unique proteins called U1A, U1 70K, and U1C. The 12 S U2 snRNP contains two unique protein subunits called U2A' and U2B" while the U2 snRNP found in the 17 S fraction contains at least nine

Table 1 Protein composition of the splicing snRNPs purified from HeLa cells[a,b]

Protein name	Molecular weight kDa	12S U1	12S U2	17S U2	20S U5	25S U4/U6–U5	12S U4/U6
G	9	●	●	●	●	●	●
F	11	●	●	●	●	●	●
E	12	●	●	●	●	●	●
D1	16	●	●	●	●	●	●
D2	16.5	●	●	●	●	●	●
D3	18	●	●	●	●	●	●
B	28	●	●	●	●	●	●
B'	29	●	●	●	●	●	●
69K	69	●	●	●	●	●	●
C	22	●					
A	34	●					
70K	70	●					
B''	28.5		●	●			
A'	31		●	●			
	35			●			
	53			●			
	60			●			
	66			●			
	92			●			
	110			●			
	120			●			
	150			●			
	160			●			
	15				●	●	
	40				●	●	
	52				●	●	
	100				●	●	
	102				●	●	
	116				●	●	
	200				●	●	
	205				●	●	
	15.5					●	
	20					●	
	27					●	
	60					● ●	●
	90					●	

[a] Adapted from Lührmann et al. (4).
[b] ● indicates protein components present in each snRNP.

additional protein components (*Table 1*). The 17 S U2 snRNP is the form which functions in the splicing reaction (26). Similarly, although a 12 S U4/U6 snRNP and both a simple 10 S and a complex 20 S U5 snRNP can be purified, it is the larger 25 S U4/U6–U5 tri-snRNP that is functional in splicing. This snRNP contains at least fifteen specific proteins in addition to two sets of the common proteins, the latter being derived from its constituent U4/U6 and U5 snRNP components (27). Several of the HeLa proteins have now been matched to yeast homologues (4, 7–9, 28, 29) and it is to be expected that many more will be in the future. However, since virtually all the studies of RNA–protein interactions have utilized vertebrate proteins, we will not discuss this topic in more detail. A final general point is that, at least *in vitro*, there seems to be no hierarchical defined order of protein association with snRNPs, as has been found in ribosome assembly (see Chapter 4). Instead, individual proteins or protein complexes can interact with their cognate RNA-binding sites independently. Nevertheless, once bound, the different proteins interact and influence each other, usually to cause mutual stabilization in the snRNP.

4. Identification of protein-binding sites on snRNAs

In this section we will proceed by first discussing the interaction between the common proteins and the snRNAs and then describe the definition of RNA-binding sites for some snRNP-specific proteins. In a later section we will return to those interactions that are best understood, that is, those between U1A, U2A′, U2B″, and their cognate RNAs and provide more structural detail.

4.1 Common protein-binding: the Sm-binding site

Comparison of the sequences and the then predicted structures of U1, U2, U4, and U5 snRNAs led Branlant *et al.* (30) to identify a conserved structural motif. They predicted that what they called domain A, a single-stranded sequence $PuA(U)_nGPu$ ($n=3$–6) flanked by two hairpin loops, would be the binding site for the common snRNP proteins. Subsequent analysis has shown that the core single-stranded sequence is indeed required for common-protein binding. Mutation of this sequence in the U1 (31), U2 (32), U4 (33), and U5 (34) snRNAs results in the loss of association with the common proteins.

This was determined *in vivo* as follows. The cytoplasm of *Xenopus laevis* oocytes contains a pool of free snRNP proteins which are competent to form snRNPs upon provision of RNA. ^{32}P-labelled wild-type and mutant U1 snRNAs can either be synthesized by *in vitro* transcription using T7 RNA polymerase and injected into the oocyte cytoplasm (31) or by co-injecting α-[^{32}P]GTP together with wild-type or mutant U snRNA genes into oocyte nuclei, where they will be transcribed (32). Assembled snRNPs can be isolated by immunoprecipitation with antibodies against any desired U snRNP protein. If the mutant RNA under study is capable of binding to the protein the ^{32}P-labelled RNA will be immunoprecipitated. Disruption of the binding site for the protein results in loss of immunoprecipitability. Similar experiments have been carried out *in vitro* using either oocyte, egg, or HeLa cell nuclear extracts (31, 35, 36) and, more recently, in a reconstitution system based on purified HeLa cell snRNP proteins and either purified or *in vitro* syn-sized snRNAs (37). The general conclusion, that the PuA(U)$_n$GPu sequence is required for common protein binding, has been upheld.

What is sufficient for this interaction is less certain. Both by microinjection (38) and *in vitro* reconstitution (37) RNAs which resemble the canonical domain A can be shown to bind the common proteins. However, the efficiency of this interaction, in oocytes at least, is low compared to that of a 'real' snRNA (34, 38). In an extensive mutational analysis of common-protein binding to U1 and U5 snRNAs in *Xenopus* oocytes, something of the complexity of the requirements for common-protein binding was revealed (34). First, it was shown that RNA structural elements in addition to those previously considered important were required for efficient association with the core proteins. Apart from the Sm 'core' sequence (PuA(U)$_n$GPu) three other elements of U1 snRNA (*Figure 2*), stem-loops III and IV and the single-stranded 5' end, were all found to be important for efficient Sm-protein binding. It was possible to show that the sequence of the single-stranded 5' end was not important and, further, that the role of this region was confined to snRNP assembly steps, since once the common proteins were associated with U1 snRNA, this region could be removed with no detectable effect on their stable association. For technical reasons, this experiment could not be carried out with the other elements of U1 RNA required for common-protein binding.

Comparing the U1 secondary structure with that of U5 snRNA (*Figure 5*) it is evident that, of the additional elements of the U1 Sm-binding site, only the 3'-terminal hairpin has a counterpart in U5. Surprisingly, however, deletion of the 3' hairpin from U5 did not reduce its ability to bind to the common proteins. Instead, in this case, the presence of the asymmetric bulge near the base of the 5' hairpin (internal loop 2) was found to be essential for efficient common-protein binding. Neither the sequence nor the precise structure of internal loop 2 were important, but its removal (to leave a base-paired helical stem extending to the A19–U58 pair) severely reduced common-protein binding.

The lack of structural similarity between the RNA elements required for efficient binding in U1 and U5 suggested that two different, unique sites were formed. This was confirmed by the observation that exchanging the Sm 'core' AAUUUUUGG sequence of U5 with that of U1 (AAUUUGUGG) gave rise to an RNA that had almost completely lost its ability to associate with the common proteins (34).

How can the same group of proteins make stable associations with these two (and considering the structures of U2 and U4 snRNAs, probably four) substantially different binding sites. The most likely answer is their flexible ability to make a subset of a large possible total of RNA–protein interactions with each snRNA to which they bind.

The extent of this interaction was uncovered by the remarkable results of experiments carried out in yeast by Jones and Guthrie (39). In this study, the effect of many mutations in the 'core' site of U5 snRNA were tested *in vivo* in *Saccharomyces cerevisiae*. Yeast carrying the wild-type U5 snRNA gene under the control of the repressible GAL1 promoter was transformed with plasmids carrying various U5 snRNA mutant genes under the control of their natural promoter. In galactose-containing media both the wild type and mutant U5 snRNAs were expressed. The wild-type gene under the control of the GAL1 promoter could however be completely shut off by transferring the cells into medium containing glucose, allowing a test of the functional properties of the mutants.

Most point mutations at the Sm site did not affect yeast growth. Only substitution of one of the three Us at the 3' end of the $(U)_6$ stretch in the UAUUUUUUGG sequence with a subset of the possible replacement nucleotides affected growth substantially. Double nucleotide substitutions generally had more severe effects, but among the U5 snRNA mutants that could support growth levels close to wild type were some where the purines flanking the 'core' site were changed to pyrimidines and others that included runs of up to 12 U residues within the U stretch! Although the fact that these experiments were carried out *in vivo* precluded an accurate assessment of the relative affinities of these RNAs for the common proteins, the fact that some level of association was maintained with the many mutant forms of U5 snRNA speaks for flexibility and probably redundancy in the Sm 'core' site–common protein interaction (38).

Little is known about the protein side of the Sm-site interaction. cDNAs encoding all of the identified HeLa cell common snRNP proteins have now been cloned (4; R. Lührmann, personal communication). With the exception of the most recently discovered 69 kDa protein (23), none of them resembles known RNA-binding proteins (R. Lührmann, personal communication). Only one, the G protein, has been implicated in a specific interaction with the Sm 'core' site RNA sequence. Heinrichs *et al.* (40) reconstituted U1 snRNP using ^{32}P-labelled U1 snRNA and protein subunits purified from HeLa cell U1 snRNPs. After UV irradiation, the core G protein was cross-linked to U1 RNA. Fingerprint analysis showed that the AUU oligonucleotide within the Sm site (AAUUUGUGG) was cross-linked to and thus is a likely site of direct physical interaction with, the core G protein. With the availability of recombinant common proteins progress in the understanding of protein–protein and protein–RNA interactions within the snRNP core domain is likely to be rapid.

4.2 Specific protein-binding sites within U1 snRNP

The U1 snRNA secondary structure contains four stem–loops (*Figure 2*). The single-stranded 5' end of the RNA, as discussed above, is required for assembly with the common proteins, but does not seem to be stably associated with any proteins. A variety of methods has been used to define stem–loops I and II of U1 as the binding sites for the U1 70K and U1A proteins, respectively. In the case of the U1 70K protein these methods included assembly of mutant and wild-type U1 RNAs in *X. laevis* oocytes followed by immunoprecipitation with anti-U1 70K antibodies (31, 35, 41), immunoprecipitation of nuclease-treated U1 snRNPs assembled *in vitro* (42, 43) or of alkali-treated U1 snRNA-U1 70K complexes (44), and probing of the accessibility of U1 RNA in U1 snRNPs prepared such as to lack either U1A, U1A plus U1C, or all three U1-specific proteins (45). All of these studies led to the conclusion that stem–loop I was required for the association of the U1 70K protein with U1 snRNA. The nucleotide sequence of loop I has been extensively conserved in evolution and its importance in interaction with U1 70K was demonstrated by the fact that changing eight of the ten loop I positions by single nucleotide mutations reduced binding of U1 snRNA to recombinant U1 70K protein (46).

The segment of the U1 70K protein that is sufficient for interaction with stem–loop I has been defined (47). A region of 111 amino acids, encompassing a conserved 80 amino acid motif (RNP motif; see section 6.1 and Chapter 6) found in other RNA-binding proteins was found to interact specifically with U1 snRNA *in vitro*, while the presence of 14 additional amino acids at the C-terminal end of this polypeptide added to the binding affinity.

U1C protein does not seem to interact directly with U1 snRNA, but rather to associate with U1 snRNP via protein–protein interaction. This was deduced from two types of experiment. First, removal of U1C from the U1 snRNP did not render any part of U1 RNA sensitive to structural probes (45). Second, point mutations in stem–loop I that reduced or abolished U1 70K binding had exactly corresponding effects on U1 C binding (41). Since U1 70K is known to interact directly with U1 RNA (see above) this suggested that U1 C requires U1 70K in order to assemble into the U1 snRNP. The first 60 residues of the U1C protein contain many hydrophilic residues, but the rest of the protein is extremely rich in proline and methionine. The N-terminal 45 amino acids of U1 C are sufficient for its incorporation into U1 snRNP (48). This protein fragment contains two histidine and three cysteine residues. One cysteine can be mutated without affecting the ability to bind U1 snRNP, but mutation of either of the other two cysteine residues or either of the two histidine residues abolishes binding. This has been interpreted as evidence that these cysteine and histidine residues may be involved in the coordination of a zinc ligand. There is no direct evidence, however, to show that this region folds around the zinc ion to form a zinc finger-like structure and the amino acid sequence in this region has no appreciable homology to the canonical zinc finger motif (see Chapter 8). Upon removal of U1C protein, U1 snRNP loses its ability to bind RNA containing a 5' splice-site sequence, while readdition of U1C protein restores this activity (49). U1C protein may therefore either interact directly with the 5' splice-site sequence or be essential for the presentation of the 5' end of U1 RNA for base pairing with the 5' splice site.

Similar methods to those used with U1 70K were also employed in conjunction with the U1A protein to determine that it interacts directly with stem–loop II of U1 snRNA (4, 50). One novel method was employed in these studies and since it was also used to identify determinants of specificity in the two different snRNA–protein interactions discussed in section 6 below, we will describe it in more detail. It is a simple assay by which the interaction between any RNA (or mutants thereof) and any protein (or mutants thereof) can potentially be examined. In this example U1 snRNA is the RNA and U1A the protein (50).

The wild-type and mutant U1 RNAs are transcribed in the presence of biotinylated uridine triphosphate in such a way that a small proportion of U residues is replaced by biotinylated-U. The U1A protein cDNA is transcribed into mRNA using T7 RNA polymerase and isotopically labelled U1A protein synthesized by *in vitro* translation of this mRNA in the presence of [^{35}S]methionine. The labelled protein is mixed with biotinylated RNA and the RNA-bound protein recovered together with the biotinylated RNA by binding to streptavidin–agarose beads and analysed by SDS–polyacrylamide gel electrophoresis.

In the analysis of the U1 snRNA–U1A interaction it was found that U1A protein failed to bind to RNA lacking stem–loop II but bound to all other U1 RNA deletion mutants. Conversely, RNA substrates containing only stem–loop II or artificial hairpin loops similar to stem–loop II only in the sequence of the loop nucleotides, did bind to U1A protein. This identified stem–loop II as the U1A-binding site and indicated that the loop sequence was critical for specific binding. It was found, however, that U1A protein can be incorporated into a U1 snRNP containing a U1 snRNA mutant that lacks stem-loop II. This indicates that protein–protein interactions can also mediate binding of U1A protein to the snRNP (31, 35, 50).

The binding of the wild-type and mutant U1A proteins to U1 snRNA (or stem-loop II) has been studied using three different methods: the biotinylated RNA–streptavidin method described above (50, 51), native gel electrophoresis, where free and protein-bound RNA is separated on the basis of the reduced electrophoretic mobility of the RNA–protein complex (52, 53), and nitrocellulose filter-binding assays, where free and protein-bound RNA is separated by the retention of the latter on a nitrocellulose filter (54, 55). The biotinylated RNA method is least quantitative but its main advantage is that a crude protein mixture such as cell extract or *in vitro* translated proteins can be used directly for the assay and the overexpression of proteins in *Escherichia coli*, which often results in the formation of insoluble proteins, is not necessary. The advantage of the native gel method is that the formation of a specific complex can be visualized. The binding of an additional factor or antibody to the complex as well as the stoichiometry of the protein and RNA in the complex can also be studied. However, the binding constant can be underestimated as protein can be depleted from the complex continuously during electrophoresis and caution should therefore be exercised when using this method to determine binding constants. The filter-binding assay is the only method which allows the measurement of the binding constant under true equilibrium conditions. However, specific RNA-protein complex formation and non-specific aggregation cannot be distinguished easily by this method and therefore careful control experiments must be carried out. These three methods have been used extensively to study this and many other protein–RNA interactions.

4.3 Specific protein-binding sites within U2 snRNP

The currently accepted secondary structure of U2 snRNA, with four stem–loop structures, is shown in *Figure 3*. The binding sites of U2A' and U2B", the two U2 snRNP-specific proteins within the 12 S U2 particle, were mapped in ways similar to those employed with the U1 snRNP proteins. Initially, various deletion mutants of U2 snRNA were synthesized *in vivo* in *Xenopus* oocytes and their ability to associate with U2A' and U2B" tested by immunoprecipitation (32, 38). It was deduced from these experiments that the integrity of hairpin IV was required for binding of the two specific proteins.

These results were confirmed by *in vitro* experiments using the biotinylated RNA selection method described above. Deletion of stem–loop I, II, or III or mutation of the Sm core sequence had no effect on the binding of recombinant U2A' and U2B" proteins. Deletion of stem–loop IV, on the other hand, abolished their binding. Conversely, a transcript consisting only of stem–loop IV bound the two proteins similarly to wild-type U2 snRNA (51).

17 S U2 snRNP has at least nine additional U2-specific proteins in addition to those found in the 12 S U2 snRNP (*Table 1*). The molecular weights of these proteins, as estimated by SDS–polyacrylamide gel electrophoresis are 35, 53, 60, 66, 92, 110, 120, 150, and 160 kDa (26, 29). These protein components have not yet been characterized in detail. However, the results of various studies suggest strongly that they bind to the 5' half of U2 snRNA.

The 5' end of U2 snRNA contains sequences complementary to the branch site of pre-mRNA and is known to play a key role in pre-mRNA splicing (see Introduction). Furthermore, recent genetic and biochemical studies have shown that two short stretches of U2 RNA (U6-I and U6-II in *Figure 3*) base pair with U6 snRNA at certain stages of the splicing cycle (reviewed in Moore *et al.* (8) and Newman (12)). To determine whether these were available for base pairing in the U2 snRNP, 2'-O-methyl or 2'-O-allyl RNA oligonucleotides complementary to various regions of U2 snRNA were used to probe the structure of U2 RNA within the U2 snRNP (56). The stem I and IIa regions as well as the branch site complementary region were found to be accessible to oligonucleotide binding, whereas the stem III and IV regions were not (*Figure 3*)

In separate experiments the availability of sequences near the 5' end of U2 RNA for base-pairing interactions was tested by digestion with RNase H in the presence of complementary deoxyoligonucleotides (26). RNase H cleaves the RNA strand of RNA–DNA hybrid duplexes. Consistent with the results of the 2'-O-alkyl-modified oligonucleotide experiments, an oligonucleotide complementary to nucleotides 1–14 of U2 snRNA was found to induce RNase H cleavage of U2 RNA. The unexpected result of this was dissociation of 17 S U2 snRNP-specific proteins (26). Surprisingly, the RNase H digestion proceeded more readily with 17 S U2 snRNPs than with 12 S U2 snRNPs, in which this region of the RNA apparently has no bound proteins. This may imply that the binding of 17 S U2 snRNP-specific proteins induces the proper display of bases for pairing with U6 snRNA. An oligonucleotide complementary to nucleotides 15–26 of U2 snRNA did not induce RNase H digestion. The interpretation of this is that either formation of a stem–loop structure or, more likely given the previous results, U2 snRNP-specific protein binding, makes this region inaccessible for oligonucleotide binding. The branch site complementary region is readily available for base pairing both in 17 S and 12 S U2 snRNPs.

The structure of the 17 S U2 snRNP has also been investigated by chemical modification of the RNA with dimethylsulphate (DMS) and kethoxal (26). DMS methylates adenosine and cytosine whereas kethoxal selectively reacts with guanosine. Chemically modified bases can be identified by primer extension using reverse transcriptase since modification renders the bases unable to form Watson–Crick base pairs and, thus prevents extension of the complementary strand by reverse transcriptase. It was found that the reactivity of C8 and C10 was higher in 17 S U2 snRNP than in 12 S U2 snRNP whereas C40, G42, and C45 were modified more efficiently in the 12 S than in the 17 S U2 snRNP. The results support the interpretation that specific proteins bind to the 5' half of U2 snRNA in the 17 S U2 snRNP and that this binding renders the bases in the 5' hairpin available for pairing with U6 snRNA (*Figure 3*). It is not currently known which of the 17 S U2 snRNP-specific proteins contact RNA directly. The location of the 17 S particle-specific proteins has been nicely confirmed by electron microscopic studies (see below).

4.4 U4/U6 snRNP

U4/U6 snRNP contains only the core proteins that are bound to the Sm site within U4 snRNA (*Figure 4*). The functional U4/U6.U5 snRNP (*Table 1*) contains a large number of specific proteins (27), but nothing is known about their interaction with either U4 or U6 or about their potential involvement in the disruption of the stable U4–U6 interaction that must take place in order to form an active spliceosome (8, 12).

4.5 Specific protein-binding sites within U5 snRNP

U5 snRNA forms two stem–loop structures on either side of the Sm-core site, the 5' of which has two asymmetric internal loops (*Figure 5*). U5 snRNPs have been purified from HeLa cell nuclear extract under various conditions. 10 S U5 snRNPs contain only the core proteins whereas the 17 S U5 snRNP contains additional proteins with apparent molecular weights of 15, 40, 52, 100, 102, 116, 200, and 205 kDa. These proteins have not yet been characterized extensively. Chemical and nuclease probing of 10 and 17 S U5 snRNPs has shown that these proteins protect a limited region of the 5' hairpin of U5 snRNA encompassing internal loop 1 and the stems flanking this loop (57, 58). Newman and Norman (59) showed that the conserved loop I in U5 snRNA plays an important role in 5' and 3' splice-site selection by base pairing with the exon sequences adjacent to the 5' and 3' splice sites in a similar way to the exon-binding loop of the group II intron.

5. Electron microscopic studies of U snRNPs

The structure of U snRNPs has also been extensively investigated by electron microscopy (*Figure 6*). U1 snRNP, the most studied, has a globular domain ~80 Å in

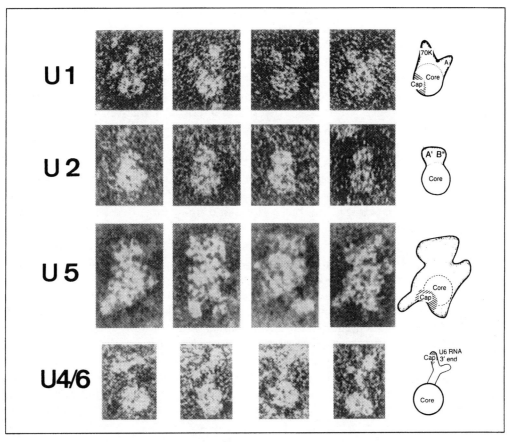

Fig. 6 Electron micrographs of U1, U2 (12S), U5, and U4/U6 snRNPs purified from HeLa cell nuclear extract. The interpretation of images is shown on the right-hand side. (Courtesy of Drs Berthold Kastner and Reinhard Lührmann.)

diameter with two protuberances 40–70 Å long and 30–40 Å wide (60) and, thus, resembles a caricature of a rabbit's head with two long ears. A similar globular domain can be seen in the other spliceosomal U snRNPs and is thought to represent the core complex consisting of the B, B', D1, D2, D3, E, F, and G proteins (snRNPs prepared in this way have lost the 69K protein). Consistent with this hypothesis, the proteinaceous parts of the 10 S form of U5 and the U4/U6 snRNP exhibit only this structure (61, 62).

As discussed above, U1 snRNP can be experimentally depleted of one or more of the specific proteins. Thus, U1 snRNPs lacking U1C, U1A, or U1 70K could also be examined by electron microscopy. One of the protuberances was lost when the U1A and U1C proteins were depleted and both protuberances disappeared when all three specific proteins were depleted. Antibodies specific to U1A and U1 70K proteins were used to identify the presence of each of the proteins in a single protuberance. The protuberance corresponding to U1 70K protein is located closer to the m_3G cap of U1 RNA (63).

The 12 S U2 snRNP, containing the core proteins, as well as U2B" and U2A', has a globular core structure 80 Å in diameter similar to that seen in U1 snRNP and a smaller domain, 60 Å wide and 40 Å long, attached to the main body. This was named the head domain and probably contains U2B" and U2A' proteins as it is lost when U2B" and U2A' proteins are depleted from U2 snRNP (61).

The 17 S U2 snRNP consists of two globular domains connected by a short filamentous structure which probably corresponds to part of U2 snRNA. One of the globular domains is thought to be the structure observed in 12 S U2 snRNP, containing the core as well as the U2A' and U2B" proteins. The second domain therefore probably contains the majority of the 17 S U2 snRNP specific proteins.

Purified U4/U6 snRNP has a globular domain 80 Å in diameter, corresponding to the core particle, from which a Y-shaped filamentous peripheral structure extends. Various experiments were carried out to identify the Y-shaped structure as U4 and U6 snRNA (62). First, antibody against the m_3G cap was added. The IgG molecule could be seen bound to the filamentous domain at a site clearly separated from the core domain, as would be expected from the structure of U4/U6 and the distance between the m_3G cap of U4 and its Sm-binding site (*Figure 4*). Second, U4/U6 snRNP was incubated with an oligonucleotide complementary to nucleotides 78–95 of U6 RNA which contained three biotinylated G nucleotides at its 5' end, then further incubated with streptavidin. The U4/U6 snRNP was fractionated on a glycerol gradient and examined. Streptavidin was observed on one arm of the Y-shaped filamentous domain, ~ 100 Å distant from the globular core domain. Third, U4/U6 snRNP was treated with RNase H in the presence of a DNA oligonucleotide complementary to nucleotides 58–84 of U4 snRNA. After this treatment the Y-shaped filamentous domain was no longer seen.

The 20 S U5 snRNP (*Figure 6*) appears to consist of two large domains which represent approximately one-third and two-thirds of the whole particle and form the head and main body, respectively (61). The 5' end of U5 snRNA has been localized to the bottom end of the main body. In summary, the electron microscope studies of U snRNPs form a useful complement to the biochemical studies described above. They help to localize some of the proteins which have not yet been studied by other methods and form the first step towards higher resolution study of the snRNP particles.

6. Structural basis for two specific RNA–protein interactions

As we have seen in the previous sections some protein-binding sites within snRNAs have been mapped using various biochemical and genetic techniques. Three of the proteins, U1 70K, U1A, and U2B", are known to bind to RNA hairpins. The core proteins and 17S U2 snRNP-specific proteins bind, at least in part, to single-stranded regions lying between secondary structure elements. The interactions of the U1A and U2B" proteins with their cognate RNA hairpins have been extensively studied and will now be discussed in more detail.

6.1 Binding of U1A and U2B" proteins

An amino acid sequence motif consisting of ~80 residues has been found in many RNA-binding proteins with diverse functions (see Chapter 6). This motif has received various names, but will be called the RNP motif here. This motif is considered to be an RNA-binding module that must have appeared relatively early in evolution. Somewhat surprisingly, the boundaries of the amino acid sequences required for RNA binding in the proteins tested often extend beyond the boundaries of the conserved motif. The RNP motif contains two particularly strongly conserved short stretches of sequence normally referred to as RNP1 and RNP2 (64; see Birney *et al.* (65) for a recent alignment of members of the family). The U1A and U2B" proteins each contain two RNP motifs which are connected by a linker polypeptide (66; *Figure 7*).

Recombinant U1A protein was found to bind U1 snRNA in the absence of any other factor (50–54). U2B" protein, in spite of its closely related amino acid sequence, did not bind specifically to U2 snRNA on its own. However, when complemented with HeLa cell nuclear extract, specific binding of U2B" protein to U2 RNA was observed. U2A' protein, the second U2-specific protein of the 12 S U2 snRNP, was an obvious candidate as a factor required for the binding of U2B" protein to U2 snRNA. U2A' protein was synthesized *in vitro* and added to the binding assay. In the presence of *in vitro* synthesized U2A' protein-specific binding of U2B" protein to U2 RNA was observed (51).

Thus, despite the fact that the amino acid sequence of U2B" protein is strikingly similar to that of U1A protein (*Figure 7*), U2B" protein binds to stem–loop IV of U2 RNA only after it forms a protein – protein complex with U2A' protein while U1A binds to stem–loop II of U1 snRNA alone. Besides the protein sequences, the RNA-binding sites for U1A protein and the U2A'–U2B" complex also share considerable similarity; the nucleotide sequence of U1 RNA loop II is **AUUGCA**CUCC whereas that of U2 RNA loop IV is **AUUGCA**GUACC (common nucleotides in bold). These similarities were exploited in studying the structural basis of the two RNA–protein interactions.

After extensive changes to the two stem–loops, it was concluded that most of the specificity of the two interactions was due to the loop sequences, with the stems playing only a minor role (51). The absence of a stem structure, however, caused a considerable reduction in the *affinity* of the U1A–U1 RNA interaction (50, 67). U2A'–U2B" binding to a specific RNA substrate in the absence of a stem structure has not been analysed to date. The predominance of the loop sequences in generating specificity was most strongly underlined by generating a U1 RNA whose loop II sequence was **AUUGCAGUACC**. This RNA bound specifically to the U2A'–U2B" complex, rather than to U1A. Conversely, a U2 RNA

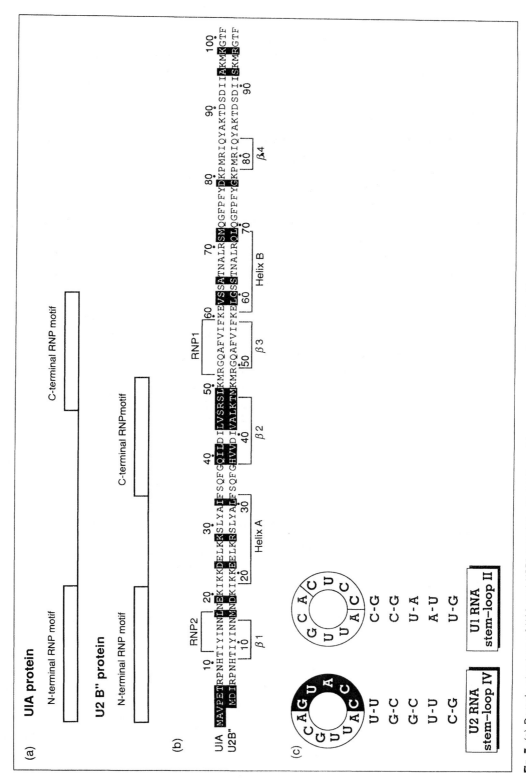

Fig. 7 (a) Domain structure of U1A and U2B" proteins, (b) amino acid sequence comparison of the N-terminal RNP motifs and the additional polypeptides necessary for RNA binding of U1A and U2B", (c) the binding sites of U1A and U2B" in U1 and U2 snRNA.

whose loop IV sequence was **AUUGCACUCC** bound to U1A protein specifically (51, 68).

On the protein side, it has already been mentioned that the U1A and U2B" proteins consist of two RNP motifs connected by polypeptides of different lengths (*Figure 7*). The sequence identity between the smallest RNA binding fragments (residues 7–101 of U1A protein and 4–98 of U2B" protein) of these proteins is 77%, with weaker similarity extending another 30 residues towards the C-terminus. The C-terminal RNP motifs of the two proteins show even higher (86%) sequence identity. An N-terminal 101 residue fragment of U1A protein retains full RNA-binding activity (50, 52–54, 69) and the corresponding fragment of U2B" protein retains interaction with U2A' protein and specific RNA-binding activity in the presence of U2A' protein (51, 70).

In order to identify amino acid residues in the two proteins that discriminate between their two cognate RNAs, chimeras of U1A and U2B" proteins were made. The residues between 38 and 49 of U1A protein and the corresponding residues in U2B" were found to be major determinants of their RNA-binding specificity (51, 68). The replacement of these residues together with two additional amino acid exchanges between the U1A and U2B" proteins was sufficient to generate a U1A protein that exhibited tight binding to U2 snRNA in concert with U2A' protein (70).

With the solution of the structure of the N-terminal RNP motif of U1A protein by both X-ray crystallography (69) and nuclear magnetic resonance (NMR) (71) the implications of these experiments became very clear. The RNP domain of U1A protein consists of a four-stranded β-sheet flanked by two α-helices which are arranged in the order β1-helix(A)-β2-β3-helix(B)-β4. The U2B"-specific amino acids required for interaction with U2A' lie on a surface formed by the lower half of β2 and helix A (as drawn in *Figure 8*), positioning U2A' adjacent to this part of the protein in the complex. It has proved more difficult to locate the RNA-binding surface of the motif precisely.

The most highly conserved stretches of sequence in the RNP motif family, RNP1 and RNP2, are located side by side in the β3- and β1-strands, respectively (*Figure 8*). Amino acid residues within RNP1 and RNP2 were mutated to determine whether they were directly involved in RNA binding (53, 69). The crystal structures of DNA–protein complexes show that backbone phosphate groups often interact with positively charged arginine and lysine side chains. In addition, hydrogen bonds between bases of DNA and protein side chains are commonly found. Therefore, Thr11, Tyr13, Asp15, and Asp16 in RNP1 were replaced individually by a non-hydrogen bond-forming amino acid of similar size. The Thr11 → Val mutation, which replaces a hydroxyl group with a methyl group, was sufficient to abolish RNA binding. Mutation of Tyr13, Asp15, and Asp16 also had a substantial effect on RNA binding, implying that these residues are directly involved in binding and possibly form hydrogen bonds with the RNA. The mutagenesis of RNP1 residues suggested that these residues were also directly involved in RNA binding. An Arg52 → Glu mutation completely abolished RNA binding whereas an Arg52 → Lys mutation had no significant effect.

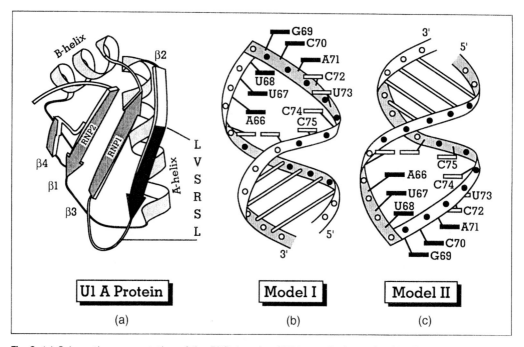

Fig. 8 (a) Schematic representation of the RNP domain of U1A protein determined by X-ray crystallography. The RNP domain consists of a four-stranded β-sheet flanked by two α-helices. The RNP1 and RNP2 sequences lie on the β3- and β1-strands, respectively. The cognate binding RNA hairpin, U1 snRNA stem–loop II, is shown in two orientations: (b) model I and (c) model II. Backbone phosphates protected from ethylation by U1A protein are shown as closed circles. Bases in the loop which are conserved between U1 snRNA stem–loop II and U2 snRNA stem–loop IV are shown in black.

How then does the surface of the four-stranded β-sheet specifically interact with RNA? The amino acid residues on the β1- and β3-strands containing RNP2 and RNP1 are conserved between U1A protein and U2B" protein, while the residues of the β2-strands differ between the two proteins. The nucleotide sequence of the 5' halves of the two loop structures to which the proteins bind is conserved, but that of the 3' halves is variable. The RNA-binding experiments with the chimeric U1A and U2B" proteins described above in combination with the point mutation analysis of the U1A motif suggested that the 5' half of the loop would most likely interact with amino acids in RNP1 and RNP2 while the 3' halves of the two loops would probably interact with specific residues in the β2-strand.

These considerations led to two possible models of interaction between the protein and RNA as shown in *Figure 8*. In model I (*Figure 8b*) the major groove side of the loop interacts with the β-sheet whereas in model II (*Figure 8c*) the minor groove site of the loop is in contact with the protein. Ethylation protection experiments were carried out to distinguish these possibilities (53). The phosphate groups of the U1 RNA stem–loop II were reactive with ethylnitrosourea in the absence of the protein but the phosphates in the 5' stem region and the 3' half of the loop were protected from ethylation by U1A protein. In model II the phosphate

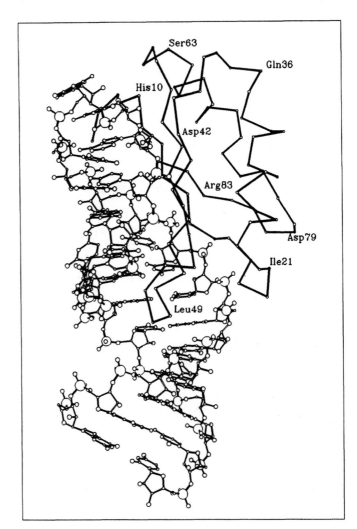

Fig. 9 A model of U1A protein complexed with its cognate RNA hairpin (53).

groups of the 5' stem are facing away from the protein and it cannot therefore account for the results of the ethylation protection experiment. In model I based on the crystal structure of U1A protein (69) and a model of stem–loop II of U1 snRNA (73) (*Figure 9*), the backbone phosphates of the 5' stem are wedged between the polypeptide loops between the β1-strand and helix A and between the β2- and β3-strands. The protein mutagenesis experiments showed that arginine and lysine residues in these loops were important for RNA binding (69). In the three dimensional (3-D) model of the complex (*Figure 9*) the bases of the RNA loop on the major groove side are seen to be well exposed and available for protein binding. The surface of the β-sheet has a right-handed twist when viewed along the β-strands (72) and this may provide a good complementary surface to the RNA loop which is extended from the right-handed double helical region and therefore has a right-handed twist. The model fits well with the available data and examination by molecular graphics reveals a good fit of the modelled RNA with the protein. However, the crystal structures of two aminoacyl–tRNA synthetases with their cognate tRNAs show that the RNA can undergo gross structural changes upon protein binding (see Chapter 3). The RNA structure incorporated into this model was predicted to be the structure of stem–loop II in free U1 snRNA (73). Study of the structure of the RNA–protein complex at high resolution will therefore be necessary before we can understand the interaction in detail.

7. A spin-off with structural feedback

The detailed investigation of the binding specificity of U1A protein for RNA directly opened an unexpected research direction. As previously discussed, the U1A protein binds to stem–loop II of U1 snRNA. The sequence of the loop, but no specific sequence of the stem is required for specific binding (50, 67). Thus, when two sequences resembling loop II of U1 snRNA, AUUGCAC, and AUUGUAC, were found in an evolutionarily conserved region of the 3' untranslated region (UTR) of the U1A mRNA, it was tempting to determine whether this region would bind U1A protein. The result was positive (74). This binding turned out to be part of an autoregulatory loop by which U1A protein limits its own production by binding to the 3' UTR sequence in the nuclear precursor of its mRNA and preventing its cleavage and polyadenylation, that is, preventing production of the mature U1A mRNA (74).

Investigation of the secondary structure of the U1A-binding sites in the 3' UTR of the human U1A pre-mRNA by a variety of methods led to the elucidation of the structure shown in *Figure 10* (55). Comparison with stem–loop II of U1 snRNA

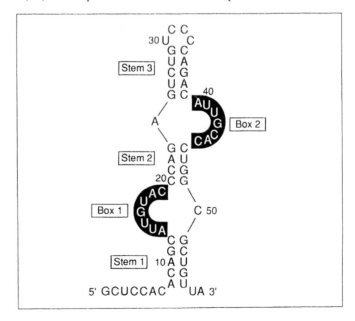

Fig. 10 The binding site for U1A protein in the 3' untranslated region (UTR) within its own mRNA. The sequences in bold type are closely related to that found in the 5' half of U1 snRNA stem–loop II.

(*Figure 7*) reveals little obvious similarity outside of the internal loop sequences in the 3' UTR structure (labelled boxes 1 and 2) that resemble the 5'-most seven nucleotides of the loop in stem–loop II of U1 snRNA. Both RNAs bind the protein with comparable affinity (K_ds of ~ $5-10 \times 10^{-11}$ M (55)). Thus, it is very likely that the AUUGCAC sequence makes an identical sequence-specific contact with U1A protein in complexes with U1A stem–loop II and the internal loop in the 3'UTR sequence. Jessen *et al.* (53) concluded that bases within the AUUGCAC sequence are in contact with amino acid side chains of RNP1 and RNP2 whereas the phosphate backbone of the 3' half of U1 RNA stem–loop II is in close contact with U1A protein. It is remarkable that the same specific-binding sequence, AUUGCAC, is displayed to U1A protein using different RNA secondary structure scaffolds. The different sequences outside these seven nucleotides are presumably used to increase protein-binding affinity by stabilizing the loop structure and providing phosphate contacts and this function may be achieved in different ways in the two cases. Comparison of the two structures will provide important insight into how nature can exploit diverse situations to generate specific RNA–protein interactions.

References

1. Berget, S. M., Moore, C., and Sharp, P. A. (1977) Spliced segments at the 5' terminus of adenovirus 2 late mRNA. *Proc. Natl Acad. Sci. USA*, **74**, 3171.
2. Chow, L. T., Gelinas, R. E., Broker, T. R., and Roberts, R. J. (1977) An amazing sequence arrangement at the 5' ends of adenovirus 2 messenger RNA. *Cell*, **12**, 1.
3. Abelson, J. (1979) RNA processing and the intervening sequence problem. *Ann. Rev. Biochem.*, **48**, 1035.
4. Lührmann, R., Kastner, B., and Bach, M. (1990) Structure of spliceosomal snRNPs and their role in pre-mRNA splicing. *Biochim. Biophys. Acta*, **1087**, 265.
5. Green, M. R. (1991) Biochemical mechanisms of constitutive and regulated pre-mRNA splicing. *Ann. Rev. Cell Biol.*, **7**, 559.
6. Guthrie, C. (1991) Messenger RNA splicing in yeast: clues to why the spliceosome is a ribonucleoprotein. *Science*, **253**, 157.
7. Rymond, B. C. and Rosbash, M. (1992) Yeast pre-mRNA splicing. In *Molecular and Cellular Biology of the Yeast* Saccharomyces: *Gene Expression.* Jones, E. W., Pringle, J. R., and Broach, J. R. (eds). Cold Spring Harbor Laboratory Press, Cold Spring Harbor, NY, Vol. 2, p. 143.
8. Moore, M. J., Query, C., and Sharp, P. A. (1993) Splicing of precursors to mRNA by the spliceosome. In *The RNA World*. Gesteland, R. F. and Atkins, J. F. (eds). Cold Spring Harbor Laboratory Press, Cold Spring Harbor, NY, p. 303.
9. Beggs, J. D. (1993) Yeast protein splicing factors involved in nuclear pre-mRNA splicing. *Mol. Biol. Rep.*, **18**, 99.
10. Mattaj, I. W., Tollervey, D., and Séraphin, B. (1993) Small nuclear RNAs in messenger RNA and ribosomal RNA processing. *FASEB.*, **7**, 47.
11. Baserga, S. J. and Steitz, J. A. (1993) The diverse world of small ribonucleoproteins. In *The RNA World*. Gesteland, R. F. and Atkins, J. F. (eds). Cold Spring Harbor Laboratory Press, Cold Spring Harbor, NY, p. 359.
12. Newman, A. J. (1993) RNA:RNA interactions in the spliceosome. *Mol. Biol. Rep.*, **18**, 85.

13. Hamm, J. and Mattaj, I. W. (1990) Monomethylated cap structures facilitate RNA export from the nucleus. *Cell*, **63**, 109.
14. Izaurralde, E., Stepinski, J., Darzynkiewicz, E., and Mattaj, I. W. (1992) A cap binding protein that may mediate nuclear export of RNA polymerase II-transcribed RNAs. *J. Cell Biol.*, **118**, 1287.
15. Mattaj, I. W. (1988) UsnRNP assembly and transport. In *Structure and Function of Major and Minor Small Nuclear Ribonucleoprotein Particles*. Birnstiel, M. L. (ed.). Springer-Verlag, Berlin, Heidelberg, p. 100.
16. Izaurralde, E. and Mattaj, I. W. (1992) Transport of RNA between nucleus and cytoplasm. *Seminars Cell Biol.*, **3**, 279.
17. Singh, R. and Reddy, R. (1989) Gamma-monomethyl phosphate: a cap structure in spliceosomal U6 small nuclear RNA. *Proc. Natl Acad. Sci. USA*, **86**, 8280.
18. Shumyatsky, G., Wright, D., and Reddy, R. (1993) Methylphosphate cap structure increases the stability of 7SK, B2 and U6 small RNAs in *Xenopus* oocytes. *Nucleic Acids Res.*, **21**, 4756.
19. Lund, E. and Dahlberg, J. E. (1992) Cyclic 2′,3′-phosphates and nontemplated nucleotides at the 3′ end of spliceosomal U6 small nuclear RNA's. *Science*, **255**, 327.
20. Guthrie, C. and Patterson, B. (1988) Spliceosomal snRNAs. *Ann. Rev. Genet.*, **22**, 387.
21. Reddy, R. and Busch, H. (1988) Small nuclear RNAs: RNA sequences, structure, and modifications. In *Structure and Function of Major and Minor Small Nuclear Ribonucleoprotein Particles*. Birnstiel, M. L. (ed.). Springer-Verlag, Berlin, Heidelberg, p. 1.
22. Bach, M., Bringmann, P., and Lührmann, R. (1990) Purification of small nuclear ribonucleoprotein particles with antibodies against modified nucleotides of small nuclear RNAs. *Methods Enzymol.*, **181**, 232.
23. Hackl, W., Fischer, U., and Lührmann, R. (1993) A 69-kD protein that associates with the Sm core domain of several spliceosomal snRNP species. *J. Cell. Biol.*, **124**, 261.
24. Lerner, M. R. and Steitz, J. A. (1979) Antibodies to small nuclear RNAs complexed with proteins are produced by patients with systemic lupus erythematosus. *Proc. Natl Acad. Sci. USA*, **76**, 5495.
25. Zieve, G. W. and Sauterer, R. A. (1990) Cell biology of the snRNP particles. *CRC Crit. Rev. Biochem. Mol. Biol.*, **25**, 1.
26. Behrens, S.-E., Tyc, K., Kastner, B., Reichelt, J., and Lührmann, R. (1993) Small nuclear ribonucleoprotein (RNP) U2 contains numerous additional proteins and has a bipartite RNP structure under splicing conditions. *Mol. Cell. Biol.*, **13**, 307.
27. Behrens, S.-E. and Lührmann, R. (1991) Immunoaffinity purification of a [U4/U6.U5] tri-snRNP from human cells. *Genes Devel.*, **5**, 1439.
28. Rymond, B. C., Rokeach, L. A., and Hoch, S. O. (1993) Human snRNP polypeptide D1 promotes pre-mRNA splicing in yeast and defines nonessential yeast Smd1p sequences. *Nucleic Acids Res.*, **21**, 3501.
29. Behrens, S.-E., Galisson, F., Legrain, P., and Lührmann, R. (1993) Evidence that the 60-kDa protein of 17 S U2 small nuclear ribonucleoprotein is immunologically and functionally related to the yeast PRP9 splicing factor and is required for the efficient formation of prespliceosomes. *Proc. Natl Acad. Sci. USA*, **90**, 8229.
30. Branlant, C., Krol, A., Ebel, J. P., Lazar, E., Haendler, B., and Jacob, M. (1982) U2 RNA shares a structural domain with U1, U4 and U5 RNAs. *EMBO J.*, **1**, 1259.
31. Hamm, J., Kazmaier, M., and Mattaj, I. W. (1987) *In vitro* assembly of U1 snRNPs. *EMBO J.*, **6**, 3479.

32. Mattaj, I. W. and De Robertis, E. M. (1985) Nuclear segregation of U2 snRNA requires binding of specific snRNP proteins. *Cell*, **40**, 111.
33. Vankan, P., McGuigan, C., and Mattaj, I. W. (1990) Domains of U4 and U6 snRNAs required for snRNP assembly and splicing complementation in *Xenopus* oocytes. *EMBO J.*, **9**, 3397.
34. Jarmolowski, A. and Mattaj, I. W. (1993) The determinants for Sm protein binding to *Xenopus* U1 and U5 snRNAs are complex and non-identical. *EMBO J.*, **12**, 223.
35. Hamm, J., Van Santen, V. L., Spritz, R. A., and Mattaj, I. W. (1988) Loop I of U1 small nuclear RNA is the only essential RNA sequence for binding of specific U1 small nuclear ribonucleoprotein particle proteins. *Mol. Cell. Biol.*, **8**, 4787.
36. Patton, J. R., Patterson, R. J., and Pederson, T. (1987) Reconstitution of the U1 small nuclear ribonucleoprotein particle. *Mol. Cell. Biol.*, **7**, 4030.
37. Sumpter, V., Kahrs, A., Fischer, U., Kornstädt, U., and Lührmann, R. (1992) In vitro reconstitution of U1 and U2 snRNPs from isolated proteins and snRNA. *Mol. Biol. Rep.*, **16**, 229.
38. Mattaj, I. W. (1986) Cap trimethylation of U snRNA is cytoplasmic and dependent on U snRNP protein binding. *Cell*, **46**, 905.
39. Jones, M. H. and Guthrie, C. (1990) Unexpected flexibility in an evolutionarily conserved protein–RNA interaction: genetic analysis of the Sm binding site. *EMBO J.*, **9**, 2555.
40. Heinrichs, V., Hackl, W., and Lührmann, R. (1992) Direct binding of small nuclear ribonucleoprotein G to the Sm site of small nuclear RNA. Ultraviolet light cross-linking of protein G to the AAU stretch within the Sm site (AAUUUGUGG) of U1 small nuclear ribonucleoprotein reconstituted *in vitro*. *J. Mol. Biol.*, **227**, 15.
41. Hamm, J., Dathan, N. A., Scherly, D., and Mattaj, I. W. (1990) Multiple domains of U1 snRNA, including U1 specific protein binding sites, are required for splicing. *EMBO J.*, **9**, 1237.
42. Patton, J. R. and Pederson, T. (1988) The M_r 70,000 protein of the U1 small nuclear ribonucleoprotein particle binds to the 5′ stem–loop of U1 RNA and interacts with Sm domain proteins. *Proc. Natl Acad. Sci. USA*, **85**, 747.
43. Patton, J. R., Habets, W., Van Venrooij, W. J., and Pederson, T. (1989) U1 small nuclear ribonucleoprotein particle-specific proteins interact with the first and second stem–loops of U1 RNA, with the A protein binding directly to the RNA independently of the 70K and Sm proteins. *Mol. Cell. Biol.*, **9**, 3360.
44. Query, C. C., Bentley, R. C., and Keene, J. D. (1989) A specific 31-nucleotide domain of U1 RNA directly interacts with the 70K small nuclear ribonucleoprotein component. *Mol. Cell. Biol.*, **9**, 4872.
45. Bach, M., Krol, A., and Lührmann, R. (1990) Structure-probing of U1 snRNPs gradually depleted of the U1-specific proteins A, C and 70k. Evidence that A interacts differentially with developmentally regulated mouse U1 snRNA variants. *Nucleic Acids Res.*, **18**, 449.
46. Surowy, C. S., Van Santen, V. L., Scheib-Wixted, S. M., and Spritz, R. A. (1989) Direct, sequence-specific binding of the human U1–70K ribonucleoprotein antigen protein to loop 1 of U1 small nuclear RNA. *Mol. Cell. Biol.*, **9**, 4179.
47. Query, C. C., Bentley, R. C., and Keene, J. D. (1989) A common RNA recognition motif identified within a defined U1 RNA binding domain of the 70K U1 snRNP protein. *Cell*, **57**, 89.
48. Nelissen. R. L. H., Heinrichs, V., Habets, W. J., Simons, F., Lührmann, R., and Van Venrooij, W. J. (1991) Zinc finger-like structure in U1-specific protein C is essential for specific binding to U1 snRNP. *Nucleic Acids Res.*, **19**, 449.

49. Heinrichs, V., Bach, M., Winkelmann, G., and Lührmann, R. (1990) U1-specific protein C needed for efficient complex formation of U1 snRNP with a 5' splice site. *Science*, **247**, 69.
50. Scherly, D., Boelens, W., Van Venrooij, W. J., Dathan, N. A., Hamm, J., and Mattaj, I. W. (1989) Identification of the RNA binding segment of human U1A protein and definition of its binding site on U1 snRNA. *EMBO J.*, **8**, 4163.
51. Scherly, D., Boelens, W., Dathan, N. A., van Venrooij, W. J., and Mattaj, I. W. (1990) Major determinants of the specificity of interaction between small nuclear ribonucleoproteins U1A and U2B" and their cognate RNAs. *Nature*, **345**, 502.
52. Lutz-Freyermuth, C., Query, C. C., and Keene, J. D. (1990) Quantitative determination that one of two potential RNA-binding domains of the A protein component of the U1 small nuclear ribonucleoprotein complex binds with high affinity to stem–loop II of U1 RNA. *Proc. Natl Acad. Sci. USA*, **87**, 6393.
53. Jessen, T.-H., Oubridge, C., Hiang Teo, C., Pritchard, C., and Nagai, K. (1991) Identification of molecular contacts between the U1A small nuclear ribonucleoprotein and U1 RNA. *EMBO J.*, **10**, 3447.
54. Hall, K. B. and Stump, W. T. (1992) Interaction of N-terminal domain of U1A protein with an RNA stem–loop. *Nucleic Acids Res.*, **20**, 4283.
55. Van Gelder, C. W. G., Gunderson, S. I., Jansen, E. J. R., Boelens, W. C., Polycarpou-Schwarz, M., Mattaj, I. W., and van Venrooij, W. J. (1993) A complex secondary structure in U1A pre-mRNA that binds two molecules of U1A protein is required for regulation of polyadenylation. *EMBO J.*, **12**, 5191.
56. Lamond, A. I., Sproat, B., Ryder, U., and Hamm, J. (1989) Probing the structure and function of U2 snRNP with antisense oligonucleotides made of 2'-OMe RNA. *Cell*, **58**, 383.
57. Black, D. L. and Pinto, A. L. (1989) U5 small nuclear ribonucleoprotein: RNA structure analysis and ATP-dependent interaction with U4/U6. *Mol. Cell. Biol.*, **9**, 3350.
58. Bach, M. and Lührmann, R. (1991) Protein–RNA interactions in 20 S U5 snRNPs. *Biochim. Biophys. Acta*, **1088**, 139.
59. Newman, A. J. and Norman, C. (1992) U5 snRNA interacts with exon sequences at 5' and 3' splice sites. *Cell*, **68**, 743.
60. Kastner, B. and Lührmann, R. (1989) Electron microscopy of U1 small nuclear ribonucleoprotein particles: shape of the particle and position of the 5' RNA terminus. *EMBO J.*, **8**, 277.
61. Kastner, B., Bach, M., and Lührmann, R. (1990) Electron microscopy of small nuclear ribonucleoprotein (snRNP) particles U2 and U5: evidence for a common structure-determining principle in the major U snRNP family. *Proc. Natl Acad. Sci. USA*, **87**, 1710.
62. Kastner, B., Bach, M., and Lührmann, R. (1991) Electron microscopy of U4/U6 snRNP reveals a Y-shaped U4 and U6 RNA containing domain protruding from the U4 core RNP. *J. Cell Biol.*, **112**, 1065.
63. Kastner, B., Kornstädt, U., Bach, M., and Lührmann, R. (1992) Structure of the small nuclear RNP particle U1: identification of the two structural protuberances with RNP-antigens A and 70K. *J. Cell Biol.*, **116**, 839.
64. Bandziulis, R. J., Swanson, M. S., and Dreyfuss, G. (1989) RNA-binding proteins as developmental regulators. *Genes Devel.*, **3**, 431.
65. Birney, E., Kumar, S., and Krainer, A. R. (1993) Analysis of the RNA-recognition motif and RS and RGG domains: conservation in metazoan pre-mRNA splicing factors. *Nucleic Acids Res.*, **21**, 5803.

66. Sillekens, P. T. G., Habets, W. J., Beijer, R. P., and van Venrooij, W. J. (1987) cDNA cloning of the human U1 snRNA-associated A protein: extensive homology between U1 and U2 snRNP-specific proteins. *EMBO J.*, **6**, 3841.
67. Tsai, D. E., Harper, D. S., and Keene, J. D. (1991) U1-snRNP-A protein selects a ten nucleotide consensus sequence from a degenerate RNA pool presented in various structural contexts. *Nucleic Acids Res.*, **19**, 4931.
68. Bentley, R. C. and Keene, J. D. (1991) Recognition of U1 and U2 small nuclear RNAs can be altered by a 5-amino-acid segment in the U2 small nuclear ribonucleoprotein particle (snRNP) B" protein and through interactions with U2 snRNP-A' protein. *Mol. Cell. Biol.*, **11**, 1829.
69. Nagai, K., Oubridge, C., Jessen, T. H., Li, J., and Evans, P. R. (1990) Crystal structure of the RNA-binding domain of the U1 small nuclear ribonucleoprotein A. *Nature*, **348**, 515.
70. Scherly, D., Dathan, N. A., Boelens, W., van Venrooij, W. J., and Mattaj, I. W. (1990) The U2B" RNP motif as a site of protein–protein interaction. *EMBO J.*, **9**, 3675.
71. Hoffman, D. W., Query, C. C., Golden, B. L., White, S. W., and Keene, J. D. (1991) RNA-binding domain of the A protein component of the U1 small nuclear ribonucleoprotein analyzed by NMR spectroscopy is structurally similar to ribosomal proteins. *Proc. Natl Acad. Sci. USA*, **88**, 2495.
72. Chothia, C. (1984) Principles that determine the structure of proteins. *Ann. Rev. Biochem.*, **53**, 531
73. Krol, A., Westhof, E., Bach, M., Lührmann, R., Ebel, J.-P., and Carbon, P. (1990) Solution structure of human U1 snRNA. Derivation of a possible three-dimensional model. *Nucleic Acids Res.*, **18**, 3803.
74. Boelens, W. C., Jansen, E. J. R., van Venrooij, W. J., Stripecke, R., Mattaj, I. W., and Gunderson, S. I. (1993) The human U1 snRNP-specific U1A protein inhibits polyadenylation of its own pre-mRNA. *Cell*, **72**, 881.

8 | Interaction of 5S RNA with TFIIIA

TOMAS PIELER

1. Introduction

Among the structural motifs which identify DNA- and RNA-binding proteins in eukaryotes, zinc finger modules are unique in defining the nucleic acid-binding domain in proteins with a demonstrated ability to form sequence-specific complexes with either RNA or DNA or even with both classes of nucleic acids (1). Transcription factor IIIA (TFIIIA) is the founding member of the zinc finger protein superfamily, which is comprised of several hundred members in vertebrates (2, 3). TFIIIA is one of the most abundant proteins in immature *Xenopus laevis* oocytes, where it is found primarily in a complex with 5S ribosomal RNA (4). The key structural feature in TFIIIA comes from nine copies of a sequence motif that has been termed zinc finger, since two pairs of invariant cysteines and histidines mediate the three-dimensional (3-D) folding of two antiparallel β-sheets and one α-helical element, mainly by coordination of zinc (*Figure 1*). The zinc finger cluster in TFIIIA is not only responsible for the specific recognition of 5S RNA, but also promotes sequence-specific binding to the internal control region of the 5S RNA gene, thereby defining the first step in the formation of an active transcription complex (5, 6). *Xenopus* oocytes contain a second abundant zinc finger protein which is also in a specific complex with 5S RNA and which is termed p43 (7). However, in contrast to TFIIIA, this protein cannot interact with the 5S RNA gene (*Figure 1*). A third function that has been ascribed to TFIIIA is in the nucleocytoplasmic transport of newly transcribed 5S RNA. In *Xenopus* oocytes, the TFIIIA–5S RNA complex (7S RNP) forms in the nucleus and migrates to the cytoplasmic compartment, where it accumulates. Without protein binding, nuclear export of 5S RNA is inhibited (8).

This chapter will discuss structural aspects of the specific recognition between 5S RNA and TFIIIA. In the absence of a final structural solution for the 7S RNP, we rely entirely on experimental results coming from chemical and enzymatic probing of the complex, as well as on systematic mutational analyses of RNA and protein components. Solution of the co-crystal structure for a (different) zinc finger protein–DNA complex has revealed basic principles of zinc finger–DNA recognition: base-specific contacts with the zinc finger α-helical element lying in the major

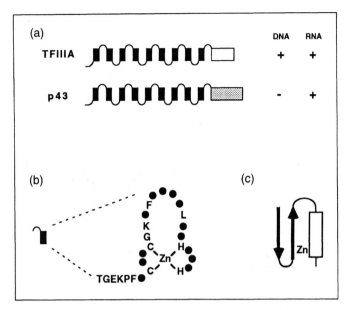

Fig. 1 Structure and nucleic acid-binding properties of TFIIIA and p43. (a) Schematic representation of the structural architecture of TFIIIA and p43; both proteins contain nine zinc finger modules in tandem array, starting from the N-terminus (indicated by filled rectangles). The C-termini differ in length and sequence (indicated by open and stippled rectangles, respectively). DNA and RNA binding activities of the two proteins are also indicated. (b) Conserved amino acid sequence elements and zinc coordination in a single zinc finger module; invariant zinc-coordinating residues as well as highly conserved residues are specified and variable positions are indicated by dots. (c) Schematic representation of common secondary–tertiary structure features in zinc finger modules; two antiparallel β-strands (arrows) and one α-helical element (rectangle) are indicated.

groove of the DNA are primary determinants of binding specificity (9). DNA recognition by TFIIIA seems to operate via similar interactions and a reasonable model for the TFIIIA–5S gene complex has been proposed (10). Starting from this situation, one central objective of this essay will be in trying to answer the question of whether zinc finger–RNA recognition may operate via similar principles to those described for zinc finger–DNA recognition or if RNA binding is achieved by basically different mechanisms.

2. Structural features of 5S ribosomal RNA

The availability of more than 700 different 5S RNA primary sequences has unequivocally established its universal secondary structure, which is shown in *Figure 2*. This structure defines five helical elements (I–V), separated by internal loops (A, B, and E) or closed by hairpin loops (C and D). Primary sequence conservation is usually higher in loops than in the centre of base-paired elements. It can be more

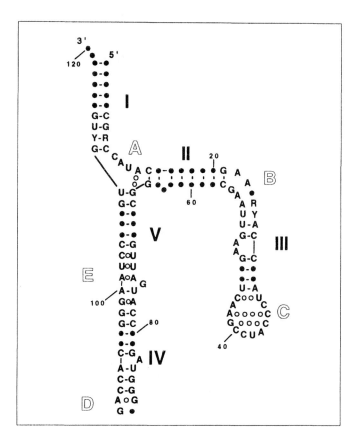

Fig. 2 Common structural features in eukaryotic 5S ribosomal RNA; conservation of primary sequence and secondary structure is indicated (Erdmann et al. (56) with modifications according to Westhof et al. (13), Baudin et al. (15), and Cheong et al. (16)). Empty circles represent non-canonical base-pair interactions.

generally assumed that conserved loop regions adopt a rigid structure, as has been proposed (11) and recently demonstrated (12) for the loop E region that joins helices IV and V by two-dimensional (2-D) nuclear magnetic resonance (NMR) analysis. Loop E includes several unusual base pairs stabilized by extensive interstrand stacking. Together with the flanking helices IV and V, formation of these non-canonical base pairs helps to build up a continuous, co-axial helical structure with local distortions. According to the model (12), G74 is in a single-base bulge conformation, lying in the major groove of the flanking base pairs. This irregular structure could serve as a specific recognition signal for TFIIIA (as discussed below). A local structural distortion in the context of regular A-type RNA helical elements is also to be expected with the single- and double-base bulges in helices II, III, and IV.

Based upon experimental information concerning the accessibility of individual nucleotides for chemical and enzymatic probes, a detailed structural model has been proposed for the two highly conserved internal loops, A and B (13–15). Very much in analogy with the loop E region, loop B, which connects helices II and III, appears to adopt a largely helical conformation, stabilized by stacking interactions bringing the two flanking helices into a co-axial configuration.

The structure of the remaining internal loop, loop A, is more difficult to elucidate. It is present at a strategic location, namely at the junction between the three helical elements I, II, and V. Its conformation must play a key role in determining alternative possibilities in the co-axial stacking of the three major structural domains in 5S RNA: (1) helix I; (2) helix II–loop B–helix III–loop C; (3) helix V–loop E–helix IV–loop D (*Figure 2*). Extensive mutational analysis of loop A combined with structural probing and model building has provided strong evidence for co-axial stacking of domain 2 with domain 3 (15). This arrangement is thought to be facilitated and stabilized by a triple interaction forming between the conserved G in the terminal base pair of helix V (G66–U109) and the conserved A13 within loop A (*Figure 2* (15)).

Of the two hairpin loops, the four-membered loop D meets the sequence consensus for the GNRA tetraloop. This conserved tetranucleotide sequence comprises > 55% of all loops present in ribosomal RNAs and its structure has been solved (Chapter 1; 16, 17). Formation of an unusual G–A base pair between the first and last nucleotide, stacking interactions and additional hydrogen bonding account for the high stability of the tetraloop structure. Extended base-pairing and stacking interactions have also been proposed to fold the 12-membered, conserved hairpin loop C into a highly organized structure (*Figure 2* (14)). None of the structural studies on 5S RNA has provided evidence for the formation of tertiary interactions which would involve loop C or any other region of the molecule.

Therefore, the 5S RNA molecule is best described as consisting of three largely independent domains (*Figure 2*). Domain 1 (I) is a helical stem that holds the 3' and 5' ends of the RNA molecule together. Domain 2 (II-B–III-C) is in an extended helical conformation with local distortions, such as bulged nucleotides in helices II and III, as well as a more flexible internal loop B region. It is closed by the rigidly structured hairpin loop C. Domain 3 (V-E–IV-D) is defined by a similar, though significantly more stable helical arm with local distortions coming from bulged nucleotides (helix IV and loop E) and from non-canonical base pairing (loop E). It is closed by the exceptionally stable tetraloop structure. The internal loop A region will be critically involved and responsible for the co-axial stacking of domain 2 with domain 3. All of these three major domains in 5S RNA appear to fold independently and there is no evidence for the formation of tertiary interactions. Such a structural configuration with no buried, internal elements should make the entire RNA chain available for protein binding.

3. RNA structural requirements for the binding of TFIIIA

The abundance of the 7S RNP, which consists of one molecule of 5S RNA complexed with one molecule of TFIIIA, in *Xenopus* oocytes makes the native particle readily available for biochemical analysis. One experimental strategy for the identification of RNA elements which are in contact with TFIIIA makes use of chemical and enzymatic probes; protein binding protects a substantial portion of the 5S RNA molecule from chemical modification or enzymatic attack. Most of the internal loop E and of the 3' strand in loop B, major portions of helices II, IV, and V, as well as the region in helix I adjacent to loop A were found to exhibit reduced accessibility to the various probes employed (*Figure 3*; 18–23). However, interpretation of these results suffers from a number of limitations; protection from chemical modification does not necessarily identify contacting residues, since these effects could equally be caused indirectly, such as by RNA structural rearrangements or by other alterations of the chemical environment. Furthermore, protection against enzymatic hydrolysis could be the result of steric hindrance and not necessarily of direct protein–RNA contacts. Therefore, these experiments tend to overestimate the region of 5S RNA which is in contact with TFIIIA. Nevertheless, taken together, chemical and enzymatic experiments on the 7S RNP clearly suggest that TFIIIA forms multiple contacts over a wide region of the surface of the 5S RNA molecule (*Figure 3*). Details of these findings will be discussed below, together with data obtained from RNA mutational analyses.

Several different experimental strategies have been developed in order to study

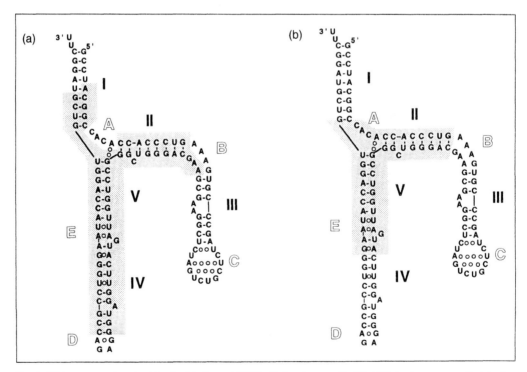

Fig. 3 RNA structure involved in the binding of TFIIIA; *X. laevis* somatic 5S RNA is drawn according to the general secondary structure for eukaryotic 5S ribosomal RNA (as in *Figure 2*). (a) Stippling indicates protection by TFIIIA against chemical modification and nuclease digestion. (b) Stippling indicates structural elements in 5S RNA which are involved in TFIIIA binding according to RNA mutational analysis.

the effects of 5S RNA sequence alterations on TFIIIA binding. Initially, the ability of various heterologous 5S RNA molecules to compete with 5S genes for binding of TFIIIA was measured as a function of 5S transcription inhibition *in vitro* (24) or via interference with the 5S DNA-footprinting activity (25). Another approach made use of the ability of free 5S RNA molecules to exchange with 5S RNA in the 7S RNP (ribonucleoprotein particle) upon co-incubation (26, 27). Finally, with the advent of a nitrocellulose filter-binding assay for 7S RNPs reconstituted from the isolated components, a quantitative binding assay was available (19). In good agreement with each other, these early studies reveal one important element in 5S RNA–TFIIIA recognition: the ability of various heterologous 5S RNA molecules to bind to TFIIIA with an affinity comparable to the *Xenopus* 5S RNA demonstrates that conserved elements of 5S RNA structure (*Figure 2*) are sufficient for TFIIIA recognition, implying that primary sequence variability in helical elements, such as in helices II, V, and IV, which are in contact with TFIIIA, is tolerated by the protein. This finding provided a first argument for RNA secondary–tertiary structure recognition by TFIIIA, rather than for a binding mode dependent entirely on primary sequence, that is, base-specific contacts.

The importance of conserved nucleotides or conserved elements of secondary structure cannot be assessed via binding studies with heterologous 5S RNA molecules. This issue has been addressed by site-directed mutagenesis of *Xenopus* 5S RNA. Binding assays employed were *in vitro* reconstitution–filter binding (28–31) and RNA exchange as above (32), as well as analysis of 7S RNP formation in the *Xenopus* oocyte; this latter assay made use of 5S gene injection followed by immunoprecipitation of 7S RNPs with antibodies directed against TFIIIA (8, 33). There is excellent agreement between these different studies with respect to a second important element of 5S RNA–TFIIIA recognition: structural alterations introduced into either helix II or helix V interfere with TFIIIA binding if they disrupt base pairing. Complementary sequence alterations, which restore base pairing, rescue binding activity.

The single base bulge present in helix II (position 63, *Figure 2*) does not appear to constitute an important signal for TFIIIA recognition. In contrast to helices II and V, structural changes in either helix III or helix IV do not or only moderately influence TFIIIA binding. For helix III this was to be expected since this part of the 5S RNA molecule is not included in the region protected against chemical modification or enzymatic hydrolysis. However, these same probing studies suggested that helix IV might be in contact with TFIIIA. If RNA–protein contacts form here, they appear not to be essential for complex formation.

The same picture emerges from extensive deletion mutagenesis of 5S RNA. A major portion of helices III and IV, as well as the adjacent hairpin loops, can be deleted without significantly altering the ability of these artificial RNA constructs to bind to TFIIIA (34). Mutational analysis also reveals some importance of helix I for 7S RNP formation (34). The involvement of this RNA element in TFIIIA binding had also been suggested by protection of nucleotides located adjacent to the internal loop A, as discussed in detail below.

Very much in contrast to the excellent agreement of results obtained from protection and mutation experiments for helices II and V, these studies disagree on the importance of loop A for TFIIIA binding. The highly conserved primary sequence of this four-nucleotide element is found to be critical for TFIIIA recognition in mutational studies (15, 19), but is not found to be protected against any of the structural probes utilized. It would therefore seem unlikely that these nucleotides are involved in the formation of direct contacts with TFIIIA, which would be mechanistically comparable to the contacts formed with helical elements or with the other internal loops. It would be more reasonable to suppose that the loop A region might primarily serve a function in determining the three dimensional (3-D) orientation of the three major structural domains in 5S RNA (as discussed above). However, the loop A element is part of a major cross-linking site between 5S RNA and TFIIIA, containing three out of the eight putative cross-linking nucleotides (35). It is also proposed to be in proximity to the one zinc finger in TFIIIA which is most critical for RNA recognition on the basis of nuclease protection data (F6, discussed below). Therefore, the final judgement on a more direct involvement of loop A in TFIIIA binding must await further experimentation.

The loop E element in 5S RNA is the other irregular structure that appears to be important for TFIIIA recognition, as can be concluded on the basis of results from mutational analysis. 2-D NMR analysis has revealed that this region is highly structured; together with the adjacent base-paired regions IV and V, it defines a continuous helical segment which is in contact with TFIIIA (as discussed above). Since we are only beginning to understand the structural details of the loop E region, it is difficult to predict the effect of sequence alterations introduced here on secondary–tertiary structure. However, several single-site mutations in loop E interfere with 7S RNP formation in the *Xenopus* oocyte system (33). It is remarkable that replacement of loop E by a canonical RNA duplex is tolerated in the *in vitro* reconstitution assay (29); this experiment clearly demonstrates that helicity within this region is one important RNA structural element for TFIIIA binding. However, a more specific recognition function for the loop E region with its characteristic irregularities should not be excluded; in support of this view is the finding that loop E and the adjacent base pairs were the only elements in 5S RNA affected in 'missing nucleoside' modification-interference experiments (23). This assay makes use of 5S RNA populations with random single-nucleoside gaps. Such gaps reduce TFIIIA binding if introduced into the loop E region, but not if they are generated in any other region of the 5S RNA molecule.

In summary, extensive use of structural probes and systematic RNA mutational analyses identify essential structural elements in 5S RNA for recognition by TFIIIA (*Figure 3*). These structures are centred around the internal loop A region, which joins the three TFIIIA recognition helices I, II, and V–loop E. Most observations made are indicative of secondary–tertiary structure recognition rather than base-specific contacts. However, the possibility of base-specific contacts should not be excluded for internal loops A and E, as well as for helix I. It is also evident that TFIIIA interacts with a broad region on the RNA surface, which is indicative of the formation of a multitude of specific contacts, perhaps reflecting interaction with the multiple zinc finger modules in the protein.

4. Functional domains of TFIIIA

On the basis of its primary sequence, *X. laevis* TFIIIA can be divided into two major domains, the N-terminal portion, containing the nine zinc finger repeats and the C-terminal portion, which is not involved in nucleic acid binding (see also *Figure 1*; 2). Sequence information for TFIIIA is also available from four other amphibian species, namely *Xenopus borealis, Rana catesbeiana, Rana pipiens,* and *Bufo americanus* (36, 37).

It is reasonable to assume that all of these proteins will bind to 5S RNA, since 7S RNPs have been isolated from the oocytes of another amphibian in addition to *X. laevis, Triturus cristatus* (38) and also from different fish, such as the teleost *Tinca tinca* (39) and from the trout (40). The five protein sequences available are closely related within the zinc finger domain (*Figure 4*); they all contain nine repeats in tandem array and corresponding zinc finger modules share all of the unusual structural characteristics which distinguish TFIIIA zinc fingers from the consensus C_2H_2-type module. Examples of distinguishing structural features are provided by unusual variations in the number of amino acids separating invariant cysteines and histidines, as well as in the number of amino acids separating the second conserved histidine of one finger from the first conserved cysteine of the next finger (H–C link).

A second oocyte-specific protein which is structurally related to TFIIIA and which is also found in a complex with 5S RNA is termed p43 (7). This protein carries an identical number of zinc finger units and most of the structural features which typify TFIIIA are present in p43. However, on the level of the TFIIIA and p43 primary sequences there is only very limited conservation beyond the invariant zinc-coordinating or hydrophobic core residues (*Figure 4*). One would expect that sequence elements conserved in TFIIIA but not in p43 might be important for DNA binding but not for RNA binding. Conversely, sequence information conserved in both TFIIIA and p43 might be significant for RNA binding.

TFIIIA has been subject to extensive mutational analysis. Zinc fingers can be treated as independently folding and independently functioning units. Deletion analysis has revealed that although all fingers appear to be involved in RNA and DNA binding, the N-terminal units (F1–F3) make a more important contribution to the overall DNA-binding activity than the remaining units (33, 41–48). It could even be demonstrated that this N-terminal finger triplet is sufficient for 5S gene recognition with an affinity that is equal to the full-length protein. Conversely, the central finger modules (F4–F6/F7) were found to govern RNA binding, being sufficient to yield high-affinity 5S RNA recognition (33, 49, 50). The remaining C-terminal units (F8–F9) are the least important in the processes of RNA and DNA recognition and they were described as being more likely to be involved in tran-

-VYKRY*I	C	SF-D	C	-A-YNKN-KL-A	H	LCK	H	F1	TFIIIA consensus
------*-	C	----	C	-A-Y-K--KL--	H	---	H		TFIIIA/p43 consensus
TGE-PFP	C	--EG	C	-KGF--L-HL-R	H	---	H	F2	
-------	C	----	C	-K-F--------	H	---	H		
TGEK---	C	----	C	-L-FTT--N---	H	--R-	H	F3	
---K---	C	----	C	---F-T------	H	----	H		
------Y-	C	-F--	C	---F-KHNQLK-	H	Q--	H	F4	
-------*	C	---*	C	---F-K---L-*	H	---	H		
T-Q-P--	C	-HEG	C	DK----PS-LKR	H	EK-	H	F5	
-------	C	---G	C	--------L--	H	---	H		
AGYP	C	-KD--	C	-FVGKTW--Y-K	H	----	H	F6	
-GY-	C	----*	C	--V--TW-----	H	---*	H		
----*	C	--	C	-R-F--K--L--	H	---	H	F7	
----*****	C	--	C	---F-----L--	H	---	H		
--ER-VY*	C	----	C	-R-YT--F-L-S	H	---F	H	F8	
-------*	C	----*	C	-------F-L--	H	----	H		
E--RPF-	C	-H--	C	GK-FAMK-SL-R	H	---	H	F9	
-------	C	-H--	C	---FAM--SL-R	H	---	H		
-------	C	--**	C	---F-----L--	H	---	H		General consensus

Fig. 4 Primary sequence comparison of corresponding zinc finger modules in TFIIIA and p43; for each of the nine zinc finger elements, primary sequence information strictly conserved in five different amphibian species is specified (top line) and compared with primary sequence information conserved in two p43 variants and these same five TFIIIA sequences (bottom line). Primary sequence information conserved in TFIIIA, but not in TFIIIA–p43, is indicative of a function in DNA binding; primary sequence information conserved for TFIIIA–p43 is indicative of a function in RNA binding. Conserved dipeptide motifs which include neither invariant zinc-coordinating residues nor conserved amino acids which contribute to the formation of the hydrophobic core are indicated by stippling.

scription complex formation, perhaps helping to bring TFIIIB and TFIIIC into their proper positions on the 5S RNA gene (48).

Footprinting experiments on RNA and DNA have allowed pairing of individual zinc finger units or functional groups of fingers with their associated RNA and DNA sequence elements. It turns out that the zinc finger modules of TFIIIA are localized over corresponding primary sequence elements in 5S RNA and 5S DNA (*Figure 5*). However, it is evident from both protein and RNA–DNA mutational analyses that the most crucial contacts involve distinct sets of fingers and distinct nucleic acid sequence elements in the 7S RNP and the TFIIIA–5S RNA gene complexes. TFIIIA mutagenesis has provided evidence for a pivotal function of F6 in RNA recognition (33, 39, 50). Inspection of our TFIIIA–p43 sequence comparison (*Figure 4*) reveals the presence of conserved, unusual primary sequence charac-

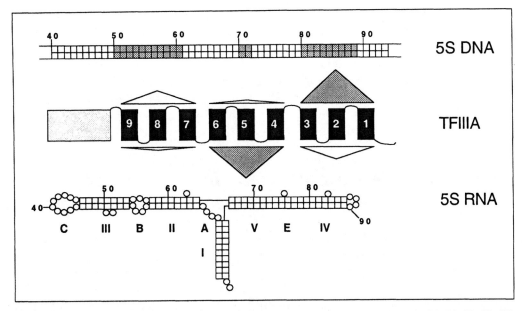

Fig. 5 Binding of zinc finger triplets in TFIIIA to 5S RNA and 5S DNA; the three triplets (F1–F3, F4–F6, F7–F9) bind to corresponding sequence elements in DNA and RNA. The sizes of the triangles above and below the zinc finger domain correspond to the relative contributions of each of the three triplets to RNA and DNA binding, respectively. Stippling of DNA elements indicates the three functional elements of the 5S gene internal control region.

teristics in finger 6: the exceptionally short H–C link (four residues instead of the usual seven) and the presence of a conserved TW motif in the finger loop region that joins the second conserved cysteine with the first conserved histidine. RNA footprint experiments localize F6 to the loop A region of the 5S RNA molecule (33, 49); this could bring the conserved, aromatic tryptophan in proximity to bases within the loop, perhaps allowing for stacking interactions, which have been proposed to be involved in the formation of other specific RNA–protein complexes. Mutagenesis of this critical tryptophan in finger 6 does indeed reduce RNA-binding activity to a significant extent (49; O. Theunissen and T. Pieler, unpublished results).

For the remainder of the zinc finger modules in TFIIIA, which all appear to be in contact with the RNA surface, we have no experimental information available on the amino acid residues that might be directly involved in RNA recognition–binding. RNA mutational studies, which favour the idea that it is the specific secondary–tertiary structure and not the sequence of 5S RNA which is the primary determinant for specific complex formation (as discussed above), predict phosphate contacts to be one major determinant in protein binding. Preliminary results from ethylation interference experiments do indeed support the notion of multiple phosphate residues in the central core of the 5S RNA molecule that are essential for TFIIIA recognition (O. Theunissen and T. Pieler, unpublished results).

In summary, the N-terminal zinc finger triplet (F1–F3) is in contact with the helix IV–loop E region; interactions are tight but not essential for RNA recognition. The middle portion of the zinc finger domain (F4–F6/F7) is involved in tight interactions with RNA helices I, II, and IV, centred around loop A. These contacts involve phosphate residues and they are essential and constitute the driving force for the stability and the specificity of the complex. The C-terminal zinc finger modules (F8–F9) are the least important for RNA binding and they are located over the loop B–helix III region of the 5S RNA molecule.

5. Other RNA-binding zinc finger proteins

A large number of zinc finger protein encoding sequences of the C2H2–TFIIIA type have been isolated from vertebrate and invertebrate species. The exact molecular function of most of these proteins remains to be determined (see El-Baradi and Pieler (3) for a review). RNA-binding activity has been attributed to several of these proteins; one, termed STP1, is a yeast protein which is involved in pre-tRNA processing (51). It was found to carry one p43–TFIIIA-type zinc finger and two additional putative zinc coordinating modules; however, direct biochemical evidence for an RNA-binding activity of this isolated zinc finger has not been presented to date.

A final example is provided by two subfamilies of structurally related *X. laevis* zinc finger proteins; members of these groups of proteins share either of two conserved sequence elements, termed the KRAB and FAR domains, respectively (52, 57). FAX (finger associated boxes) and KRAB (Krüppel associated box) are located at the amino-terminus of these proteins connected to structurally divergent zinc finger clusters. These zinc finger clusters exhibit specific RNA homopolymer-binding activity; like many other specific RNA-binding proteins they will form a complex with poly(U) and poly(G) but not with other RNA homopolymers (52–54). The RNA-binding activities of these proteins may be regulated by phosphorylation (54, 55). However, the *in vitro* RNA homopolymer-binding activities described are not likely to reflect the physiological, *in vitro* RNA-binding specificities of KRAB and FAR zinc finger proteins, which therefore remain to be determined.

References

1. Mattaj, I. W. (1993) RNA recognition: a family matter? *Cell*, **73**, 837.
2. Miller, J., McLachlan, A. D., and Klug, A. (1985) Repetitive zinc-binding domains in the protein transcription factor IIIA from *Xenopus* oocytes. *EMBO J.*, **4**, 1609.
3. El-Baradi, T. and Pieler, T. (1991) Zinc finger proteins: what we know and what we would like to know. *Mech. Devel.*, **35**, 155.
4. Picard, B. and Wegnez, M. (1979) Isolation of a 7S particle from *Xenopus laevis* oocytes: a 5S RNA–protein complex. *Proc. Natl Acad. Sci. USA*, **76**, 241.
5. Engelke, D. R., Ng, S. Y., Shastry, B. S., and Roeder, R. G. (1980) Specific interaction of a purified transcription factor with an internal control region of 5S RNA genes. *Cell*, **19**, 717.

6. Pelham, H. R. B. and Brown, D. D. (1980) A specific transcription factor that can bind either the 5S RNA gene or 5S RNA. *Proc. Natl Acad. Sci. USA*, **77**, 4170.
7. Joho, K. E., Darby, M. K., Crawford, E. T., and Brown, D. D. (1990) A finger protein structurally similar to TFIIIA that binds exclusively to 5S RNA in *Xenopus*. *Cell*, **6**, 293.
8. Guddat, U., Bakken, A. H., and Pieler, T. (1990) Protein-mediated nuclear export of RNA: 5S rRNA containing small RNPs in *Xenopus* oocytes. *Cell*, **60**, 619.
9. Pavletich, N. P. and Pabo, C. O. (1991) Zinc finger–DNA recognition: crystal structure of a Zif268–DNA complex at 2.1 Å. *Science*, **252**, 809.
10. Clemens, K. R., Liao, X., Wolf, V., Wright, P. E., and Gottesfeld, J. M. (1992) Definition of the binding sites of individual zinc fingers in the transcription factor IIIA–5S RNA gene complex. *Proc. Natl Acad. Sci. USA*, **89**, 10822.
11. Andersen, J., Delihas, N., Hanas, J. S., and Wu, C. W. (1984) 5S RNA structure and interaction with transcription factor A. 1. Ribonuclease probe of the structure of 5S RNA from *Xenopus laevis* oocytes. *Biochemistry*, **23**, 5752.
12. Wimberley, B., Varani, G., and Tinoco, I. (1993) The conformation of loop E of eukaryotic 5S ribosomal RNA. *Biochemistry*, **32**, 1078.
13. Westhof, E., Romby, P., Romaniuk, P. J., Ebel, J. P., Ehresmann, C., and Ehresmann, B. (1989) Computer modeling from solution data of spinach chloroplast and of *Xenopus laevis* somatic and oocyte 5S rRNAs. *J. Mol. Biol.*, **207**, 417.
14. Brunel, C., Romby, P., Westhof, E., Romaniuk, P. J., Ehresmann, B., and Ehresmann, C. (1990) Effect of mutations in domain 2 on the structural organization of oocyte 5S rRNA from *Xenopus laevis*. *J. Mol. Biol.*, **215**, 103.
15. Baudin, F., Romaniuk, P. J., Romby, P., Brunel, C., Westhof, E., Ehresmann, B., and Ehresmann, C. (1991) Involvement of 'hinge' nucleotides of *Xenopus laevis* 5S rRNA in the RNA structural organization and in the binding of transcription factor TFIIIA. *J. Mol. Biol.*, **218**, 69.
16. Cheong, C., Varani, G., and Tinoco, I. (1990) Solution structure of an unusually stable RNA hairpin, 5'GGAC(UUCG)GUCC. *Nature*, **346**, 680.
17. Heus, H. A. and Pardi, A. (1991) Structural features that give rise to the unusual stability of RNA hairpins containing GNRA loops. *Science*, **253**, 191.
18. Pieler, T. and Erdmann, V. A. (1983) Isolation and characterization of a 7 S RNP particle from mature *Xenopus laevis* oocytes. *FEBS Lett.*, **157**, 283.
19. Romaniuk, P. J. (1985) Characterization of the RNA binding properties of transcription factor IIIA of *Xenopus laevis* oocytes. *Nucleic Acids Res.*, **13**, 5369.
20. Huber, P. W. and Wool, I. G. (1986) Identification of the binding site on 5S rRNA for the transcription factor IIIA: proposed structure of a common binding site on 5S rRNA and on the gene. *Proc. Natl Acad. Sci. USA*, **83**, 1593.
21. Christiansen, J., Brown, R. S., Sproat, B. S., and Garret, R. A. (1987) *Xenopus* transcription factor IIIA binds primarily at junctions between double helical stems and internal loops in oocyte 5S RNA. *EMBO J.*, **6**, 463.
22. Sands, M. S. and Bogenhagen, D. F. (1991) Two zinc finger proteins from *Xenopus laevis* bind the same region of 5S RNA but with different nuclease protection patterns. *Nucleic Acids Res.*, **19**, 1797.
23. Darsillo, P. and Huber, P. W. (1991) The use of chemical nucleases to analyse RNA–protein interactions. *J. Biol. Chem.*, **266**, 21075.
24. Pieler, T., Erdmann, V. A., and Appel, B. (1984) Structural requirements for the interaction of 5S rRNA with the eukaryotic transcription factor IIIA. *Nucleic Acids Res.*, **12**, 8393.

25. Hanas, J. S., Bogenhagen, D. F., and Wu, C. W. (1984) Binding of *Xenopus* transcription factor A to 5S RNA and to single stranded DNA. *Nucleic Acids Res.*, **12**, 2745.
26. Andersen, J., Delihas, N., Hanas, J. S., and Wu, C. W. (1984) 5S RNA structure and interaction with transcription factor A. 2. Ribonuclease probe of the 7S particle from *Xenopus laevis* immature oocytes and RNA exchange properties of the 7S particle. *Biochemistry*, **23**, 5759.
27. Andersen, J. and Delihas, N. (1986) Characterization of RNA–protein interactions in 7S ribonucleoprotein particles from *Xenopus laevis* oocytes. *J. Biol. Chem.*, **261**, 2912.
28. Baudin, F. and Romaniuk, P. J. (1989) A difference in the importance of bulged nucleotides and their parent base pairs in the binding of transcription factor IIIA to *Xenopus* 5S RNA and 5S RNA genes. *Nucleic Acids Res.*, **17**, 2043.
29. Romaniuk, P. J. (1989) The role of highly conserved single-stranded nucleotides of *Xenopus* 5S RNA in the binding of transcription factor IIIA. *Biochemistry*, **28**, 1388.
30. You, Q. and Romaniuk, P. J. (1990) The effects of disrupting 5S RNA helical structures on the binding of *Xenopus* transcription factor IIIA. *Nucleic Acids Res.*, **18**, 5055.
31. You, Q., Veldhoen, N., Baudin, F., and Romaniuk, P. J. (1991) Mutations in 5S DNA and 5S RNA have different effects on the binding of *Xenopus* transcription factor IIIA. *Biochemistry*, **39**, 2495.
32. Sands, M. S. and Bogenhagen, D. F. (1987) TFIIIA binds to different domains of 5S RNA and the *Xenopus borealis* 5S RNA gene. *Mol. Cell. Biol.*, **7**, 3985.
33. Theunissen, O., Rudt, F., Guddat, U., Mentzel, H., and Pieler, T. (1992) RNA and DNA binding zinc fingers in *Xenopus* TFIIIA. *Cell*, **71**, 679.
34. Bogenhagen, D. F. and Sands, M. S. (1992) Binding of TFIIIA to derivatives of 5S RNA containing sequence substitutions or deletions defines a minimal TFIIIA binding site. *Nucleic Acids Res.*, **20**, 2639.
35. Baudin, F., Romby, P., Romaniuk, P. J., Ehresmann, B., and Ehresmann, C. (1989) Crosslinking of transcription factor TFIIIA to ribosomal 5S RNA from *X. laevis* by trans-diaminedichloroplatinum (II). *Nucleic Acids Res.*, **17**, 10035.
36. Gaskins, C. J. and Hanas, J. S. (1990) Sequence variations in transcription factor IIIA. *Nucleic Acids Res.*, **18**, 2117.
37. Gaskins, C. J., Smith, J. F., Ogilvie, M. K., and Hanas, J. S. (1992) Comparison of the sequence and structure of transcription factor IIIA from *Bufo americanus* and *Rana pipiens*. *Gene*, **120**, 197.
38. Kloetzel, P. M., Whitfield, W., and Sommerville, J. (1981) Analysis and reconstruction of an RNP particle which stores 5S RNA and tRNA in amphibian oocytes. *Nucleic Acids. Res.*, **9**, 605.
39. Denis, H., Picard, B., Le Maire, M., and Clerot, J. C. (1980) Biochemical research on oogenesis. The storage particles of the teleost fish *Tinca tinca*. *Devel. Biol.*, **77**, 218.
40. Welfle, H. and Misselwitz, R. (1987) Nucleoprotein particles from trout oocytes. *Studia Biophys.*, **122**, 99.
41. Smith, D. R., Jackson, I. J., and Brown, D. D. (1984) Domains of the positive transcription factor specific for the *Xenopus* 5S RNA gene. *Cell*, **37**, 645.
42. Fiser-Littell, R. A., Duke, A. L., Yanchick, J. S., and Hanas, J. S. (1988) Deletion of the N-terminal region of *Xenopus* transcription factor IIIA inhibits specific binding to the 5S RNA gene. *J. Biol. Chem.*, **263**, 1607.
43. Vrana, K. E., Churchill, M. E. A., Tullius, T. D., and Brown, D. D. (1988) Mapping functional regions of transcription factor TFIIIA. *Mol. Cell. Biol.*, **8**, 1684.
44. Christensen, J. H., Hansen, P. K., Lillelund, O., and Thogersen, C. (1991) Sequence-

specific binding of the N-terminal three-finger fragment of *Xenopus* transcription factor IIIA to the internal control region of a 5S RNA gene. *FEBS Lett.*, **281**, 181.
45. Liao, X., Clemens, K. R., Tennant, L., Wright, P. E., and Gottesfeld, J. M. (1992) Specific interaction of the first three zinc fingers of TFIIIA with the internal control region of the *Xenopus* 5S RNA gene. *J. Mol. Biol.*, **223**, 857.
46. Smith, J. F., Hawkins, J., Leonard, R. E., and Hanas, J. S. (1991) Structural elements in the N-terminal half of transcription factor IIIA required for factor binding to the 5S RNA gene internal control region. *Nucleic Acids Res.*, **19**, 6871.
47. Darby, M. K. and Joho, K. E. (1992) Differential binding of zinc fingers from *Xenopus* TFIIIA and p43 to 5S RNA and the 5S RNA gene. *Mol. Cell. Biol.*, **12**, 3155.
48. Del Rio, S. and Setzer, D. R. (1993) The role of zinc fingers in transcriptional activation by transcription factor IIIA. *Proc. Natl Acad. Sci. USA*, **89**, 168.
49. Clemens, K. R., Wolf, V., McBryant, S. J., Zhang, P., Liao, X., Wright, P. E., and Gottesfeld, J. M. (1993) Molecular basis for specific recognition of both RNA and DNA by a zinc finger protein. *Science*, **260**, 530.
50. Rollins, M. B., Del Rio, S., Galey, A. L., Setzer, D. R., and Andrews, M. (1993) Role of TFIIIA zinc fingers *in vivo*: analysis of single-finger function in developing *Xenopus* embryos. *Mol. Cell. Biol.*, **13**, 4776.
51. Wang, S. S., Stanford, D. R., Silvers, C. D., and Hopper, A. K. (1992) STP1, a gene involved in pre-tRNA processing, encodes a nuclear protein containing zinc finger motifs. *Mol. Cell. Biol.*, **12**, 2633.
52. Klocke, B., Köster, M., Hille, S., Bouwmeester, T., Böhm, S., Pieler, T., and Knöchel, W. (1994) The FAR domain defines a new *X. laevis* zinc finger protein subfamily with specific RNA homopolymer binding activity. *Biochimica et Biophysica Acta*, **1217**, 81.
53. Köster, M., Kühn, U., Bouwmeester, T., Nietfeld, W., El-Baradi, T., Knöchel, W., and Pieler, T. (1991) Structure, expression and *in vitro* functional characterization of a novel RNA binding zinc finger protein from *Xenopus*. *EMBO J.*, **10**, 3087.
54. Andreazzoli, M., De Luccini, S., Costa, M., and Barsacchi, G. (1993) RNA binding properties and evolutionary conservation of the *Xenopus* multifinger protein Xfin. *Nucleic Acids Res.*, **21**, 4218.
55. Van Wijk, I., Burfeind, J., and Pieler, T. (1992) Nuclear transport and phosphorylation of the RNA binding *Xenopus* zinc finger protein XFG 5-1. *Mech. Devel.*, **39**, 63.
56. Erdmann, V. A., Wolters, J., Pieler, T., Digweed, M., Specht, T., and Ulbrich, N. (1988) Evolution of organisms and organelles as studied by computer and biochemical analysis by ribosomal 5S RNA structure. *Ann. NY Acad. Sci.*, **503**, 103.
57. Bellefroid, E. J., Poncelet, D. A., Lecocq, P. J., Revelant, O., and Martial, J. A. (1991). The evolutionarily conserved Krüppel-associated box domain defines a subfamily of eukaryotic multifingered proteins. *Proc. Natl Acad. Sci. USA*, **88**, 3608.

9 | Control of human immunodeficiency virus gene expression by the RNA-binding proteins tat and rev

JONATHAN KARN, MICHAEL J. GAIT,
MARK J. CHURCHER, DEREK A. MANN, IVAN MIKAÉLIAN,
and CLARE PRITCHARD

1. Introduction

Replication of the human immunodeficiency virus (HIV) is controlled by two viral proteins, the *trans*-activator protein (tat) and the regulator of virion expression (rev). These regulatory proteins play complementary roles in the HIV life cycle: tat stimulates transcription from the viral long-terminal repeat (LTR), whereas rev is required for the efficient cytoplasmic expression of the mRNAs encoding the structural proteins of the virus. This system of dual regulation is illustrated in *Figure 1*. In permissive cells, transcription of the integrated proviral genome is initiated by host cell transcription factors (1–3). Initially, a low level of small, multiply spliced mRNAs encoding tat and rev are produced (4). tat synthesized during the primary round of transcription establishes a 'positive feedback loop' which boosts tat and rev mRNA production 200–1000-fold (5–7). Once a critical rev threshold is surpassed, HIV gene expression switches away from the production of the mRNAs for the regulatory proteins towards the production of the unspliced or partially spliced mRNAs encoding the virion proteins (8–10).

In contrast to previously studied retroviral regulatory proteins which act through DNA elements, both tat and rev are RNA-binding proteins that exert their effects through specific *cis*-acting viral RNA regulatory sequences. In this chapter, we describe our recent experiments and ideas about the mechanism of action of both tat and rev and their interactions with target RNAs.

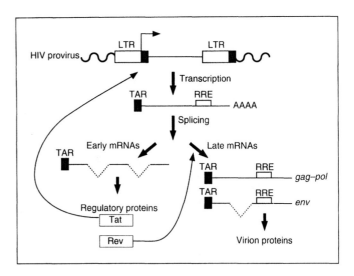

Fig. 1 Regulatory circuits in HIV gene expression. Early gene expression produces the regulatory proteins tat and rev from doubly spliced mRNAs. Initially, tat stimulates transcription from the viral long-terminal repeat (LTR) by interacting with the *trans*-activation response region (TAR). Once tat and rev levels are sufficiently high, rev then binds to the rev-response element (RRE) and stabilizes the mRNAs for the virion proteins. Diagram adapted from Cullen (113).

2. Stimulation of transcription by tat
2.1 TAR is functional as an RNA transcript

The activity of tat requires the *trans*-activation responsive region (TAR), a regulatory element of 59 nucleotides (nt) located immediately downstream of the initiation site for transcription (11–14). Because of its position in the HIV genome, each viral mRNA carries a copy of TAR at its 5′ end.

The location of TAR within a transcribed region was surprising and raised the possibility that tat could be acting through an RNA element rather than a DNA element. By contrast to enhancer elements, the TAR element is only functional when it is placed downstream of the start of HIV transcription and in the correct orientation and position. Further evidence that the TAR element functioned as RNA came from the observation that TAR RNA forms a highly stable, nuclease-resistant, stem–loop structure (13). *Figure 2* shows the apical portion of the TAR RNA stem–loop. Point mutations which disrupt base pairing in the upper TAR RNA stem invariably abolish *tat*-stimulated transcription (1, 14, 15).

Berkhout *et al.* (16) have provided convincing evidence that the TAR RNA sequence must be correctly folded during transcription in order for *trans*-activation to occur. In this experiment, an antisense sequence which was designed to compete for the formation of the TAR RNA hairpin–loop structure was inserted either upstream or downstream of TAR. When the antisense sequence was placed on the 5′ side of TAR, TAR RNA was unable to form a normal stem–loop structure and, as a result, the tat response was inhibited. By contrast, incorporation of the antisense sequence on the 3′ side of TAR allowed normal *trans*-activation, since the TAR RNA present on nascent transcripts was able to adopt its normal structure prior to the transcription of the downstream antisense sequences.

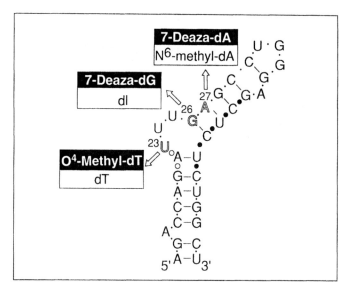

Fig. 2 Structure of TAR RNA and sequence requirements for tat binding. The top of the TAR RNA stem–loop structure is drawn to indicate bending of the stem induced by the bulged nucleotides U23–U25 and the structure of the apical loop. Three bases in the stem known to be critical for tat binding (U23, G26, and A27) are highlighted. Synthetic RNA duplexes carrying O^4-methyl-dT, 7-deaza-dG, or 7-deaza-dA at these positions show a greater than 10-fold loss of tat affinity (filled boxes). By contrast dT, dI, and N^6-methyl-dA are well tolerated at these positions (open boxes). Phosphate residues important for tat binding are indicated by the large circles between the bases. The open circles denoting P21 and P22 are sites where ethylation by ethylnitrosourea (ENU) or substitution by methylphosphonates inhibits tat binding. Ethylation of phosphates on the opposite strand (filled circles) also inhibits tat binding.

2.2 tat is an RNA-binding protein

The genetic evidence described above strongly suggested that tat recognizes TAR RNA by direct binding. In 1989 we provided the first demonstration that recombinant tat expressed in *Escherichia coli* could bind specifically to bases in the stem of the TAR RNA (17). However, it was some time before our conclusions were generally accepted. One reason for this was that earlier attempts to demonstrate specific binding of tat to TAR RNA were unsuccessful due to difficulties in preparing the protein (18, 19). Furthermore, it was assumed that tat must be interacting with the apical loop sequences of TAR RNA (19) since, at that time, mutations in this region were the only point mutations in TAR that were known to block *trans*-activation (14). These criticisms were answered by Roy *et al.* (20) and Dingwall *et al.* (21). When mutations in TAR RNA that show reduced binding to tat *in vitro* are introduced into the HIV LTR there is always a corresponding reduction in the tat responses (20, 21). Ironically, the role played by the loop sequences in *trans*-activation is still indeterminate. The most likely function for this sequence is to serve as an RNA-binding site for cellular co-factors of tat (22, 23), but other roles for the sequence, including a role as a DNA element involved in transcriptional initiation, have not been ruled out.

The binding site for tat includes a U-rich trinucleotide bulge found in the upper stem of TAR RNA (*Figure 2*). Residues essential for tat recognition include the first residue in the bulge, U23 and the two base pairs immediately above the bulge, G26–C39 and A27–U38 (20, 21, 24, 25). The two base pairs below the bulge make a comparatively small contribution to tat-binding specificity and are probably not points of direct contact with the protein (24, 25). The other residues in the bulge, C24 and U25, appear to act predominantly as spacers and may be replaced by other nucleotides or even by non-nucleotide linkers (26).

Figure 3 (top panel) shows an example of a competition-binding experiment similar to the experiments used to define the tat-binding site on TAR RNA (24). In

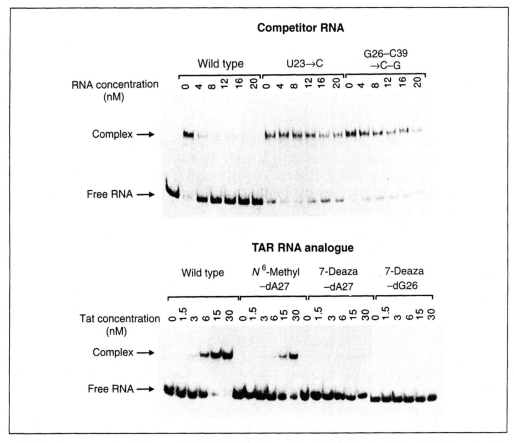

Fig. 3 Gel-mobility shift assays for tat binding to TAR RNA in which free RNA was fractionated from the tat–TAR RNA complex by electrophoresis on non-denaturing polyacrylamide gels. Top panel: competition-binding experiment. Complexes were formed between 2 nM ^{35}S-labelled TAR RNA, 0–20 nM unlabelled competitor TAR RNA and 5 nM tat protein. TAR RNAs carrying the U23 → C and G26–C39 → C–G mutations competed with more than 10-fold lower affinity than the wild-type sequence. Bottom panel: binding of chemically modified TAR RNA to tat protein. Synthetic RNA duplexes were prepared carrying the N^6-methyl-dA27, 7-deaza-dA27, or 7-deaza-dG26 substitutions. Binding reactions contained 10 nM ^{32}P-labelled TAR RNA analogue and 0–30 nM recombinant tat protein.

the absence of competitor RNA, tat formed stoichiometric complex with TAR RNA (a one-to-one molar ratio of tat protein to TAR RNA in the complex) and had a binding constant of ~3 nM. Unlabelled TAR RNA acts as an effective competitor with a $D_{1/2}$ (the concentration of competitor required to reduce binding by 50%) of ~3 nM. Unlabelled TAR RNAs carrying mutations in the binding site, such as the U23 → C and G26–C39 → C–G mutations were poor competitors for protein binding and there was virtually no loss of radioactivity from the tat–TAR complex in the presence of 20 nM competitor RNA ($D_{1/2} > 20$ nM).

Experiments using hybrid proteins have also provided evidence that one role for TAR RNA is to act as a 'loading site' that brings tat in proximity with the transcription machinery (27, 28). For example, fusion proteins containing sequences from tat and bacteriophage R17 coat protein can stimulate transcription from HIV LTRs when the TAR RNA sequence is replaced entirely by an RNA stem–loop structure carrying the R17 operator sequence (28). As in the case of tat recognition of TAR RNA, binding of the fusion protein to its RNA target appears to be direct, since mutations in the R17 RNA operator sequence that reduce its affinity to coat protein produced a corresponding decrease in *trans*-activation by the tat–R17 fusion protein *in vivo*.

2.2.1 Recognition of specific bases in the major groove of TAR RNA

The mutations in TAR described above have defined residues that are important to the interaction with tat protein. In order to identify some of the functional groups that are recognized by the protein, we have used synthetic chemistry to prepare model RNA structures (29).

Weeks & Crothers (34) postulated that the U-rich bulge creates a distortion that opens the otherwise narrow major groove of the TAR RNA duplex. This important suggestion was based on the observation that both G26 and A27 show increased reactivity to diethylpyrocarbonate (DEPC) when one or more bulged nucleotides are inserted into the TAR RNA stem. Our functional group mutagenesis studies (*Figures 2* and *3*) have provided strong evidence that tat recognizes bases in the major groove of the TAR RNA duplex (29). For example, the G26 residue has been replaced by either 7-deaza-dG, which removes a potential hydrogen-bond contact from the major groove or by dI (hypoxanthine), which removes a potential hydrogen-bond contact from the minor groove. The 7-deaza-dG substitution resulted in a complete loss of binding (*Figure 3*, bottom). By contrast, the dI substitution was well tolerated at G26. At the A27–U38 base pair, 7-deaza-dA substitution reduced tat binding by more than 20-fold, whereas the N^6-methyl-dA27 substitution was well tolerated and allowed nearly normal levels of tat binding (*Figure 3*, bottom). Thus, hydrogen bonding at the N^7 position of both G26 and A27 is essential for tat recognition.

Recently some of the phosphate contacts between tat and TAR RNA have been mapped by chemical modification-interference experiments (24, 31). Ethylation of P22 (between G21 and A22), P23, and five phosphates (P36–P40) located on the strand opposite the U-rich bulge caused substantial interference of tat binding.

Similarly, removal of the negative charge at the P22 or at P23, by methylphosphonate incorporation, also strongly interferes with tat binding (*Figure 2*).

2.2.2 Specific recognition of TAR RNA requires more than the arginine-rich sequence

Although there is now general agreement about the identity of the bases in TAR RNA that are involved in tat binding, it remains unclear which amino acid side chains in the tat protein participate in RNA recognition. Several groups have proposed that an arginine-rich sequence located towards the C-terminus of tat (residues 48–57) acts as an independent RNA-binding domain (30–35), but our data are inconsistent with this hypothesis. The arginine-rich peptides not only bind 100-fold less tightly to TAR RNA than tat protein, but they also have a significantly reduced binding specificity (24). For example, short basic region peptides (such as ADP3, residues 48–72) show less than a 2-fold difference in affinity between wild-type TAR RNA and TAR sequences carrying the U23 → C and G26-C39 → C-G mutations (24). By contrast, peptides such as ADP-1 (residues 37–72) that contain residues from the conserved 'core' region of lentivirus *trans*-activators (36), had a higher affinity for TAR RNA and were able to discriminate between TAR RNA mutants with a specificity which more closely resembled that of the tat protein (24).

Short basic peptides derived from tat appear to be unstructured and only assume the configurations that are imposed by the RNA-binding site (32, 37–39). By contrast to the protein, the basic peptides display a high affinity for duplex RNA sequences that carry one or more bulged residues. For example, monomeric complexes between peptides and TAR are only obtained when TAR is truncated and the competing binding sites normally present on the lower stem are removed (24, 30, 34). Thus, one can plausibly argue that the basic sequence in tat detects structural perturbations of the RNA helix non-specifically, while the flanking residues in tat are responsible for site-specific recognition. For example, the core region could help to orientate the basic region in the major groove. Alternatively, essential amino acid residues in the core region, such as Lys 41 (40), could form specific contacts with TAR RNA.

As discussed in Chapter 10, Frankel and colleagues have argued that the only specific contact between tat and TAR RNA involves a single arginine residue (31, 32, 41–43). Obviously, a single arginine would be unable to interact with all of the diverse functional groups comprising the tat-binding site unless TAR RNA undergoes extensive conformational rearrangements. One possible rearrangement, including the formation of a base triple between U23 and A27–U38 has been proposed by Puglisi *et al.* (Chapter 10; 41, 42) on the basis of nuclear magnetic resonance (NMR) studies of complexes between TAR and a derivative of arginine, argininamide. It is still unclear whether a structure resembling the TAR–argininamide complex is present in tat–TAR complexes. In quantitative terms the two binding reactions are very different. tat binding is achieved at nanomolar concentrations, whereas the free argininamide binds only at millimolar concentrations (43). Furthermore, the tat protein binds with high affinity to TAR RNA carrying N^6-methyl-A27, a modified base that should be unable to participate in the formation of a base triple with U23. Until a structure for the tat protein–TAR RNA complex is solved, the number and type of contacts that exist between the protein and the RNA will remain ambiguous.

2.3 *Trans*-activation mechanism

How does tat stimulate transcription? In the absence of tat, the majority of the viral transcripts are short RNA molecules ranging in size from 60 to 80 nt (44–47). Nuclear run-off experiments have shown that the majority of RNA polymerases initiating transcription in the absence of tat stall near the promoter (48, 49). Following addition of tat, there is a dramatic increase in the density of RNA polymerases found downstream of the promoter (45, 48–53).

The observations described above have led to the working hypothesis for the mechanism of action of tat that is illustrated in *Figure 4*. According to this proposal, the binding of cellular initiation factors allows RNA polymerase to initiate transcription from the 5'LTR of the HIV provirus. However, by contrast to cellular promoters, the RNA polymerase engaged by the HIV LTR is believed to be intrinsically unstable. Soon after transcription through TAR, the polymerase either pauses or falls off the template. If tat is present in the cell, it can associate with TAR RNA and the transcribing RNA polymerase. It seems likely that this reaction involves not only the binding of tat to TAR RNA but also the recruitment of cellular co-factors. The modified transcription complex is then able to transcribe the remainder of the HIV genome efficiently. As pointed out by Peterlin and colleagues (15, 28, 45) the antitermination mechanism envisioned in *Figure 4* may be analogous to the mechanism used by the bacteriophage λN protein (54). However, in the λN system antitermination takes place at a specific ρ-dependent site, whereas in the HIV system, tat may function as a generalized elongation factor.

2.3.1 *Trans*-activation *in vitro*

The recent development of cell-free transcription systems that respond to tat is now allowing direct tests of the *trans*-activation model (50, 52, 53, 55, 56). When used in standard cell-free systems, the HIV LTR behaves as a highly active promoter even in the absence of tat. This high level of tat-independent transcription is short-lived. After 30 min, most transcription in the reaction becomes tat dependent (52, 53, 56). Alternatively, extracts that have been pre-incubated in the presence of ATP (53, 56) or sodium citrate (50) or prepared in the presence of 100 μM $ZnSO_4$ (55), have been used to detect tat-dependent transcription.

The experiments shown in *Figure 5* demonstrate that the tat-stimulated RNA polymerase has an intrinsic antitermination activity (55). A synthetic terminator sequence (τ) consisting of a stable RNA stem–loop structure followed by a tract of nine uridine residues was placed ~ 200 nt downstream of the start of transcription. The presence of τ caused ~ 30% of the transcribing polymerases to disengage

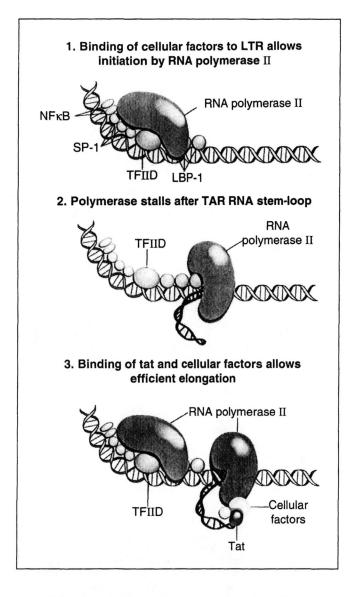

Fig. 4 Model for control of transcriptional elongation by tat. Step 1: cellular transcription factors binding to the proviral LTR allow RNA polymerase II to be engaged by the promoter. Step 2: after transcription through TAR RNA, the RNA polymerase pauses. Step 3: tat, cellular TAR binding factors, and possibly other co-factors form a complex together with the RNA polymerase and the TAR RNA present on the nascent chain. In the absence of tat, these factors are not recruited and the unstable RNA polymerase usually disengages from the template within several hundred nucleotides of the end of TAR. Once the polymerase has moved away from the TAR region, the promoter becomes accessible and a new RNA polymerase is able to initiate transcription.

prematurely from the template. Addition of recombinant tat protein (*Figure 5*) to the reaction stimulated the production of run-off product (ρ) by more than 25-fold.

Stimulation of transcription by tat in the cell-free system requires a functional TAR element. Templates carrying the ΔU23–25 mutation in TAR (a mutation that abolishes specific tat binding *in vitro*) were stimulated less than 3-fold by the addition of tat (*Figure 5*, left). Studies using an extensive series of templates carrying mutations in TAR have shown that tat-dependent *trans*-activation *in vitro* has the same sequence requirements seen *in vivo*. Templates carrying mutations

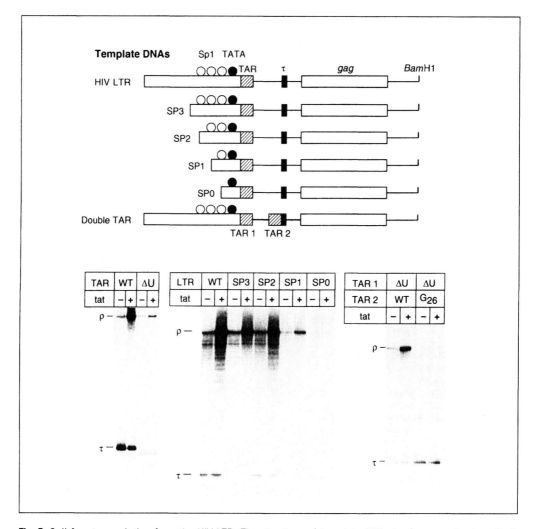

Fig. 5 Cell-free transcription from the HIV LTR. The structure of template DNAs is shown on the top. Each plasmid carried synthetic terminators (τ) inserted ~230 nt downstream of the start of transcription and were linearized by cleavage at the *Bam*H1 site (or an adjacent *Xba*I site) located ~650 nt downstream of the start of transcription. Plasmids SP3, SP2, SP1, and SP0 carried 5' deletions in the HIV LTR. These deletions removed all the LTR sequences upstream of the third, second, or first Sp1 sites or the TATA box, respectively. The double TAR plasmid carried a duplicated TAR sequence inserted immediately upstream of the terminator site. Transcription reactions were performed essentially as described (52, 53, 55) in the presence of 0 (−) or 200 ng (+) recombinant tat protein. Left panel: templates carried an intact LTR and either wild-type TAR or the ΔU23−25 mutation. Addition of tat protein increased the synthesis of the run-off product (ρ) ~25-fold from the wild-type template, but less than 3-fold from the template carrying the ΔU23−25 mutation. The level of transcription products ending at the terminator (τ) remained relatively constant in the presence or absence of tat. Centre panel: templates carried either the intact LTR or the SP3, SP2, SP1 and SP0 deletions. In each case, tat was able to stimulate transcription through to the run-off product (ρ) by 5−25-fold. Right panel: templates carrying TAR duplications are able to respond to tat even when the first TAR is inactivated by mutation. A template carrying the ΔU23−25 mutation in TAR 1 and a wild-type sequence in TAR 2 produced ~10-fold more run-off product (ρ) after addition of tat. When TAR 2 was inactivated by the G26−C39 → C−G mutation, a less than 2-fold response to tat was observed.

that reduce tat affinity for TAR as well as mutations of the apical loop of TAR RNA show poor responses to tat in the cell-free system (55).

Biochemical evidence suggests that TAR itself can also act as a terminator (46, 47, 57, 58). In the cell-free system, the major termination sites correspond to regions of the transcripts that can fold into stable RNA hairpin–loop structures, such as TAR RNA, and other stem–loop structures found before the start of the *gag* gene, as well as the artificial terminator (55).

2.3.2 Control of elongation, but not initiation, by tat

Can tat stimulate initiation as well as elongation? Mathews and colleagues (48, 59) have obtained data from nuclear run-off experiments that suggest that tat can stimulate initiation 2–5-fold. Furthermore, when tat is fused to a DNA-binding domain it can activate promoters carrying the appropriate DNA-binding site (60, 61). These observations have led to a variety of suggestions that tat functions primarily as an initiation factor. For example, Jeang and colleagues (62, 63) have proposed that tat binds directly to the cellular transcription factor Sp1. In another model, Cullen (64) has suggested that tat selectively stimulates initiation by a class of RNA polymerases which are able to elongate more efficiently past blocks near the HIV LTR than ordinary RNA polymerases. A limitation of all these models is that they don't specify how tat enters the system through TAR. The models seem to imply that tat dissociates from TAR RNA and then acts on later rounds of transcription, whereas it seems much more likely that tat remains with the elongating RNA polymerase after forming a stable complex with TAR RNA. Fortunately, formal tests of these models are now possible.

Several recent experiments using the cell-free systems suggest that tat acts exclusively at the level of elongation (50, 52, 53, 55, 56). Most obviously, RNase protection experiments show that tat stimulates elongation without increasing the overall levels of transcripts initiating on template DNAs (50, 53, 56). In addition, the only element required to generate a tat response in the *in vitro* systems appears to be TAR. Hybrid promoters containing the adenovirus major late promoter fused to HIV TAR are fully functional (56).

tat is even able to boost transcription when initiation rates are experimentally reduced by removal of the sites for upstream factors. In *Figure 5* (centre), the transcriptional activity of template DNAs carrying three, two, one, or no Sp1 sites were compared in the cell-free system. As Sp1 sites were removed there was a progressive decrease in the level of tat-independent transcription. However, for each template, including templates that contain no upstream elements, tat is able to stimulate the production of full-length RNA. The data suggest that although the levels of both tat-independent and tat-dependent transcription are limited by the mutations in the promoter, tat provides a complementary activity due to its ability to stimulate elongation.

2.3.3 Introduction of tat at a distance

If tat is able to associate with the TAR RNA produced by the elongating RNA polymerase, then it should also be possible to introduce tat when TAR is placed considerably downstream of the promoter. In an early experiment, Selby et al. (15) found that maximal promoter activity required TAR to be located immediately downstream of the start of transcription. When TAR was displaced by 88 nt, 20% of the tat-stimulated transcriptional activity was obtained. This experiment has been interpreted to mean that TAR needs to be near the start of transcription in order to allow tat to interact with the upstream elements in the promoter (61). However, it is also possible that the insertion mutations used by Selby et al. (15) inhibited transcription by disrupting DNA elements, such as the binding sites for the leader-binding protein (LBP) known to be present in this region of TAR (1, 65).

To test whether tat could be introduced at a distance, under conditions where the sequences near the start of HIV transcription were unaltered, we prepared templates carrying duplicated TAR elements (Figure 5). The first TAR carried the $\triangle$U23–25 mutation and is therefore unable to bind tat efficiently. The second TAR carried either an intact tat-binding site or, as a control, the G26–C39 → C–G mutation in the tat-binding site. As shown in Figure 5 (right) when a wild-type sequence is present in the second TAR, tat is able to stimulate transcription efficiently.

In conclusion, the cell-free systems are providing strong support for the *trans*-activation model outlined in *Figure 4*. The experimental measurements suggesting that tat stimulates initiation could simply be a consequence of the proximity of TAR and the start of transcription. For example, it is possible that the pausing of RNA polymerase near TAR could block the access of the next enzyme to the promoter. It remains a challenge for the future to devise biochemical experiments that will allow the identification of the components of tat-modified transcription complexes.

3. Control of late mRNA expression by rev

3.1 The rev-response and INS elements

The HIV genome (*Figure 6*) encodes nine genes, but at least 30 different mRNA transcripts are produced by splicing using the six major splice acceptors and two major splice donor sequences (66, 67). During mutagenesis experiments on the *tat* gene, Sodroski et al. (8) noticed that mutations near the two coding exons of *tat* dramatically reduced the ability of HIV to express the gag and env proteins. These mutations could not be complemented by the *tat* gene, suggesting that a second regulatory gene, now known as the *rev* gene, also governs HIV expression.

Only the short fully spliced viral mRNAs appear in the cytoplasm of rev-minus cells (9, 68, 69). It is now known that both positive and negative regulatory sequences required for rev function are present in the regions of the *gag* and *env* mRNAs which are removed by splicing (Figure 6). A *cis*-acting sequence, called the rev-responsive element (RRE), acts as a positive control element that is absolutely required for rev activity. The RRE is located within the *env* reading frame (70–72). The rev-dependent mRNAs also carry negative regulatory elements, now called

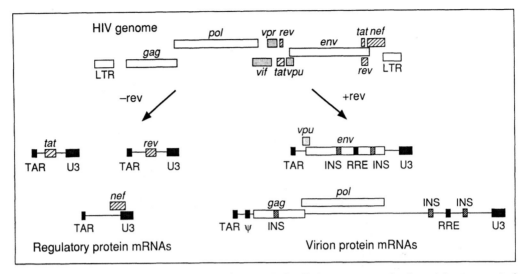

Fig. 6 Structure of HIV genome and the major viral mRNAs. Early gene expression is mainly composed of 1.8 kb transcripts containing doubly spliced mRNAs encoding the regulatory proteins tat, rev, and nef. Each mRNA begins with the TAR sequence at the 5' end and the U3 region of the viral LTR at its 3' end. Late gene expression includes the 4.3 kb mRNA for *vpu–env*. Messengers for *vpr*, *vif*, and the first exon of *tat* are also ~4.3 kb (not shown). The 9.2 kb virion RNA also acts as the mRNA for *gag–pol*. Both the *vpu–env* and *gag–pol* mRNAs carry the rev-response element (RRE) and instability (INS) elements. These regulatory elements, required for rev activity, are removed from the early mRNAs by splicing.

instability (INS) sequences (70, 73, 74). In the absence of either rev or a functional RRE sequence rev-dependent mRNAs are highly unstable in the cytoplasm. Their cytoplasmic levels rise dramatically if rev is added in *trans* (73, 75) or when the INS sequences are inactivated by a series of mutations (74).

The mechanism of action of rev is still unknown. Messenger RNA precursors carrying heterologous 5' or 3' splice sites may become rev responsive by the addition of RRE sequences into an intron (68, 76–78). However, it seems unlikely that rev has a direct effect on splicing, since rev can stimulate production of *gag* mRNA even when there are no splice donors or acceptors present (75).

If rev does not control splicing, how can it affect the ratios of fully and partially spliced viral mRNAs? In the absence of rev there is a large pool of the unspliced or partially spliced mRNAs that appear in the nucleus, but only doubly spliced mRNAs appear in the cytoplasm. After addition of rev, the larger mRNAs can be found in the cytoplasm as well as in the nucleus (68, 79, 80). Although these observations have been interpreted to mean that rev exerts a direct effect on mRNA transport, the ability of rev to selectively stabilize mRNAs carrying INS sequences in the cytoplasm and the nucleus could also account for these results (69, 75, 79, 81).

3.2 rev binds directly to the RRE

As shown in *Figure 7*, the RRE RNA forms a complicated stem–loop structure (71). The RRE RNA is specifically bound by rev *in vitro* (82–85), but the binding reaction

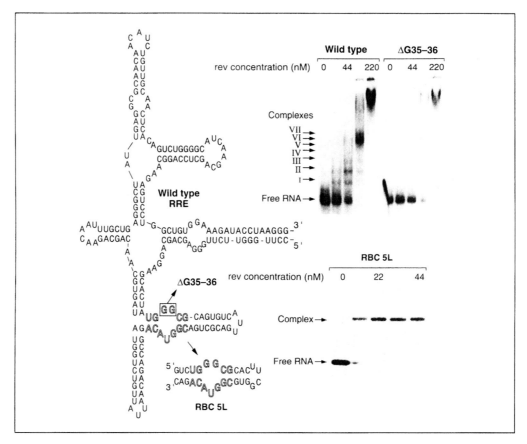

Fig. 7 Binding of rev to synthetic RNA binding sites and RRE RNA transcripts. The predicted secondary structures for a 238 nt fragment of the RRE and a short synthetic sequence, RBC5L, are shown on the left. Residues in the high-affinity site shown to be essential for rev recognition (86) by mutagenesis experiments are highlighted. Top panel: RRE transcripts (238 nt) carrying either the wild-type sequence or the mutation ΔG35–36 were tested for rev binding by gel-mobility shift assays. Reactions contained 25 nM ^{32}P-labelled RNA and up to 220 nM of recombinant rev protein. Note that the characteristic stepwise assembly of rev on the RRE is blocked by the mutation in the high-affinity binding site. Bottom panel: binding of rev to a minimal synthetic structure. Binding reactions contained 10 nM ^{32}P-labelled RNA and between 0 and 44 nM of recombinant rev protein. Note that even in the presence of excess rev only a single complex is formed.

is complex and involves an initial interaction with a high-affinity site followed by the cooperative addition of rev monomers to lower affinity sites (83). The simplest way to visualize this process is by electrophoretic mobility-shift assays. As shown in the top panel of *Figure 7*, the rev-binding reaction produces a series of complexes. There is a progressive increase in the formation of the lowest mobility complexes as the molar ratio of rev to RRE RNA increases (83, 86–88).

3.2.1 High-affinity binding by rev at the 'bubble' structure

The high affinity rev-binding site was mapped by constructing artificial stem–loop structures carrying fragments of the RRE (86). The binding site contains an unusual purine-rich 'bubble' containing bulged GG and GUA residues that was not predicted in the original models for the RRE structure (68). The identification of the 'bubble' sequence as a high-affinity binding site is confirmed by chemical footprinting data (89, 90), as well as by the selection of high-affinity binding sites for rev from pools of randomly generated mutants (91).

In the RRE, the 'bubble' is located towards the base of a stem–loop near a three-way junction (*Figure 7*). rev does not appear to recognize features of the junction and this sequence can be replaced by any short stretch of duplex RNA. An example of one of these artificial binding sites, RBC5L, is shown in *Figure 7* (bottom). RBC5L is able to bind a monomer of rev with a binding constant of ~ 2 nM. Even in the presence of a 10-fold molar excess of rev, only a single complex is formed.

3.2.2 Non-Watson–Crick base pairs stabilize the 'bubble'

Recognition of the high-affinity binding site by rev shows strong parallels to the recognition of TAR RNA by tat. Both proteins recognize base pairs in a distorted major groove of duplex RNA and both proteins make contacts with the adjacent phosphates. However, the nature of the distortion in the 'bubble' is clearly different from that introduced by the bulge in TAR RNA.

The 'bubble' structure is believed to be stabilized by non-Watson–Crick G–A and G–G base pairs formed between the bulged residues (86, 91, 92). Using a strategy based on our experience with tat binding to TAR RNA, we investigated the nature of the non-Watson–Crick base pairs in the 'bubble' by functional group replacement (*Figure 8*, top). Bartel *et al.* (91) have shown that the G36–G59 base pair can be replaced by an isosteric A–A base pair. Replacement of each of the G residues in the 'bubble' by dI (hypoxanthine), either individually or simultaneously, produced no significant loss in rev affinity. However substitution of O^6-methyl-G for either G35 or G36 resulted in a substantial loss of rev binding. By contrast, replacement of A61 by dI resulted in substantial loss of binding to rev, but N^6-methyl-dA is tolerated at this position. Thus, none of the exocyclic amino groups of bulged G residues is likely to be involved in intrachain base pairing or in rev binding, whereas the exocyclic amino group of A61 is important. These data are consistent with chemical interference data showing that modification of G and A residues by dimethylsulphate (DMS) (89) or by DEPC (89, 90) inhibits the binding of rev.

The functional group replacement studies limit the possible base-pairing schemes for the G36–G59 and G35–A61 pairs to the structures shown in *Figure 9*. The G36–G59 pair in the bubble appears to be an *anti–syn* configuration since the exocyclic amino groups of G36 and G59 are not essential for rev binding, and N^7-deaza-dG is mostly tolerated at G35 and fully tolerated at G36. The G35–A61 base pair is probably in the *anti–anti* configuration, since inosine is tolerated at G35 and both N^6-methyl-dA and N^7-deaza-dA are tolerated at A61.

3.2.3 Recognition of functional groups in the major groove by rev

Surprisingly, we have not obtained any evidence for direct contacts between rev and the bulged residues in the bubble. Each of functional groups on the

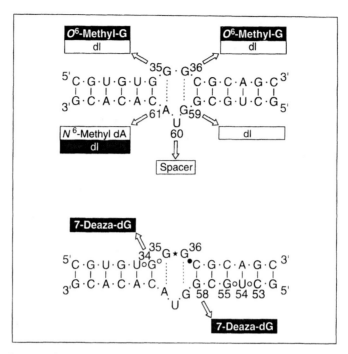

Fig. 8 The effect of chemical modifications on the rev-binding site. Top: analysis of non-Watson–Crick base pairs. Synthetic RNA duplexes carrying O^6-methyl-G at positions G35 and G36 show a greater than 10-fold loss of rev affinity (filled boxes). By contrast dI is well tolerated at these positions (open boxes) and at G59. A triple substitution of dI at positions 35, 36, and 59 binds rev with essentially wild-type affinity. At position A61, N^6-methyl-dA is tolerated, but dI inhibits rev binding. Bottom: critical contacts for rev binding. Substitution of 7-deaza-dG at positions G34 or G58 reduces rev affinity by more than 10-fold. By contrast, 7-deaza-dG is better tolerated at G35 and G36. The open circles denoting P33, P34, P53, and P54 are sites where ethylation by ENU or substitution by both diastereoisomers of methylphosphonates inhibits rev binding. At P36 (filled circle) only one diastereoisomer of methylphosphonate blocks rev binding, whereas at P35 (star) methylphosphonate substitutions are tolerated.

bulged nucleotides is either non-essential or appears to be involved in base pairing.

The RRE 'bubble' appears to be highly structured. The only unpaired base in the 'bubble' appears to be U60, which acts as a spacer. Base substitutions at position 60 are freely tolerated (86, 92). U60 can even be replaced by an abasic propyl linker sequence without interfering with rev-binding activity (*Figure 8*). Model building suggests that there is no significant bending at the 'bubble', but the non-Watson–Crick pairs plus the presence of the U60 residue allows sufficient opening of the major groove for specific recognition to take place at G34 and other flanking base residues. Consistent with this model, functional group replacements have demonstrated that the base pairs in the RRE RNA stem near the 'bubble' are recognized in the major groove (*Figure 8*, bottom). Replacement of either G34 or G58 by N^7-deaza-dG results in a substantial drop in rev affinity. Thus, recognition of the high-affinity site involves contacts made primarily, if not exclusively, with the flanking Watson–Crick base pairs.

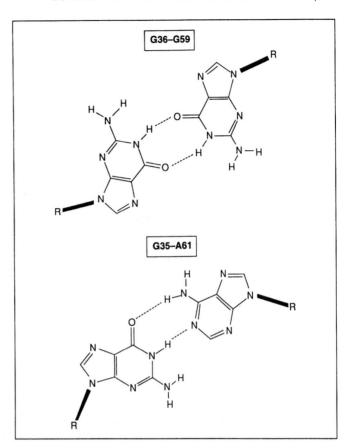

Fig. 9 Structure of non-Watson–Crick base pairs found in the RRE 'bubble'. Top: the G36–G59 pair in the bubble appears to be an *anti–syn* configuration. Evidence that the exocyclic amino groups do not participate in forming the base pair comes from experiments showing that simultaneous substitutions of G36 and G59 with inosines are tolerated (92). Bottom: the G35–A61 base pair is probably in the *anti–anti* configuration as shown, since inosine is tolerated at G35 and both N^6-methyl-dA and N^7-deaza-dA are tolerated at A61.

3.2.4 Contacts with the phosphate backbones

Short RNA duplexes with less than six flanking base pairs are unable to bind rev with high affinity, suggesting that the region of duplex RNA covered by rev must extend beyond the 'bubble' structure (92). Similarly, ethylation interference experiments have shown that rev is positioned asymmetrically with respect to the 'bubble' and makes contacts with phosphates located on both strands of the duplex (P33–34 and P53–54) (89).

We have used methylphosphonate substitution experiments to define the nature of the phosphate contacts made by rev more precisely. Methylphosphonate can interfere with rev binding either by neutralizing the charge on the phosphate, removing a hydrogen-bond contact, or by steric interference. We found that at P33, P34, P53, and P54 either diastereoisomer of the methylphosphonate blocked binding. These positions are therefore likely to be sites where salt bridges are formed between rev and the phosphate backbone. By contrast, at P36 only one diastereoisomer of methylphosphonate inhibited binding. Inhibition in this case is therefore due to the loss of a hydrogen-bonding contact or due to a steric clash. At P35 both methylphosphonate diastereoisomers were tolerated, suggesting in agreement with the base modification data described above, that rev protein does not make direct contact with this region of the 'bubble'.

3.2.5 rev binds preferentially to double-stranded regions of the RRE RNA

Ordered assembly of rev only takes place when a functional 'bubble' sequence is present on the RRE. When RRE RNAs carrying mutations in the 'bubble' sequence are used in gel-shift assays, the characteristic stepwise pattern of complex formation is not observed and only non-specific complexes are seen at high rev concentrations (*Figure 7*, top).

The regular nature of the secondary structure for the RRE (*Figure 7*) suggests that after initial recognition of the 'bubble' on one arm, there is an ordered recognition of the other five arms, but the precise details of rev assembly on the RRE are not yet understood. It is clear, however, that between six and eight rev monomers are able to bind to the 234 nt RRE RNA (92, 93). In the absence of rev, the RRE RNA is susceptible to nuclease digestion, but in the presence of rev, protected fragments of increasing size are observed. Analysis of these patterns has shown that the sites of nuclease cleavage are restricted to the apices of the loops present in the RRE RNA structure. The simplest interpretation of the data is that rev binds preferentially to the double-stranded regions of the RRE RNA.

3.2.6 Protein interactions in rev oligomerization

The *in vitro* binding experiments described above strongly suggest that rev binding to the high-affinity site within the RRE RNA may be considered to be the nucleation event for a complicated assembly process, analogous to the packaging of TMV and other RNA viruses by their coat proteins (94). However, unlike the viral packaging reactions, rev neither completely protects the RNA from nuclease digestion, nor denatures double-stranded regions in the RRE RNA.

At very high protein concentrations the RRE RNA is packaged into rod-like ribonucleoprotein filaments (86, 95). The rev protein can also aggregate in low-salt buffers (50 mM NaCl) and form large filaments ~ 140 Å wide and up to 15 000 Å long (86, 95). Although these structures may not be of physiological significance, the observations demonstrate that protein–protein interactions, as well as protein–RNA interactions allow assembly of multiple rev molecules on the RRE RNA. Similarly mutations throughout rev are known to produce defects in the formation of multiple complexes with RRE RNA (71, 88, 96, 97).

3.3 Activity of rev *in vivo*

3.3.1 An intact 'bubble' is necessary for rev activity

Is the RRE RNA packaged *in vivo*? A series of mutations was introduced into the RRE and incorporated into a reporter plasmid in order to test whether both nucleation and rev oligomerization takes place *in vivo*. As shown in *Figure 10*, the reporter plasmid used is a defective HIV prophage carrying a deletion of the *pol*, *vif*, and *vpr* genes. The *nef* gene and 3'LTR have also been replaced by a polyadenylation site derived from SV40, while the *rev* gene has been inactivated by filling in the unique *Bam*H1 site found in its second exon. When transfected into cells, this plasmid produces *tat* and *rev* mRNAs which are easily detected by RNase protection assays using a probe overlapping their second exons (*Figure 11*, left). Only low levels of the *gag* and *env* mRNAs are detected in the absence of rev, but these increase dramatically when rev is supplied.

In agreement with previous studies, rev activity in this system requires the high-affinity rev-binding site (72, 84, 91, 98–100). For example, deletion of G35 and G36 in the 'bubble' abolishes rev-dependent *trans*-activation (*Figure 11*, left).

3.3.2 An extended stem 1 structure is required for full rev activity

The RRE has previously been defined as a 234 nt RNA. However, we found that the sequences as originally defined by Malim *et al.* (68) showed reduced activity in the reporter gene systems. On the basis that the flanking sequences might be contributing to the RRE structure additional regions of possible secondary structure immediately 3' and 5' of the mapped RRE region were sought. Surprisingly, stem I could be extended by nearly twice the length of the stem I region defined by Malim *et al.* (68). The extended stem I (*Figure 10*) does not form a perfect duplex and it is periodically interspersed with purine-rich 'bubbles', including one at the junction of the sequence analysed by Malim *et al.* (68). It is tempting to speculate that these purine-rich 'bubbles' are stabilized by non-Watson–Crick G–G and G–A base pairs similar to those found at the high-affinity rev-binding site (91, 92).

In order to test whether the complete sequence of this proposed stem is required for rev activity, a series of RREs carrying truncations in stem I was tested in the reporter system (*Figure 10*). As shown in *Figure 11*, the magnitude of the rev response decreased as stem I was shortened. Plasmids carrying truncation 5, which deletes most of stem I, were as inactive as plasmids carrying the ΔG35–36 mutation in the high-affinity binding site. The RREs carrying truncation 3 (218 nt) are similar in size to the RRE studied by Malim *et al.* (68) and showed intermediate responses to rev.

The mutagenesis data are consistent with the view that rev packages double-stranded regions in the RRE following a nucleation event at the 'bubble'. This model nicely explains a variety of apparently contradictory results in the literature. For example, it is has been reported that mutations that disrupt double-stranded regions in the RRE invariably inactivate rev function, yet most of the individual stem structures in the RRE can be deleted without impairing RRE activity (72, 84, 98, 100, 101). However, packaging of RRE RNA *in vivo* is probably more complicated than the simple binding reaction seen *in vitro*. For example, it seems likely that hnRNA-binding proteins, which preferentially bind to single-stranded RNA but which can also catalyse the unwinding of double helical RNA (102), could act as competitors of rev packaging of the RRE and help regulate rev responsiveness.

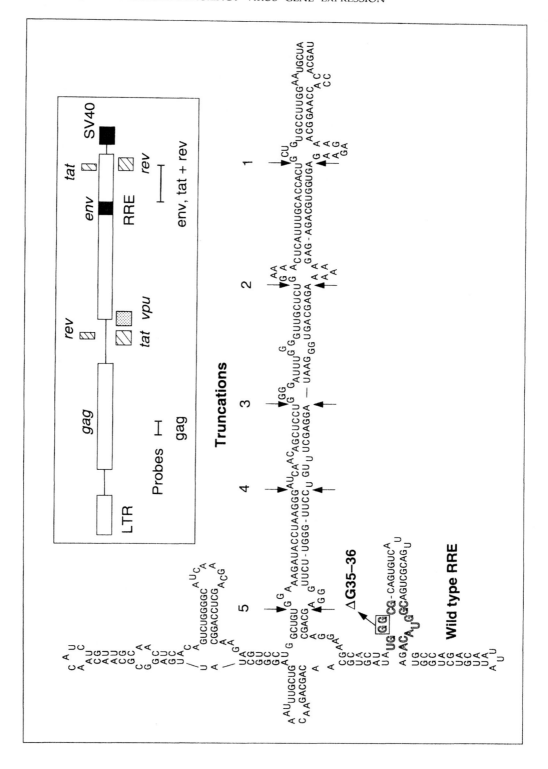

Fig. 10 Structure of rev-responsive reporter gene constructs and complete RRE. Insert: the test plasmid is a defective HIV provirus derived from the infectious molecular clone HIV-1_{NL4-3} (114). The *pol, vif,* and *vpr* genes were removed by a deletion between the *Bal*I and *Eco*RI sites of the provirus. In addition, a polyadenylation derived from SV40 placed downstream of the HIV sequences replaces the *nef* gene (starting at the *Xho*I site) and the 3′ LTR of the original provirus. To inactivate rev function, a frame-shift mutation was introduced at the *Bam*H1 site in the second exon of *rev*. Shown below are the positions of two probes used in the RNase protection experiments (*Figure 11*). Bottom: the secondary structure of the complete RRE from the HIV-1_{NL4-3} virus (114) is shown folded according the prediction programme of Zuker (115). Similar structures can be predicted for other HIV-1 strains including HIV-1_{SF-2} and HIV-1_{CSF-1}. The high-affinity binding site is highlighted. Arrows indicate the 3′ and 5′ ends of the truncations studied in *Figure 11*. The RRE sequence as originally defined by Malim *et al.* (68) corresponds to truncation 3.

4. Biological implications
4.1 Principles of RNA recognition

Studies of the interactions of tat and rev with RNA have provided new insights into the chemistry of nucleic acid recognition. There are many examples of RNA-binding proteins that recognize bases displayed in single-stranded loop and bulge structures (103). By contrast, both tat and rev associate with functional groups on Watson–Crick base pairs that are exposed within the major groove of a distorted double-stranded RNA. This new principle of RNA recognition is likely to extend to many other protein–RNA interactions.

There is undoubtedly a subtle interplay between the RNA structures and both the tat and rev proteins. The data on both tat binding to TAR and rev binding to the RRE suggest that there are numerous contacts made between the amino acid side chains of tat and functional groups displayed in pre-existing RNA-binding sites. However, conformational changes in the RNA after binding by the proteins have not been ruled out. In addition to base-specific contacts, both proteins make phosphate contacts on both strands of the flanking RNA duplexes.

tat and rev share an arginine-rich basic domain. There is little obvious sequence specificity in this region and the basic domains of the two proteins can be interchanged or even replaced by random sequences (32). It seems likely that the basic domain inserts into the distorted RNA helices found at the recognition sites for tat and rev. However, this interaction appears to represent only one of several types of contacts made between the RNA and the regulatory proteins. The complete proteins are needed to achieve full RNA recognition specificity.

4.2 Thresholds and kinetics controlling HIV gene expression

How relevant are the *in vitro* measurements of tat and rev binding to the physiological situation? For both proteins there is a strict correlation between their ability to specifically recognize RNA sequences and the efficiency of the *in vivo* response. The RNA-binding properties of the two proteins also make it likely that the distinct kinetic phases of HIV transcription are a reflection of their intracellular levels. Assuming a diameter of 5–10 μm for the nucleus of a typical human cell, the presence of only 40–320 molecules of tat in the nucleus would correspond to a concentration of 1 nM. At these concentrations, ~30% of the nascent transcripts

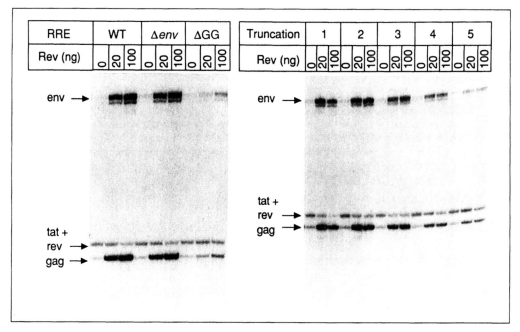

Fig. 11 Effect of mutations in the RRE on rev activity *in vivo*. Test plasmids carrying RRE mutations (500 ng) were co-transfected into HeLa cells together with up to 100 ng of pF31, a plasmid that expresses rev under the control of the CMV immediate early gene promoter. Levels of RNA synthesis were determined by RNase protection assays 48 h after transfection. ^{32}P-Labelled RNA transcripts corresponding to the antisense strands of the regions shown in *Figure 10* were simultaneously hybridized to cytoplasmic RNA obtained from the transfected cells. After digestion with an excess of RNase T$_1$, the protected fragments were detected by gel electrophoresis. Under the chosen digestion conditions, the unannealed probe is completely degraded. The 5' probe is specific for the *gag* mRNA, whereas the 3' probe detects simultaneously the second exons of the *tat* and *rev* genes as well as the partially spliced mRNAs for *vpu–env* and the unspliced mRNA for *gag*. In the absence of rev, both the *gag* and *env* sequences are poorly expressed. As rev is added, the levels of these mRNAs increase 10–20-fold. rev activity requires an intact high-affinity binding site (left). There is a reduced response to the ΔG35–36 mutation (ΔGG), but by contrast, deletion of envelope sequences flanking the end of the stem I duplex (Δ*env*) had no effect on rev activity. Truncations of stem I (right) produced a progressive decline in rev responses.

from a single integrated provirus would be expected to be bound by tat. Since tat is functional as a monomer and has a catalytic effect on transcription, these calculations explain how a strong stimulation of transcription can be achieved early during the virus growth cycle. Once rev concentrations rise above 10 nM oligomerization of rev on the RRE is expected to occur and late mRNA synthesis can begin.

The HIV growth cycle may also include a latent stage where viral gene expression is silent. Little or no virus production is detected in infected but non-cycling T cells obtained from the peripheral blood of patients, but stimulation of these cells with mitogens dramatically increases virus production (104). Although it has been suggested that rev might regulate latency directly (10, 105), any effect of rev on latency must be indirect since rev is not constitutively expressed in quiescent HIV-infected T cells. On the other hand, there is good evidence that viral transcription is limited in cells containing low levels of NF-κB (2, 3, 106, 107). Thus, the cellular DNA-binding transcription factors can act as 'molecular switches' that allow HIV to make the transition from a latent state to lytic growth. The rate of virus growth is further moderated by the regulatory proteins. If the LTR activity is too weak to allow the production of necessary levels of tat or there are insufficient tat levels to allow the threshold production of rev, virus replication will be aborted.

4.3 Future challenges

Important themes for anti viral drug development are also emerging from studies of the HIV regulatory proteins. It is possible to inhibit virus replication by over-expression of TAR RNA and RRE RNA 'decoy' sequences (108–111) or dominant mutants of rev (71, 112). This suggests that it may eventually be possible to inhibit virus replication by a 'gene therapy' strategy aimed at interfering with regulatory protein function. A more immediate prospect is to look for small molecules that inhibit tat and rev activity. The biochemical assays for binding and *trans*-activation lend themselves nicely to the screening of compounds. There may even be some prospects for drug design emerging from the study of the regulatory proteins, since the binding sites for both tat and rev appear to involve relatively small regions of RNA and the three-dimensional (3-D) structures of the binding sites should soon be known.

Acknowledgements

We thank Gabriele Varani and Jo Butler for helpful discussions and the present and former members of the HIV group at LMB for their collaboration.

References

1. Garcia, J. A., Harrich, D., Soultanakis, E., Wu, F., Mitsuyasu, R., and Gaynor, R. B. (1989) Human immunodeficiency virus type 1 LTR TATA and TAR region sequences required for transcriptional regulation. *EMBO J.*, **8**, 765.
2. Jones, K., Kadonaga, J., Luciw, P., and Tjian, R. (1986) Activation of the AIDS retrovirus promoter by the cellular transcription factor, Sp1. *Science*, **232**, 755.
3. Nabel, G. and Baltimore, D. A. (1987) An inducible transcription factor activates expression of human immunodeficiency virus in T cells. *Nature*, **326**, 711.
4. Kim, S., Byrn, R., Groopman, J., and Baltimore, D. (1989) Temporal aspects of DNA and RNA synthesis during human immunodeficiency virus infection: evidence for differential gene expression. *J. Virol.*, **63**, 3708.
5. Arya, S. K., Guo, C., Josephs, S. F., and Wong-Staal, F. (1985) *Trans*-activator gene of human T-lymphotropic virus type III (HTLV-III). *Science*, **229**, 69.
6. Sodroski, J., Patarca, R., Rosen, C., Wong-Staal, F., and Haseltine, W. A. (1985) Location of the *trans*-acting region on the genome of human T-cell lymphotropic virus type III. *Science*, **229**, 74.

7. Sodroski, J. G., Rosen, C. A., Wong-Staal, F., Salahuddin, S. Z., Popovic, M., Arya, S., Gallo, R. C., and Haseltine, W. A. (1985) *Trans*-acting transcriptional regulation of human T-cell leukemia virus type III long terminal repeat. *Science*, **227**, 171.
8. Sodroski, J., Goh, W. C., Rosen, C. A., Dayton, A., Terwilliger, E., and Haseltine, W. A. (1986) A second post-transcriptional activator gene required for HTLV-III replication. *Nature*, **321**, 412.
9. Feinberg, M. B., Jarrett, R. F., Aldovini, A., Gallo, R. C., and Wong-Staal, F. (1986) HTLV-III expression and production involve complex regulation at the levels of splicing and translation of viral RNA. *Cell*, **46**, 807.
10. Pomerantz, R. J., Trono, D., Feinberg, M. B., and Baltimore, D. (1990) Cells non-productively infected with HIV-1 exhibit an aberrant pattern of viral RNA expression: a molecular model for latency. *Cell*, **61**, 1271.
11. Cullen, B. R. (1986) *Trans*-activation of human immunodeficiency virus occurs via a bimodal mechanism. *Cell*, **46**, 973.
12. Rosen, C. A., Sodroski, J. G., and Haseltine, W. A. (1985) The location of *cis*-acting regulatory sequences in the human T cell lymphotropic virus type III (HTLV-III/LAV) long terminal repeat. *Cell*, **41**, 813.
13. Muesing, M. A., Smith, D. H., and Capon, D. J. (1987) Regulation of mRNA accumulation by a human immunodeficiency virus *trans*-activator protein. *Cell*, **48**, 691.
14. Feng, S. and Holland, E. C. (1988) HIV-1 tat *trans*-activation requires the loop sequence within TAR. *Nature*, **334**, 165.
15. Selby, M. J., Bain, E. S., Luciw, P., and Peterlin, B. M. (1989) Structure, sequence and position of the stem–loop in TAR determine transcriptional elongation by tat through the HIV-1 long terminal repeat. *Genes Devel.*, **3**, 547.
16. Berkhout, B., Silverman, R. H., and Jeang, K.-T. (1989) Tat *trans*-activates the human immunodeficiency virus through a nascent RNA target. *Cell*, **59**, 273.
17. Dingwall, C., Ernberg, I., Gait, M. J., Green, S. M., Heaphy, S., Karn, J., Lowe, A. D., Singh, M., Skinner, M. A., and Valerio, R. (1989) Human immunodeficiency virus 1 tat protein binds *trans*-activation-responsive region (TAR) RNA *in vitro*. *Proc. Natl Acad. Sci. USA*, **86**, 6925.
18. Frankel, A. D., Bredt, D. S., and Pabo, C. O. (1988) Tat protein from human immunodeficiency virus forms a metal-linked dimer. *Science*, **240**, 70.
19. Marciniak, R. A., Garcia-Blanco, M. A., and Sharp, P. A. (1990) Identification and characterization of a HeLa nuclear protein that specifically binds to the *trans*-activation response (TAR) element of human immunodeficiency virus. *Proc. Natl Acad. Sci. USA*, **87**, 3624.
20. Roy, S., Delling, U., Chen, C.-H., Rosen, C. A., and Sonenberg, N. (1990) A bulge structure in HIV-1 TAR RNA is required for tat binding and tat-mediated *trans*-activation. *Genes Devel.*, **4**, 1365.
21. Dingwall, C., Ernberg, I., Gait, M. J., Green, S. M., Heaphy, S., Karn, J., Lowe, A. D., Singh, M., and Skinner, M. A. (1990) HIV-1 tat protein stimulates transcription by binding to a U-rich bulge in the stem of the TAR RNA structure. *EMBO J.*, **9**, 4145.
22. Sheline, C. T., Milocco, L. H., and Jones, K. A. (1991) Two distinct nuclear transcription factors recognize loop and bulge residues of the HIV-1 TAR RNA hairpin. *Genes Devel.*, **5**, 2508.
23. Wu, F., Garcia, J., Sigman, D., and Gaynor, R. (1991) Tat regulates binding of the human immunodeficiency virus *trans*-activating region RNA loop-binding protein TRP-185. *Genes Devel.*, **5**, 2128.

24. Churcher, M., Lamont, C., Dingwall, C., Green, S. M., Lowe, A. D., Butler, P. J. G., Gait, M. J., and Karn, J. (1993) High affinity binding of TAR RNA by the human immunodeficiency virus tat protein requires amino acid residues flanking the basic domain and base pairs in the RNA stem. *J. Mol. Biol.*, **230**, 90.
25. Delling, U., Reid, L. S., Barnett, R. W., Ma, M. Y.-X., Climie, S., Sumner-Smith, M., and Sonenberg, N. (1992) Conserved nucleotides in the TAR RNA stem of human immunodeficiency virus type 1 are critical for tat binding and *trans*-activation: model for TAR RNA tertiary structure. *J. Virol.*, **66**, 3018.
26. Sumner-Smith, M., Roy, S., Barnett, R., Reid, L. S., Kuperman, R., Delling, U., and Sonenberg, N. (1991) Critical chemical features in *trans*-acting-responsive RNA are required for interaction of human immunodeficiency virus type 1 tat protein. *J. Virol.*, **65**, 5196.
27. Southgate, C., Zapp, M. L., and Green, M. R. (1990) Activation of transcription by HIV-1 tat protein tethered to nascent RNA through another protein. *Nature*, **345**, 640.
28. Selby, M. J. and Peterlin, B. M. (1990) *Trans*-activation by HIV-1 tat via a heterologous RNA binding protein. *Cell*, **62**, 769.
29. Hamy, F., Asseline, U., Grasby, J., Iwai, S., Pritchard, C., Slim, G., Butler, P. J. G., Karn, J., and Gait, M. J. (1993) Hydrogen-bonding contacts in the major groove are required for human immunodeficiency virus type-1 tat protein recognition of TAR RNA. *J. Mol. Biol.*, **230**, 111.
30. Weeks, K. M. and Crothers, D. M. (1991) RNA recognition by tat-derived peptides: interaction in the major groove? *Cell*, **66**, 577.
31. Calnan, B. J., Tidor, B., Biancalana, S., Hudson, D., and Frankel, A. D. (1991) Arginine-mediated RNA recognition: the arginine fork. *Science*, **252**, 1167.
32. Calnan, B. J., Biancalana, S., Hudson, D., and Frankel, A. D. (1991) Analysis of arginine-rich peptides from the HIV tat protein reveals unusual features of RNA–protein recognition. *Genes Devel.*, **5**, 201.
33. Cordingley, M. G., LaFemina, R. L., Callahan, P. L., Condra, J. H., Sardana, V. V., Graham, D. J., Nguyen, T. M., LeGrow, K., Gotlib, L., Schlabach, A. J., and Colonno, R. J. (1990) Sequence specific interaction of tat and tat peptides with the TAR sequence of HIV-1 *in vitro*. *Proc. Natl Acad. Sci. USA*, **87**, 8985.
34. Weeks, K. M., Ampe, C., Schultz, S. C., Steitz, T. A., and Crothers, D. M. (1990) Fragments of the HIV-1 tat protein specifically bind TAR RNA. *Science*, **249**, 1281.
35. Weeks, K. M. and Crothers, D. M. (1992) RNA binding assays for tat-derived peptides: implications for specificity. *Biochemistry*, **31**, 10281.
36. Derse, D., Carvalho, M., Carroll, R., and Peterlin, B. M. (1991) A minimal lentivirus tat. *J. Virol.*, **65**, 7012.
37. Tan, R. and Frankel, A. D. (1992) Circular dichroism studies suggest that TAR RNA changes conformation upon specific binding of arginine or guanidine. *Biochemistry*, **31**, 10288.
38. Loret, E. P., Georgel, P., Johnson, W. C. J., and Ho, P. S. (1992) Circular dichroism and molecular modeling yield a structure for the complex of human immunodeficiency virus type 1 *trans*-activation response RNA and the binding region of tat, the *trans*-acting transcriptional activator. *Proc. Natl Acad. Sci. USA*, **89**, 973.
39. Loret, E. P., Vives, E., Ho, P. S., Rochat, H., Rietschoten, J. V., and Johnson, W. C. J. (1991) Activating region of HIV-1 tat protein: vacuum UV circular dichroism and energy minimization. *Biochemistry*, **30**, 6013.

40. Tiley, L. S., Brown, P. H., and Cullen, B. R. (1990) Does the human immunodeficiency virus tat *trans*-activator contain a discrete activation domain? *Virology*, **178**, 560.
41. Puglisi, J. D., Tan, R., Calnan, B. J., Frankel, A. D., and Williamson, J. R. (1992) Conformation of the TAR RNA–arginine complex by NMR spectroscopy. *Science*, **257**, 76.
42. Puglisi, J. D., Chen, L., Frankel, A. D., and Williamson, J. R. (1993) Role of RNA structure in arginine recognition of TAR RNA. *Proc. Natl Acad. Sci. USA*, **90**, 3680.
43. Tao, J. and Frankel, A. D. (1992) Specific binding of arginine to TAR RNA. *Proc. Natl Acad. Sci. USA*, **89**, 2723.
44. Toohey, M. G. and Jones, K. A. (1989) In vitro formation of short RNA polymerase II transcripts that terminate within the HIV-1 and HIV-2 promoter-proximal downstream regions. *Genes Devel.*, **3**, 265.
45. Kao, S.-Y., Calman, A. F., Luciw, P. A., and Peterlin, B. M. (1987) Anti-termination of transcription within the long terminal repeat of HIV-1 by tat gene product. *Nature*, **330**, 489.
46. Kessler, M. and Mathews, M. B. (1992) Premature termination and processing of human immunodeficiency virus type 1-promoted transcripts. *J. Virol.*, **66**, 4488.
47. Ratnasabapathy, R., Sheldon, M., Johal, L., and Hernandez, N. (1990) The HIV-1 long terminal repeat contains an unusual element that induces the synthesis of short RNAs from various mRNA and snRNA promoters. *Genes Devel.*, **4**, 2061.
48. Laspia, M. F., Rice, A. P., and Mathews, M. B. (1989) HIV-1 tat protein increases transcriptional initiation and stabilizes elongation. *Cell*, **59**, 283.
49. Laspia, M. F., Rice, A. P., and Mathews, M. B. (1990) Synergy between HIV-1 tat and adenovirus E1a is principally due to stabilization of transcriptional elongation. *Genes Devel.*, **4**, 2397.
50. Kato, H., Sumimoto, H., Pognonec, P., Chen, C.-H., Rosen, C. A., and Roeder, R. G. (1992) HIV-1 tat acts as a processivity factor *in vitro* in conjunction with cellular elongation factors. *Genes Devel.*, **6**, 655.
51. Feinberg, M. B., Baltimore, D., and Frankel, A. D. (1991) The role of tat in the human immunodeficiency virus life cycle indicates a primary effect on transcriptional elongation. *Proc. Natl Acad. Sci. USA*, **88**, 4045.
52. Marciniak, R. A., Calnan, B. J., Frankel, A. D., and Sharp, P. A. (1990) HIV-1 tat protein *trans*-activates transcription *in vitro*. *Cell*, **63**, 791.
53. Marciniak, R. A. and Sharp, P. A. (1991) HIV-1 tat protein promotes formation of more processive elongation complexes. *EMBO J.*, **10**, 4189.
54. Greenblatt, J., Nodwell, J. R., and Mason, S. W. (1993) Transcriptional antitermination. *Nature*, **364**, 401.
55. Graeble, M. A., Churcher, M. J., Lowe, A. D., Gait, M. J., and Karn, J. (1993) Human immunodeficiency virus type 1 transactivator protein tat, stimulates transcriptional read-through of distal terminator sequences *in vitro*. *Proc. Natl Acad. Sci. USA*, **90**, 6184.
56. Laspia, M. F., Wendel, P., and Mathews, M. B. (1993) HIV-1 tat overcomes inefficient transcriptional elongation *in vitro*. *J. Mol. Biol.*, **232**, 732.
57. Bengal, E. and Aloni, Y. (1991) Transcriptional elongation by purified RNA polymerase II is blocked at the *trans*-activation-responsive region of human immunodeficiency virus type 1 *in vitro*. *J. Virol.*, **65**, 4910.
58. Sheldon, M., Ratnasabapathy, R., and Hernandez, N. (1993) Characterization of the inducer of short transcripts, a human immunodeficiency virus type 1 transcriptional element that activates the synthesis of short RNAs. *Mol. Cell. Biol.*, **13**, 1251.

59. Kessler, M. and Mathews, M. B. (1991) Tat transcription of the human immunodeficiency virus type 1 promoter is influenced by basal promoter activity and the simian virus 40 origin of DNA replication. *Proc. Natl Acad. Sci. USA*, **88**, 10018.
60. Southgate, C. D. and Green, M. R. (1991) The HIV-1 tat protein activates transcription from an upstream DNA-binding site: implications for tat function. *Genes Devel.*, **5**, 2496.
61. Berkhout, B., Gatignol, A., Rabson, A. B., and Jeang, K.-T. (1990) TAR-independent activation of the HIV-1 LTR: evidence that tat requires specific regions of the promoter. *Cell*, **62**, 757.
62. Jeang, K.-T., Chun, R., Lin, N. H., Gatignol, A., Glabe, C. G., and Fan, H. (1993) In vitro and in vivo binding of human immunodeficiency virus type 1 tat protein and Sp1 transcription factor. *J. Virol.*, **67**, 6224.
63. Berkhout, B. and Jeang, K.-T. (1992) Functional roles for the TATA promoter and enhancers in basal and tat-induced expression of the human immunodeficiency virus type 1 long terminal repeat. *J. Virol.*, **66**, 139.
64. Cullen, B. R. (1993) Does HIV-1 tat induce a change in viral initiation rights? *Cell*, **73**, 417.
65. Kato, H., Horikoshi, M., and Roeder, R. G. (1991) Repression of HIV-1 transcription by a cellular protein. *Science*, **251**, 1476.
66. Schwartz, S., Felber, B. K., Benko, D. M., Fenyo, E.-M., and Pavlakis, G. N. (1990) Cloning and functional analysis of multiply spliced mRNA species of human immunodeficiency virus type 1. *J. Virol.*, **64**, 2519.
67. Arrigo, S. J., Wietsman, S., Zack, J. A., and Chen, I. S. Y. (1990) Characterization and expression of novel singly spliced RNA species of human immunodeficiency virus type 1. *J. Virol.*, **64**, 4585.
68. Malim, M. H., Hauber, J., Le, S.-Y., Maizel, J. V., and Cullen, B. R. (1989) The HIV-1 rev *trans*-activator acts through a structured target sequence to activate nuclear export of unspliced viral mRNA. *Nature*, **338**, 254.
69. Arrigo, S. J. and Chen, I. S. Y. (1991) Rev is necessary for translation but not cytoplasmic accumulation of HIV-1 *vif*, *vpr*, and *env/vpu* 2 RNAs. *Genes Devel.*, **5**, 808.
70. Rosen, C. A., Terwilliger, E., Dayton, A. I., Sodrowski, J. G., and Haseltine, W. A. (1988) Intragenic *cis*-acting art-responsive sequences of the human immunodeficiency virus. *Proc. Natl Acad. Sci. USA*, **85**, 2071.
71. Malim, M. H., Bohnlein, S., Hauber, J., and Cullen, B. R. (1989) Functional dissection of the the HIV-1 rev *trans*-activator: derivation of a *trans*-dominant repressor of rev function. *Cell*, **58**, 205.
72. Dayton, E. T., Powell, D. M., and Dayton, A. I. (1989) Functional analysis of CAR, the target sequence for the rev protein of HIV-1. *Science*, **246**, 1625.
73. Hadzopoulou-Cladaras, M., Felber, B. K., Cladaras, C., Athanassopoulos, A., Tse, A., and Pavlakis, G. N. (1989) The rev/art protein of human immunodeficiency virus type 1 affects viral mRNA and protein expression via a *cis*-acting sequence in the *env* region. *J. Virol.*, **63**, 1265.
74. Schwartz, S., Campbell, M., Nasioulas, G., Harrison, J., Felber, B. K., and Pavlakis, G. N. (1992) Mutational inactivation of an inhibitory sequence in human immunodeficiency virus type 1 results in rev-independent *gag* expression. *J. Virol.*, **66**, 7176.
75. Schwartz, S., Felber, B. K., and Pavlakis, G. N. (1992) Distinct RNA sequences in the *gag* region of human immunodeficiency virus type 1 decrease RNA stability and inhibit expression in the absence of rev protein. *J. Virol.*, **66**, 150.

76. Kjems, J., Frankel, A. D., and Sharp, P. A. (1991) Specific regulation of mRNA splicing *in vitro* by a peptide from HIV-1 rev. *Cell*, **67**, 169.
77. Hammarskjold, M.-L., Heimer, J., Hammarskjold, B., Sangwan, I., Albert, L., and Rekosh, D. (1989) Regulation of human immunodeficiency virus *env* gene expression by the rev gene product. *J. Virol.*, **63**, 1959.
78. Chang, D. A. and Sharp, P. A. (1989) Regulation by HIV rev depends upon recognition of splice site. *Cell*, **59**, 789.
79. Malim, M. H. and Cullen, B. R. (1993) Rev and the fate of pre-mRNA in the nucleus: implications for the regulation of RNA processing in eukaryotes. *Mol. Cell Biol.*, **13**, 6180.
80. Emerman, M., Vazeaux, R., and Peden, K. (1989) The rev gene product of the human immunodeficiency virus affects envelope specific RNA localization. *Cell*, **57**, 1155.
81. Felber, B. K., Hadzopoulou-Cladaras, M., Cladaras, C., Copeland, T., and Pavlakis, G. N. (1989) Rev protein of human immunodeficiency virus type 1 affects the stability and transport of the viral mRNA. *Proc. Natl Acad. Sci. USA*, **86**, 1495.
82. Daly, T. J., Cook, K. S., Gary, G. S., Maione, T. E., and Rusche, J. R. (1989) Specific binding of HIV-1 recombinant rev protein to the *rev*-responsive element *in vitro*. *Nature*, **342**, 816.
83. Heaphy, S., Dingwall, C., Ernberg, I., Gait, M. J., Green, S. M., Karn, J., Lowe, A. D., Singh, M., and Skinner, M. A. (1990) HIV-1 regulator of virion expression (rev) protein binds to an RNA stem–loop structure located within the *rev*-response element region. *Cell*, **60**, 685.
84. Malim, M. H., Tiley, L. S., McCarn, D. F., Rusche, J. R., Hauber, J., and Cullen, B. R. (1990) HIV-1 structural gene expression requires binding of the *rev trans*-activator to its RNA target sequence. *Cell*, **60**, 675.
85. Zapp, M. L. and Green, M. R. (1989) Sequence-specific binding by the HIV-1 rev protein. *Nature*, **342**, 714
86. Heaphy, S., Finch, J. T., Gait, M. J., Karn, J., and Singh, M. (1991) Human immunodeficiency virus type-1 regulator of virion expression, rev, forms nucleoprotein filaments after binding to a purine-rich 'bubble' located within the *rev*-responsive region of viral RNA. *Proc. Natl Acad. Sci. USA*, **88**, 7366.
87. Kjems, J., Brown, M., Chang, D. D., and Sharp, P. A. (1991) Structural analysis of the interaction between the human immunodeficiency virus rev protein and the *rev* response element. *Proc. Natl Acad. Sci. USA*, **88**, 683.
88. Malim, M. H. and Cullen, B. R. (1991) HIV-1 structural gene expression requires the binding of multiple rev monomers to the viral RRE: implications for HIV-1 latency. *Cell*, **65**, 241.
89. Kjems, J., Calnan, B. J., Frankel, A. D., and Sharp, P. A. (1992) Specific binding of a basic peptide from HIV-1 *rev*. *EMBO J.*, **11**, 1119.
90. Tiley, L. S., Malim, M. H., Tewary, H. K., Stockley, P. G., and Cullen, B. R. (1992) Identification of a high-affinity RNA-binding site for the human immunodeficiency virus type 1 rev protein. *Proc. Natl Acad. Sci. USA*, **89**, 758.
91. Bartel, D. P., Zapp, M. L., Green, M. R., and Szostak, J. W. (1991) HIV-1 rev regulation involves recognition of non-Watson–Crick base pairs in viral RNA. *Cell*, **67**, 529.
92. Iwai, S., Pritchard, C., Mann, D. A., Karn, J., and Gait, M. J. (1992) Recognition of the high affinity binding site in rev-response element RNA by the human immunodeficiency virus type-1 rev protein. *Nucleic Acids Res.*, **20**, 6465.

93. Cook, K. S., Fisk, G. J., Hauber, J., Usman, N., Daly, T. J., and Rusche, J. R. (1991) Characterization of HIV-1 rev protein: binding and stoichiometry and minimal RNA substrate. *Nucleic Acid Res.*, **19**, 1577.
94. Turner, D. R., McGuigan, C. J., and Butler, P. J. G. (1989) Assembly of hybrid RNAs with tobacco mosaic virus coat protein: evidence for incorporation of disks in 5'-elongation along the major RNA tail. *J. Mol. Biol.*, **209**, 407.
95. Wingfield, P. T., Stahl, S. J., Payton, M. A., Venkatesan, S., Misra, M., and Steven, A. J. (1991) HIV-1 rev expressed in recombinant *Escherichia coli*: purification, polymerization and conformational properties. *Biochemistry*, **30**, 7527.
96. Malim, M. H., McCarn, D. F., Tiley, L. S., and Cullen, B. R. (1991) Mutational definition of the human immunodeficiency virus type 1 rev activation domain. *J. Virol.*, **65**, 4248.
97. Zapp, M. L., Hope, T. J., Parslow, T. G., and Green, M. R. (1991) Oligomerization and RNA binding domains of the type 1 human immunodeficiency virus rev protein: a dual function of an arginine-rich binding motif. *Proc. Natl Acad. Sci. USA*, **88**, 7734.
98. Dayton, E. T., Konings, D. A. M., Powell, D. M., Shapiro, B. A., Butini, L., Maizel, J. V., and Dayton, A. I. (1992) Extensive sequence-specific information throughout the CAR/RRE, the target sequence of the human immunodeficiency virus type 1 *rev* protein. *J. Virol.*, **66**, 1139.
99. Holland, S. M., Chavez, M., Gerstberger, S., and Venkatesan, S. (1992) A specific sequence with a bulged guanosine residue(s) in a stem–bulge–stem structure rev-responsive element RNA is required for *trans*-activation by human immunodeficiency virus type 1 rev. *J. Virol.*, **66**, 3699.
100. Olsen, H. S., Nelbrock, P., Cochrane, A. W., and Rosen, C. A. (1990) Secondary structure is the major determinant for interaction of HIV rev protein with RNA. *Science*, **247**, 845.
101. Cochrane, A. W., Chen, C.-H., and Rosen, C. A. (1990) Specific interaction of the human immuodeficiency virus rev protein with a structured region in the *env* mRNA. *Proc. Natl Acad. Sci. USA*, **87**, 1198.
102. Dreyfuss, G. (1986) Structure and function of nuclear and cytoplasmic ribonucleoprotein particles. *Ann. Rev. Cell Biol.*, **2**, 459.
103. Nagai, K. (1992) RNA–protein interactions. *Curr. Opin. Struct. Biol.*, **2**, 131.
104. Fauci, A. S. (1988) The human immunodeficiency virus: infectivity and mechanisms of pathogenesis. *Science*, **239**, 617.
105. Pomerantz, R. J., Seshamma, T., and Trono, D. (1992) Efficient replication of human immunodeficiency virus type 1 requires a threshold level of rev: potential implications for latency. *J. Virol.*, **66**, 1809.
106. Jones, K. A., Luciw, P. A., and Duchange, N. (1988) Structural arrangements of transcription control domains within the 5'-untranslated leader regions of the HIV-1 and HIV-2 promoters. *Genes Devel.*, **2**, 1101.
107. Jakobovits, A., Rosenthal, A., and Capon, D. J. (1990) *Trans*-activation of HIV-1 LTR-directed gene expression by tat requires protein kinase C. *EMBO J.*, **9**, 1165.
108. Graham, G. J. and Maio, J. J. (1990) RNA transcripts of the human immunodeficiency virus *trans*-activation response element can inhibit action of the viral *trans*-activator. *Proc. Natl Acad. Sci. USA*, **87**, 5817.
109. Lisziewicz, J., Rappaport, J., and Dhar, R. (1991) Tat-regulated production of multimerized TAR RNA inhibits HIV-1 gene expression. *New Biol.*, **3**, 82.
110. Sullenger, B. A., Gallardo, H. F., Ungers, G. E., and Gilboa, E. (1990) Overexpression

of TAR sequences renders cells resistant to human immunodeficiency virus replication. *Cell*, **63**, 601.
111. Sullenger, B. A., Gallardo, H. F., Ungers, G. E., and Gilboa, E. (1991) Analysis of *trans*-acting response decoy RNA-mediated inhibition of human immunodeficiency virus type 1 *trans*-activation. *J. Virol.*, **65**, 6811.
112. Hope, T. J., Klein, N. P., Elder, M. E., and Parslow, T. G. (1992) *Trans*-dominant inhibition of human immunodeficiency virus type 1 rev occurs through formation of inactive protein complexes. *J. Virol.*, **66**, 1849.
113. Cullen, B. R. (1991) Human immunodeficiency virus as a prototypic complex retrovirus. *J. Virol.*, **65**, 1053.
114. Adachi, A., Gendelman, H. E., Koenig, S., Folks, T., Willey, R., Rabson, A., and Martin, M. A. (1986) Production of acquired immunodeficiency syndrome-associated retrovirus in human and nonhuman cells transfected with an infectious molecular clone. *J. Virol.*, **59**, 284.
115. Zuker, M. (1989) On finding all suboptimal foldings of an RNA molecule. *Science*, **244**, 48.

10 | Using peptides to study RNA–protein recognition

ALAN D. FRANKEL

1. Introduction

In biology, as in other sciences, the study of a particular system often is made simpler if the system is first reduced to small components. In structural biology, studies of large macromolecules can be hampered if, for example a molecule has too many flexible segments that prevent it from crystallizing or if it exceeds the practical size limit for nuclear magnetic resonance (NMR), currently ~20–30 kDa. To overcome these problems, structural biologists often try to identify small, stable subdomains that retain the essential characteristics of the larger system but behave well under NMR conditions or crystallize. Some types of proteins, such as transcription factors, seem particularly well-suited to this approach because many are constructed from small structural and functional domains that can be studied as isolated modules (1). The structures of several DNA-binding domains were solved only after the domains were localized to short polypeptides, typically <100 amino acids, that bound to DNA with the same specificity as the intact proteins.

While reductionist approaches can help overcome critical technical difficulties, they also can reveal characteristics of a system that are not necessarily obvious in a larger context. For example, it is hard to imagine how physicists could learn about the behaviour of fundamental subatomic particles without first dissecting atoms into smaller and smaller units. This chapter describes how small model peptides have been used to help simplify structural studies of RNA–protein recognition and how these systems have highlighted unusual characteristics of RNA structure and recognition. As with any biological system studied *in vitro*, many experimental tests must be applied to ensure that an isolated model system accurately reflects the behaviour of the entire system *in vivo*. This chapter will focus primarily on two model systems derived from the human immunodeficiency virus (HIV) in which short peptides have been identified that bind specifically to their RNA sites and will discuss some remaining questions about RNA recognition in these systems.

2. The arginine-rich domain in RNA-binding proteins

The three-dimensional (3-D) structures of more than 30 protein–DNA complexes have been determined by X-ray crystallography and NMR. These structures have revealed a wide range of interactions between amino acids and bases, between amino acids and the DNA backbone, and unusual DNA conformations that all contribute to sequence-specific recognition (2–4). Because many DNA-binding proteins can be grouped into families based on conserved sequence and structural motifs (2–4), it has been particularly useful to compare at the structural level how different family members bind to their sites. While no obvious amino acid–base 'recognition code' has emerged, some common features have been seen. For example, α-helices from closely related helix–turn–helix proteins can bind in the DNA major groove with similar orientations and use some of the same amino acid–base interactions for recognition (5).

Structural studies of RNA–protein complexes are less far advanced, in part because families of related proteins have only recently been identified (6). Currently, there are two tRNA synthetase–tRNA co-crystal structures (7, 8) and an NMR structure of the HIV *trans*-activation responsive element (TAR) RNA bound to arginine (9), all from unrelated families. While it is too early to make extensive comparisons or generalizations about RNA recognition (10), at least one feature common to all three complexes is that the RNA undergoes a large conformational change upon binding.

If studies of DNA–protein recognition provide any precedent, the identification of at least nine families of RNA-binding proteins (6) should be of great help in guiding structural studies of RNA–protein recognition. Indeed, the RNA-binding domains in at least two of the nine families already have been localized to small peptides that are being studied at the structural level. The first domain, the ribonucleoprotein (RNP) domain, contains a highly conserved 80 amino acid motif whose crystal and NMR structures have been determined in the absence of RNA (see Chapters 6 and 7). The second domain, the arginine-rich domain, contains an 8–20 amino acid motif particularly rich in arginine residues. This domain first was identified by Lazinski *et al.* (11) in bacteriophage λ and P22 N antiterminator proteins, ribosomal proteins, viral coat proteins, and the HIV tat and rev proteins. Because the arginine-rich RNA-binding domain is localized to such a short region and because it contains so many basic amino acids, it has been possible in several cases to identify very short peptides that bind RNA with high affinity and specificity. Studies using peptide–RNA models of the tat–RNA and rev–RNA complexes will be described in this chapter. These studies have identified contributions to specificity made by particular amino acids, by peptide structure, and by RNA structure and emphasize some of the advantages of reductionist approaches.

3. The tat–TAR interaction
3.1 Identification of TAR

The HIV tat protein is a transcriptional activator that operates largely by increasing the efficiency of transcriptional elongation from the viral promoter (see Chapter 9;

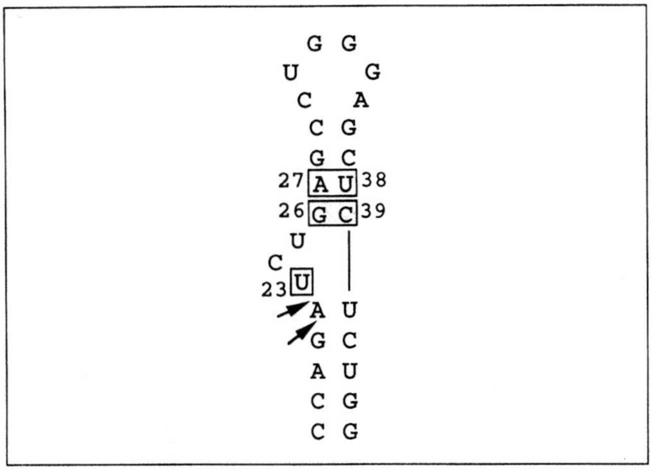

Fig. 1 Secondary structure of TAR RNA showing nucleotides (boxed) and phosphates (arrows) important for tat protein, tat peptide, and arginine binding, determined from mutagenesis and chemical modification experiments (see text).

```
              G G
            U     G
            C     A
            C  G
            G  C
        27 [A  U] 38
        26 [G  C] 39
            U   |
          C
        23[U]
         ↗A  U
         ↗G  C
          A  U
          C  G
          C  G
```

(12)). tat was first identified as a transcriptional activator by mutational analysis of the HIV genome and by analysis of cDNA clones (13, 14). The *cis*-acting region that mediates tat responsiveness, TAR, was identified by analysing deletion mutants of the HIV promoter (15). This early analysis mapped TAR to nucleotides −17 to +80 in the viral promoter (+1 is the transcription start site) and subsequent deletion and substitution mutants localized TAR further to nucleotides +18 to +44 (16). The unusual location of TAR within the transcribed region and the requirement of maintaining its location and orientation (17) suggested that TAR might not act as a 'typical' DNA enhancer element but, rather, might be an RNA element. Indeed, secondary structure predictions and ribonuclease mapping experiments (18) showed that TAR could fold into a stable RNA hairpin containing a three-nucleotide bulge and a six-nucleotide loop (see *Figure 1*). Proof that TAR functions as an RNA element was provided by extensive mutational analyses which demonstrated that base pairing within the stems of the hairpin was essential for tat activity (17, 19, 20) and by a clever experiment in which the TAR hairpin was disrupted on the nascent transcript, leading to loss of tat function (21).

3.2 Identification of a tat RNA-binding peptide

Given that the promoter element responsible for tat function appeared to be an RNA element, it seemed plausible that tat might interact directly with the TAR hairpin. In early binding assays, recombinant tat protein was shown to bind tightly to nucleic acids (22) and in later assays, tat was shown to bind TAR RNA with higher affinity than antisense TAR RNA or TAR DNA (23). While devising a purification scheme for the full-length 86 amino acid tat protein, Weeks *et al.* (24) observed that a fragment spanning amino acids 49–86 was readily proteolysed from the intact protein (*Figure 2*). Remarkably, they found that this short isolated fragment bound TAR with similar affinity and specificity as intact tat. Working

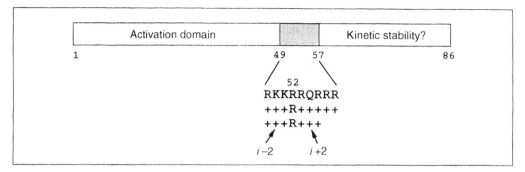

Fig. 2 Schematic representation of the tat protein highlighting the arginine-rich RNA-binding domain. Plus signs indicate positions at which lysine or arginine are allowed.

with synthetic peptides, this group and others (25–27) further localized the TAR-binding region to the nine amino acids encompassing the arginine-rich region of tat, previously implicated as a possible RNA-binding domain based on its similarity to the λN protein (11).

Besides its small size, the tat RNA-binding peptide is unusual in at least two other respects. First, circular dichroism (CD) experiments indicate that the peptide is unstructured in the absence of RNA (27), yet it binds specifically to TAR. Upon binding, the peptide may become ordered, although the CD spectra do not suggest an obvious secondary structure. Second, the amino acid sequence required for specific binding is extraordinarily flexible; the nine amino acid sequence can be reversed, scrambled, or simplified to homopolymers of nine arginines or eight lysines and one arginine and still retain full binding specificity (27–29). Thus, both the CD results and loose sequence requirements give the impression that a disordered peptide can bind to an RNA in a specific way, implying that it is largely the structure of the RNA, rather than the peptide, that determines the location and orientation of binding.

3.3 Determinants of specificity in TAR RNA

The regions of the TAR hairpin responsible for tat function can be divided roughly into two parts. The region around the three-nucleotide bulge provides the direct binding site for tat and is described below in detail. The six-nucleotide loop also is important for tat function *in vivo* (26, 30), however, the loop appears to bind a cellular protein rather than interact directly with tat. Potential roles of loop-binding proteins are discussed briefly below and in Chapter 9.

The results from *in vivo* and *in vitro* mutagenesis and chemical modification studies have established a virtually complete map of the functional groups in TAR that determine tat protein- and tat peptide-binding specificity (*Figure 1*). The important groups are all located in or directly adjacent to the bulge. Within the bulge, U23 is the only nucleotide whose identity is important for binding (26, 31, 32). Substitution of U23 with C reduces binding, suggesting that the N3 imido and C4 carbonyl groups may be important and chemical modification and substitution experiments directly implicate a hydrogen-bond donor at the N3 position (33, 34). The identity of the other two bulge nucleotides is not important, however, the length of backbone in the bulge must be equivalent to at least two nucleotides and may be replaced by a chemical spacer (32, 33, 35). In the upper stem adjacent to the bulge, the identity of the G26–C39 and A27–U38 base pairs is important. No nucleotide substitutions are tolerated at these positions (32, 35, 36) and methylation of G26 by dimethylsulphate or carbethoxylation of A27 by diethylpyrocarbonate strongly interferes with binding (24). tat binding to synthetic RNAs containing modified bases further implicated the N7 group of G26 and the N7 group of A27 as important hydrogen-bond acceptors (34). No other base substitution or modification has been found to decrease tat-binding affinity significantly.

In addition to base-specific determinants, at least two phosphate groups on the RNA backbone appear to be important for recognition (*Figure 1*). Ethylation of phosphates between G21 and A22 and between A22 and U23 strongly interferes with binding (28) as does methylphosphonate substitution of the phosphate between A22 and U23 (34). Additional phosphate groups on the strand opposite the bulge may be in the vicinity of the bound tat protein (36), however, these phosphates do not appear to contribute to the binding specificity of the peptides (28). Substitution of several ribonucleotides in TAR with deoxyribonucleotides has not implicated any 2'-OH group in binding (33, 34).

Because base substitutions and chemical modifications may disrupt the folding of TAR, it is not possible using these data alone to determine which groups directly contact tat and which contribute indirectly to binding specificity by stabilizing the RNA structure. However, it is obvious that the bulge region is important. Potential recognition features of the bulge will be interpreted below with respect to specificity determinants in the tat peptide and the NMR structure of TAR.

3.4 Determinants of specificity in the tat peptide

Having narrowed the RNA-binding domain of tat to just nine amino acids, it was possible to determine which amino acids in the peptide contributed to TAR-binding specificity. As described above, it has been shown that the sequence requirements of the peptide were quite flexible and that a homopolymer of nine arginines bound to TAR with the same specificity as the wild-type tat peptide (RKKRRQRRR). Because it also was known that a high charge density is needed for binding (27, 29, 37), a homopolymer of nine lysines was tested for binding. Unlike the homoarginine peptide, the homolysine peptide did not bind TAR specifically, indicating that arginine residues were important for recognition (28). Using a reductionist approach, Calnan *et al.* (28) then substituted arginine residues back into the homolysine peptide one at a time and found that a single arginine, with three to four basic amino acids on each side, was all that was needed for full binding specificity both *in vitro* and *in vivo* (*Figure 2*).

Because only one arginine in the tat peptide appeared to determine TAR-binding specificity and because previous experiments with a *Tetrahymena* group I self-splicing intron showed that RNA can contain specific arginine-binding sites (38), it was reasoned that arginine, even as the free amino acid, might bind specifically to TAR. Indeed, RNA-binding experiments using an arginine affinity column and tat peptide competition experiments demonstrated that free arginine binds to TAR with the same specificity as the tat peptide, although with much lower affinity (39). The same functional groups in TAR that determine tat-binding specificity (*Figure 1*) also determine arginine-binding specificity. By measuring binding affinities of arginine analogues, it was shown that the guanidinium moiety of the arginine side chain alone is responsible for recognition (39) and by taking the reductionist approach to the extreme, it was shown that even guanidine can bind specifically to the arginine-binding site in TAR (40).

Clearly, arginine binding alone cannot determine the full affinity and specificity of the tat-TAR interaction. What other factors contribute to binding and which contribute to affinity and which to specificity?

3.4.1 Electrostatics

Arginine as the free amino acid binds to TAR with a K_d in the millimolar range (39, 40) whereas the tat peptides or tat protein bind in the nanomolar range (24, 27, 41). This difference of six orders of magnitude in affinity may be explained largely, if not entirely, by additional electrostatic interactions provided by basic amino acids flanking the one required arginine. The *in vivo* activities of mutant proteins indicate that at least three positively charged amino acids are required on each side of the arginine to achieve full function (28, 29, 42), corresponding to an increase in RNA-binding affinity of at least four orders of magnitude over arginine alone (42). These six charged residues presumably form ionic interactions with RNA phosphates, consistent with the six ion pairs in the peptide–RNA complex estimated from the salt dependence of binding (43). Not only is the number of charged residues important in raising the overall affinity, but their precise distribution and density influence both binding affinity and specificity. For example, RNA-binding affinity drops approximately 5-fold when charged residues at the $i-2$ or $i+2$ positions (relative to the arginine) are removed from a simplified seven amino acid peptide containing three lysines on each side of one arginine (*Figure 2*), compared to removing charges at other positions (42). The specificity of the interaction, as determined by the relative binding constants of wild-type and mutant TAR RNAs, is higher when arginine is placed within the context of charged amino acids in a peptide than for arginine alone (42), perhaps reflecting electrostatic interactions that stabilize the bound RNA conformation (see section 3.5). Thus, while a single arginine binds to TAR using the same RNA structural determinants as the tat peptides or tat protein, the ability of arginine to discriminate between TAR RNA and other RNAs can be enhanced by its surrounding environment.

3.4.2 Peptide structure

As described above, CD experiments and the sequence flexibility of the tat peptide suggest that its conformation is unstructured in the absence of RNA. In the intact protein, this region may also be unstructured given that it displays similar sequence flexibility to the peptide (27, 28). While it cannot be excluded that the RNA-binding domain adopts a particular conformation in the context of the protein, it seems unlikely that so many different sequences would preserve the entire set of amino acid contacts needed to stabilize its conformation. A related RNA-binding peptide from the bovine immunodeficiency virus (BIV) tat protein binds to its RNA site with much higher affinity and specificity than HIV tat (L. Chen and A. Frankel, in press) but has a sequence indicative of a random coil, further suggesting that peptides need not be structured to bind to specific RNA sites. In contrast, a peptide from HIV rev binds specifically to RNA only when in an α-helical conformation (described below). From the perspective of specificity, a lack of defined structure can contribute to specificity just as a requirement for structure can, if, for example the peptide must change its conformation upon binding to conform to the structure of the ligand. In this case, it will cost binding energy to unfold the peptide. Thus, arginine-binding specificity for TAR may be increased not only because arginine is presented within an array of basic amino acids but perhaps also because it is in an unstructured peptide context. It is quite common for DNA-binding proteins to use flexible unstructured regions to make specific DNA contacts (44); the tat–TAR interaction may be analogous.

3.4.3 Kinetic stability

While the thermodynamics of binding establish the affinity and specificity of the tat–TAR interaction, kinetic considerations can also influence the apparent binding specificity and may play a role in determining the extent of TAR-site occupancy, particularly *in vivo*. Gel-shift experiments with tat peptides of various lengths have shown that while the nucleotides in TAR required for specific binding are the same for each peptide, the half-life of each complex can differ during gel electrophoresis, leading to apparent differences in the ability to discriminate between TAR and TAR mutants (45). Longer peptides containing the arginine-rich region followed by additional amino acids at its C-terminus (spanning amino acids 49–86 of tat; *Figure 2*) form more stable complexes and as a consequence show better discrimination than peptides containing only the arginine-rich region. This is consistent with the observation that TAR complexes with the shorter peptides are seen only when electrophoresed at low temperature (27). This type of kinetic instability has also been observed in DNA sequence-specific binding of a GCN4 DNA-binding peptide (45). These studies raise the possibility that, in the context of intact tat, regions outside the arginine-rich domain may increase the kinetic stability of the tat–TAR complex, either directly by forming additional interactions with the RNA or indirectly by stabilizing the bound structure of the arginine-rich domain.

3.4.4 Location of the RNA-binding domain within the protein

The experiments described above provide evidence that specificity is influenced by the position of arginine within the surrounding basic residues, however, the location of the entire RNA-binding domain within the protein also influences *in vivo* activity. The first 48 amino acids of tat comprise the activation domain, followed by the arginine-rich RNA-binding domain (*Figure 2*). When only four amino acids are inserted between these domains, tat activity decreases significantly, suggesting that the spacing between the domains is important for function (46). This is supported by even more precise mutations in which activity was shown to decrease progressively as the single required arginine at position 52 was moved one position at a time further from the activation domain (42). Why is this spacing important? The simplest interpretation is that it reflects the need for an interaction between tat and another protein bound at the loop to stabilize the complex (*Figure 3*; see below). Several experiments support this interpretation. If the tat activation domain is delivered to the RNA without the need for TAR, namely by fusing the activation domain to the bacteriophage R17 coat protein and providing an R17-binding site in place of TAR, then the spacing between the activation and RNA-binding domains is no longer critical (46). Obviously, the TAR loop is no longer required in this context, suggesting that the R17–RNA interaction is tight enough on its own to support tat function (47). The spacing between the bulge in TAR (the tat-binding site) and the loop (the cellular protein-binding site) is critical (48), consistent with the interpretation that communication between these regions may be mediated through RNA-bound protein–protein interactions. Given that the spacing between the RNA-binding and activation domains is important only when the TAR loop is required, it seems reasonable that part of the activation domain may interact with the loop-binding protein. This is supported by the finding that mutations throughout the activation domain decrease the ability of tat, fused to a

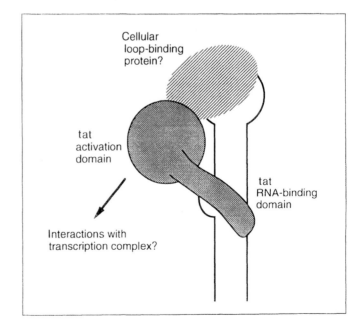

Fig. 3 Possible arrangement of the tat–TAR complex, stabilized by interactions with a protein bound at the loop. The activation domain of tat presumably also interacts with the cellular transcription complex to enhance transcriptional activity.

heterologous protein, to bind to TAR sites positioned to allow function of the heterologous protein (49).

3.4.5 Loop-binding proteins

Although binding of the arginine-rich domain of tat to the TAR bulge is necessary for *in vivo* activity, the interaction is not sufficient and appears to require an additional protein(s) bound to the loop to stabilize binding. Several proteins have been identified that bind to the loop with specificities that correlate with loop mutations known to abolish tat function (50–52). The functional importance of these proteins has not yet been established, but there is indirect evidence suggesting that one or more relevant proteins may be encoded on human chromosome 12. tat is known to function poorly in rodent cells unless chromosome 12 is provided, but the need for the chromosome is eliminated if tat is delivered to the RNA using a heterologous RNA–protein interaction that no longer requires the TAR loop (53, 54). Furthermore, mutations in the loop do not further decrease the level of tat activity observed in rodent cells when chromosome 12 is absent (55). Because the direct interaction of tat with the TAR bulge has relatively modest affinity ($K_d \sim$ 6 nM) and specificity (typically in the order of 40-fold discrimination between TAR and TAR mutants) (27, 32, 43), it seems reasonable that tat might cooperatively enhance its RNA-binding affinity and specificity through additional protein–protein interactions *in vivo* (*Figure 3*).

3.4.6 Specificity of the arginine–TAR complex

Given that other interactions are likely to exist *in vivo*, what is the evidence that the tat peptides or free arginine provide reasonable models of the tat–TAR interaction? The strongest evidence comes from the observation that the determinants in TAR required for tat protein, tat peptide, and arginine binding (*Figure 1*) are essentially identical and that *in vitro* binding specificities show an excellent correlation with *in vivo* activities. A few key comparisons are as follows: detailed mutagenesis of TAR has identified U23 in the bulge and the G26–C39 and A27–U38 base pairs above the bulge as the only nucleotides required for tat protein (26, 36), tat peptide (32), and arginine (39, 40, 56) binding. Analysis of the NMR structure of a TAR–arginine complex (section 3.5) explains the need for each of these nucleotides. Chemical interference experiments have identified two phosphates that are important for tat peptide (28) and arginine (39) binding and at least one of these has been shown to be important for tat protein binding (34). The NMR structure also explains why these phosphates contribute to binding. The reported binding constants for tat peptide and tat protein are similar, in the range of ~ 0.1–10 nM, depending on binding conditions (26, 27, 32, 41). The sequence flexibility of the arginine-rich peptides observed for TAR binding *in vitro* also is observed for tat protein function *in vivo*, for example scrambling the sequence of the nine amino acid RNA-binding domain, replacing it with arginines, or replacing it with one arginine embedded in eight lysines all have binding affinities and specificities similar to the wild-type peptide and activities identical to wild-type tat *in vivo* (27, 28, 42). Thus, while it cannot be excluded that other parts of tat outside the arginine-rich domain contribute to TAR-binding specificity, particularly *in vivo* (see above), the precise correspondence between specificity determinants in the RNA and the peptide suggest that arginine–TAR and peptide–TAR complexes accurately reflect many important aspects of the interaction.

What are the advantages of having reduced the major specificity determinant in the tat–TAR interaction to a guanidinium-binding site at the TAR bulge? As suggested in the introduction, it is easier technically to determine the structure of a small complex by NMR and the structure of a TAR–arginine complex is presented below. Several additional advantages can be seen in this case. Because TAR contains only one specific arginine-binding site, a peptide containing just one arginine or arginine itself binds to TAR in only one way. Peptides containing several arginines, such as the wild-type peptide, are able to bind to TAR in multiple ways in the absence of additional cellular factors or other parts of the protein. Arginine–TAR complexes do not aggregate whereas peptide–TAR complexes can form cross-linked species containing multiple RNAs bridged by a single peptide. Arginine NMR resonances are easily assigned to free arginine or to a simple peptide containing just one arginine. ^{15}N-Labelled arginine is commercially available and can be used to study the complex using heteronuclear NMR methods.

Beyond technical considerations, the finding of a single arginine-binding site in TAR raises several conceptual issues about RNA–protein recognition. It is clear from work on the *Tetrahymena* ribozyme that RNA alone is capable of adopting a specific arginine-binding conformation (38), however, arginine forms a different interaction with TAR and is part of a known RNA–protein interaction. Given the prevalence of the arginine-rich motif in RNA–binding proteins, it may be anticipated that similar arginine–RNA interactions will be found in other systems. Having dissected the interaction down to a specific contact between arginine and TAR, described in detail below and additional electrostatic effects on specificity, it should be possible to quantitatively establish the thermodynamic and kinetic contributions of each type of interaction to binding. Perhaps most intriguing is the dominant role that RNA structure plays in the arginine–TAR interaction, suggesting that one may view the tat–TAR interaction as a case in which the ligand (RNA) recognizes the protein. The concept that RNA structure can play an active role in a biological process, as demonstrated so elegantly by the discovery of catalytic RNAs, may be extended to RNA–protein recognition.

3.5 NMR structure of a TAR–arginine complex

How does TAR form a specific arginine-binding site? Two models were proposed initially based on the biochemical and mutagenesis data. In the major groove binding model, it was suggested that the bulge in TAR served, in part, to widen the major groove of the RNA helix, allowing access to hydrogen-bonding groups that would otherwise be inaccessible in the deep and narrow groove of an A-form helix (32). This was supported by the accessibility to diethylpyrocarbonate of the

N7 group of A27 in the helix above the bulge. In the 'arginine fork' model, it was suggested that an unusual conformation of the RNA backbone might be formed at the junction between the bulge and lower stem, positioning two adjacent phosphates to form a set of hydrogen bonds to the guanidinium group of arginine (28). This was supported by phosphate ethylation interference of peptide binding. The major groove binding and arginine fork models are not mutually exclusive and, indeed, features of both appear to be correct.

Because many nucleotide and phosphate groups in and around the bulge (*Figure 1*) are required for binding a single guanidinium group, it was expected that some groups would bind directly to arginine while others would contribute indirectly to binding by establishing a specific conformation of TAR. Models for free TAR and a TAR–arginine complex determined by NMR (9) are shown (*Figure 4*). The structure of free TAR is not very well defined by NMR constraints, however, it is clear that TAR contains upper and lower A-form helical stems as predicted and that the two uridines in the bulge and probably the cytosine are at least partially stacked between the helices. The presence of these bulge nucleotides probably causes the RNA to kink (57). Bases in the six-nucleotide loop also appear to stack (58, 59), but the structure of the loop is not well defined and the loop is not directly involved in tat or arginine binding. It is apparent from the free TAR structure alone that the set of groups in the bulge region responsible for arginine binding are far apart in space and little rationale can be proposed to explain specific binding.

In contrast to free TAR, the structure of the TAR–arginine complex (9), which is much better defined by the NMR data, seems to explain why each group in TAR is needed for specific arginine binding. Comparison of the two structures shows that the bulge region undergoes a rather large conformational change upon arginine

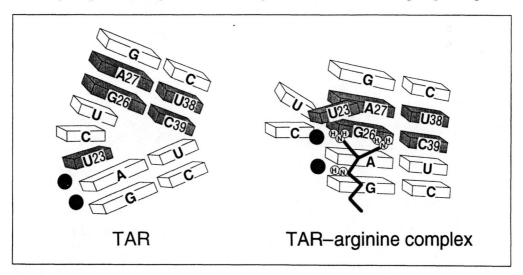

Fig. 4 Schematic representations of the structures of free TAR and TAR bound to argininamide (a tight-binding analogue of arginine) determined by NMR. Important nucleotides are shaded and two important phosphates are shown as black circles.

binding, consistent with changes observed by CD spectroscopy (40). The changes in structure upon arginine binding are restricted to the bulge region. The three bulge nucleotides, partially stacked between the stems in the unbound structure, become unstacked in the complex allowing the two helical stems to stack co-axially. The 5'-most uridine in the bulge (U23), an essential nucleotide for binding, becomes positioned in the major groove above the bulge and appears to form a base triple with the A27–U38 base pair (*Figure 5*), also essential nucleotides for binding. The other two bulge nucleotides, whose identities are not important, appear to be looped out into solution, although it is clear that the length of the bulge backbone must be equivalent to at least two nucleotides to allow U23 to reach the upper major groove, consistent with mutagenesis data on the bulge size. As a consequence of the change in bulge structure, the two phosphates whose modification interferes with binding are brought close to G26 in the major groove, apparently forming a binding pocket for the arginine guanidinium group in which hydrogen bonds may be formed with the N7 and O6 acceptors of guanine and with phosphate oxygen accepters (*Figure 5*). This type of arginine–guanine arrangement was initially proposed by Seeman *et al.* (60) and is frequently observed in protein–DNA complexes (2, 4). The position of arginine is reasonably well constrained in the structure by the observation of five resonances between the arginine δ-proton and protons in the RNA. An additional consequence of the bulge rearrangement is that several phos-

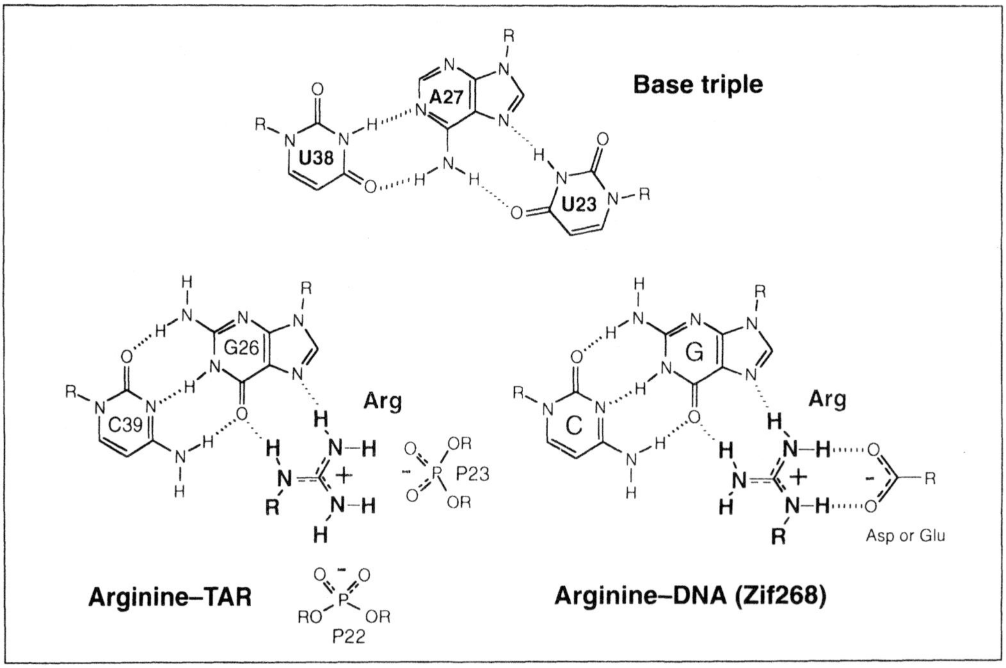

Fig. 5 Summary of the base triple and arginine–guanine–phosphate interactions in the TAR–arginine NMR model. An arginine–DNA interaction seen in the Zif268 protein–DNA co-crystal structure is shown for comparison.

phates are brought relatively close together, probably forming unfavourable electrostatic interactions. It is possible that in the context of the peptide, charged residues surrounding the arginine help neutralize these interactions, thus explaining the enhanced binding specificity when arginine is in a peptide. Nevertheless, the final structures of TAR, either bound to peptide or bound to free arginine, are the same (9).

While the TAR-arginine NMR model accounts for all the important chemical groups identified by the biochemical and mutagenesis experiments, the data provided no direct evidence for specific hydrogen bonds in the arginine contact or base triple because no resonances were observed between the interacting groups. Additional evidence for the base triple was provided by an experiment in which the proposed U23-A27-U38 interaction was replaced by an isosteric C23-G27-C38 base triple (56). NMR data indicate that the TAR-arginine structure with this mutant is essentially identical to the structure with wild-type TAR whereas mutation of either U23 or the A27-U38 base pair alone results in an RNA unable to bind arginine or adopt the bound structure. Thus, formation of the base triple appears to correlate well with specific arginine binding. Methylation of the N6 group of A27 has a small effect on binding (34), suggesting that one of the hydrogen bonds in the base triple may contribute relatively little to stabilizing the bound TAR structure. There have been no further tests of the proposed arginine-guanine-phosphate interaction (*Figure 5*), however it is interesting that possibly analogous interactions are observed in DNA-protein complexes. In the Zif268 zinc finger-DNA co-crystal structure (61), arginine guanidinium groups have been seen to hydrogen bond to O6 and N7 groups on guanines in the major groove, with additional hydrogen bonds formed between the guanidinium group and carboxyl groups on aspartic or glutamic acid residues located elsewhere in the protein (*Figure 5*). The positioning of these acidic residues serves to orient the arginine side chain to bind specifically to guanine. Perhaps the structure of the TAR RNA backbone serves a similar role.

Studies of the tat-TAR complex have illustrated several points about RNA-protein recognition, many having been seen in larger systems. The interaction highlights how important the conformation of the RNA can be for recognition, in this case perhaps being more defined than the protein conformation. Groups specifically positioned on both the backbone and bases can be directly involved in binding. Conformational changes in the RNA, also observed in tRNA-tRNA synthetase complexes (7, 8), can be crucial for correctly positioning interacting groups in the complex. Some of these conformational changes may be a consequence of different electrostatic interactions between the free and bound forms and may be used thermodynamically to drive protein binding. Bulges appear to provide a considerable amount of information for recognition.

1. Nucleotides in a bulge have exposed groups available to interact either with a protein or with other groups on the RNA.

2. Bulges widen the major groove of adjacent helices (32, 62), exposing rich arrays of hydrogen-bond donors and acceptors for protein-RNA or RNA-RNA interactions.

3. Bulges create non-helical geometries of the backbone that may form complementary surfaces for protein binding and may position phosphate or hydroxyl groups for specific hydrogen-bonding interactions (28).

Many RNA-binding proteins require bulge structures for recognition, such as rev (described below) and the R17 coat protein (63) and it will be interesting to see which features contribute to specificity in each case (see Turner (64) for a recent review on the structures of bulges).

4. The rev–RRE interaction

Structural information about the rev–RRE interaction is not yet as detailed as the tat–TAR interaction, however, comparisons of these two peptide–RNA models already provide an expanded perspective on RNA recognition by the arginine-rich motif.

4.1 Identification of RRE

The HIV rev protein is a post-transcriptional regulator of HIV gene expression that induces accumulation of unspliced or incompletely spliced viral mRNAs in the cytoplasm (see Chapter 9; (65)). The unspliced mRNAs encode the viral structural proteins which are translated and assembled into virions. The precise mechanism of rev action is unclear, however, it is believed to activate a transport pathway and/or inhibit the efficiency of mRNA splicing. rev was discovered unexpectedly by mutagenesis of the *tat* gene; several tat mutants appeared to give unusual patterns of spliced and unspliced mRNA expression but the phenotype actually resulted from mutations in the overlapping *rev* gene (66). The *cis*-acting region that mediates rev responsiveness, named RRE (for rev-response element), was localized by deletion and mutation analysis to a rather complex 234 nucleotide (nt) RNA structure encoded in the *env* gene (67). As described below, further analyses identified a single high affinity rev-binding site within the RRE (*Figure 6*). Binding to this site appears to nucleate binding of additional rev molecules to adjacent sites and oligomeric binding appears to be necessary for rev function.

4.2 Identification of a rev RNA-binding peptide

Several studies initially demonstrated that recombinant rev protein binds specifically to RRE RNA (68–73). rev contains an arginine-rich region that appeared to be a reasonable candidate for an RNA-binding domain (11) and mutagenesis studies confirmed that the region is necessary for specific RNA binding (74–76). When the arginine-rich domain was fused to heterologous proteins, the fusion proteins could function through RRE sites (77, 78), suggesting that the domain is sufficient for specific RNA recognition. Given the successful use of peptides in examining the tat–TAR interaction, peptides were synthesized spanning the arginine-rich domain of rev (amino acids 34–50; *Figure 7*) and were found to bind specifically to the RRE (79). Thus, as for tat, a short peptide appears to provide a reasonable model of the rev RNA-binding domain, but unlike tat, rev does not appear to require additional factors to bind with high affinity and specificity to the RRE *in vivo*. Interestingly, the rev peptide inhibits splicing of RRE-containing RNAs *in vitro* (80, 81), perhaps reflecting part of the *in vivo* function of rev. The determinants of rev–RRE recognition show similarities and differences to tat–TAR recognition and are discussed below.

4.3 Determinants of specificity in RRE RNA

Enzymatic and chemical footprinting of the RRE suggested that rev may bind to at least five sites in the large, 234 nt multihairpin RNA structure (82). Further deletion analyses showed that a smaller domain containing just two hairpins was sufficient for high affinity rev binding *in vitro* (70, 73) and multiple tandem copies of the smaller element were shown to be sufficient for rev function *in vivo* (83). More detailed mutagenesis and chemical interference studies further localized the rev-binding site to a single hairpin (*Figure 6*) called stem–loop IIB (79, 84–87).

Nucleotides and backbone groups in IIB important for rev protein and rev peptide binding have been identified by mutagenesis, chemical modification, and

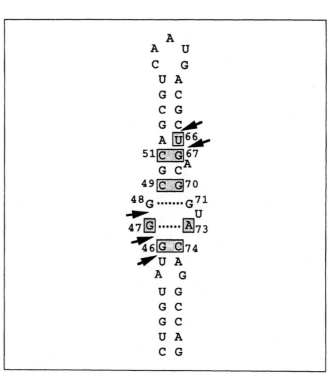

Fig. 6 Secondary structure of the RRE IIB hairpin showing nucleotides (boxed) and phosphates (arrows) important for rev protein and rev peptide binding, determined from mutagenesis, chemical modification, and *in vitro* selection experiments (see text).

in vitro selection experiments (summarized in Figure 6). It is clear that the determinants of specificity are all located in and around an asymmetric internal loop and a single adenine bulge. Mutations in the internal loop that decrease rev binding include substitution of G47, G48, G71, or A73 or deletion of G71 or A73 (79, 85, 87). The identity of U72 does not appear to be very important although deletion of the nucleotide severely reduces binding (85). Substitution of the G46–C74 or C49–G70 base pairs immediately surrounding the loop is poorly tolerated although some mutations have larger effects than others (for example, substituting G46–C74 with A–U decreases binding less than a C–G substitution) (85, 87). Multiple substitutions of G46, G47, and G48 also decrease binding (88). Interestingly, when rev-binding RNAs were selected from a partially randomized sequence pool, G48 and G71 showed a covariation suggesting that they might form a homopurine base pair across the loop; the putative G–G pair could be substituted by an isosteric A–A pair (87). Nucleotides G47 and A73 were conserved in the selection experiment, suggesting that they also might form a non-Watson–Crick pair across the loop. Chemical modification interference experiments (79, 84) and binding of chemically substituted RNAs (89) identified similar nucleotide determinants as the substitution and selection experiments and suggested that, as for TAR recognition, RRE recognition probably occurs in the major groove. For example, modification of several purine N7 groups in the loop by diethylpyrocarbonate or substitution of G46 or G47 with N7-deaza-dG strongly interfered with binding (79, 84, 89).

While important specificity determinants include nucleotides in the asymmetric internal loop, several backbone groups and the region around the single adenine bulge also appear to contribute to binding specificity. Ethylation interference experiments identified five phosphates, three located near the base of the loop and two adjacent to the bulge, that are important for binding (79). These phosphates are located next to guanine bases, reminiscent of the arginine–guanine–phosphate arrangement observed in TAR. Several deoxyribonucleotide substitutions in the loop decrease binding, suggesting that 2'-OH groups also may play a role in recognition (89). The nucleotide determinants near the adenine bulge have not been as fully defined as the loop determinants, however the selection experiment identified three invariant nucleotides adjacent to the bulge, the C51–G67 base pair and U66 (Figure 6), that may be important (87). Chemical modification of G67 and U66 also interferes with binding, further suggesting that they may be involved in recognition (79, 84).

4.4 Determinants of specificity in the rev peptide

As in the tat–TAR interaction, an isolated peptide spanning the arginine-rich domain of rev binds to the RRE IIB hairpin with affinity and specificity similar to the full-length protein (78, 79). However unlike tat, the 17-amino acid rev peptide (Figure 7) does not allow much sequence flexibility and adopts a defined structure. The rev peptide forms a relatively stable monomeric α-helix in aqueous solution (~50% α-helix at 4°C), as judged by CD spectroscopy, provided that the ends of

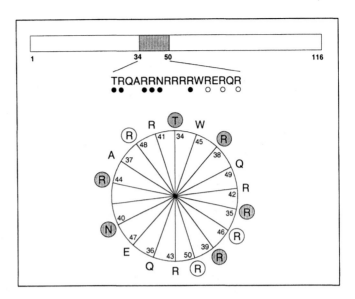

Fig. 7 Schematic representation of the rev protein highlighting the arginine-rich RNA-binding domain. Dark circles indicate six amino acids that are important for specific RRE binding and open circles indicate three additional arginines that contribute more weakly to binding. A helical projection of the RNA-binding domain is shown.

the peptide are modified to stabilize the helix macrodipole electrostatically (78). It is relatively unusual for peptides of this length to form stable helices and it is even more surprising given the high density of positively charged side chains that might be expected to interact unfavourably in a helical conformation. Helix formation is required for specific recognition of IIB RNA; destabilizing the helix causes a proportional decrease in specific, but not non-specific, RNA-binding affinity (78). Amino acids that contribute to binding specificity were mapped by substituting alanines at each position in the peptide (78). Mutation of six amino acids, one threonine, one asparagine, and four arginines, strongly decreased specific RNA-binding affinity whereas mutation of three additional arginines showed a less dramatic effect on binding (*Figure 7*). Amino acid positions important for binding do not map strictly to one side of the α-helix, suggesting that the RNA may partially surround the peptide helix.

What is the evidence that the rev peptide–RRE interaction accurately reflects the rev–RRE interaction? Footprinting and chemical interference experiments suggest that the rev protein and peptide interact with identical sites on the 234 nt RRE and on the IIB hairpin (79, 82). Similarly, mutations that decrease protein-binding affinity also decrease peptide-binding affinity (78, 79). RNA-binding affinities are similar for the peptide and protein (78, 90). Amino acid substitutions shown to decrease specific RNA-binding affinity of the rev peptide *in vitro* also decrease activity of the intact protein *in vivo* (74–76). Fusions between the rev peptide and the human T-cell leukaemia virus (HTLV)-I rex- or HIV tat-activation domains allow these proteins to function through RRE or IIB RNA-binding sites, demonstrating that the peptide is sufficient for high-affinity binding *in vivo* (77, 78). Furthermore, *in vivo* binding correlates with peptide α-helical content, as observed *in vitro* (78).

Although the arginine-rich domain is sufficient for high-affinity binding to the RRE and the specificity determinants are the same as for the intact protein, do other factors contribute to rev RNA-binding specificity *in vivo*? Clearly, as for tat, the *in vitro* binding experiments cannot rule out the possibility that other parts of the protein outside the arginine-rich domain contribute to RRE-binding specificity, however the correspondence between peptide and protein specificity determinants again suggests that the peptide–RNA complex reflects many important aspects of the interaction. One feature of the interaction not mimicked by the peptide is the possible effect of oligomerization. It is clear that rev function *in vivo* requires binding of several rev molecules to the RRE (76, 83, 91), apparently nucleated by a single high-affinity binding site and it is possible that cooperative protein–protein interactions enhance the specificity of binding to the multi-site RRE. It is not yet clear whether rev forms a discrete multimer in the absence of RNA, however, a recent study has shown that a C-terminal truncation mutant eliminates much of the oligomeric behaviour observed *in vitro* but is still functional *in vivo* (90). The precise arrangement of protein molecules on the RRE does not appear to be critical for function because rev is active when fused to the R17 coat protein and delivered to the RNA using multiple copies of the R17 RNA-binding site (92, 93). In addition to cooperative interactions between rev molecules, it is possible that additional cellular proteins may be important for RNA recognition. However, because no sequence-specific determinants have been identified in the RRE outside the high-affinity rev-binding site and because heterologous RNA-binding sites can be used by rev to function, it seems unlikely that other RRE-specific proteins are needed to enhance rev-binding specificity.

4.5 Model of a rev peptide–RRE complex

How might an isolated α-helical rev peptide bind to RNA? While there currently are only limited structural data, several features of recognition may reasonably be inferred from the biochemical and genetic data. As with tat, binding appears to occur primarily in the major groove near a distorted helical region of the RNA. Binding of an α-helix in the major groove is reminiscent of DNA–protein recognition in which side chains from 'recognition helices' often make specific contacts with bases in the major groove, a feature common among proteins from very different structural families (2–4). Typically, the orientation of the helix is determined by the surrounding protein framework which docks against the DNA using defined sets of backbone contacts. The helical orientation can differ widely among proteins even within the same structural family. It is unclear how a single α-helix from rev can orient itself in the absence of the rest of the protein, however it seems likely that contacts to the backbone will play a key role. These might, for example involve arginine interactions with RNA helix–bulge junctions (28), perhaps similar to that seen in the arginine–TAR complex (9). The details of the specific rev–RRE amino acid–base and amino acid–backbone contacts will require further structural studies.

What can be inferred about the structure of the RRE IIB-binding site? The gross architecture of the IIB hairpin may be viewed as analogous to TAR in that two helical stems are separated by a non-helical region (*Figure 6*). Specific peptide binding occurs in the asymmetric internal loop and single-nucleotide bulge region. As with the tat-TAR interaction, rev peptide binding causes a change in RNA conformation, as detected by CD (R. Tan and A. Frankel, in preparation) and NMR (J. Battiste, R. Tan, A. Frankel, and J. Williamson, in press) spectroscopy. Preliminary gel-shift experiments suggest that IIB RNA may be slightly bent (R. Tan and A. Frankel, in preparation), as observed for TAR (57). Conformational changes in the multihairpin RRE structure also have been observed upon rev protein binding (82, 94). From the NMR data, it appears that the entire region between the two helical stems may be largely unstructured (all bases are unpaired) in the absence of peptide. Upon peptide binding, base pairs are formed, including two predicted non-Watson–Crick pairs, leaving only two bulge nucleotides (A68 and U72) unpaired in the complex. Nucleotides throughout the entire region appear to be largely stacked. The identities of the non-Watson–Crick G–G and G–A pairs (*Figure 8*) are unambiguously established by the NMR data and were predicted correctly from covariation in the *in vitro* selection experiment (87) and from extensive chemical substitution data (89). The conformation of an asymmetric internal loop from 5S ribosomal RNA, determined by NMR (95), also contains stacked non-Watson–Crick base pairs and a single bulged guanine located in the major groove. This loop forms a binding site for ribosomal protein L5 and transcription factor IIIA (TFIIIA) and it is believed that the bulged G and distorted backbone conformation may provide important recognition features (see Chapter

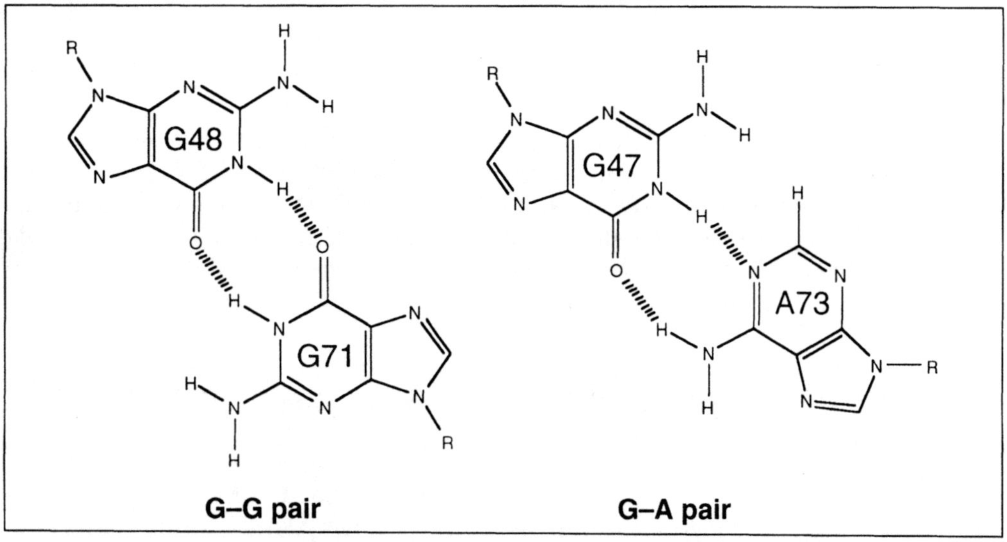

Fig. 8 Two non-Watson–Crick base pairs observed in the rev peptide–RRE IIB RNA complex by NMR. Both pairs were predicted by phylogenetic covariation and chemical substitution experiments (see text).

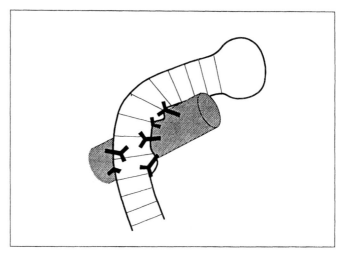

Fig. 9 Schematic arrangement of the α-helical rev peptide bound to IIB RNA. Binding appears to occur in the major groove and several amino acids probably contact specific groups on the bases and backbone. The RNA may be bent or kinked.

8). The precise effects of the G–G and G–A base pairs on the 3-D structure of RRE IIB RNA and on rev recognition remain to be determined. The general importance of non-Watson–Crick pairs in RNA structure has been reviewed recently (96).

While the structural details are still very sketchy, a preliminary model for the overall arrangement of the rev peptide–IIB RNA complex may be proposed (*Figure 9*). The α-helical rev peptide binds in the RNA major groove, spanning the asymmetric internal loop and bulge region and presents an array of at least six amino acids for interactions with the bases and backbone. Contacts at multiple points on the RNA backbone, perhaps mediated by arginines, may serve to position the helix in a defined orientation. A slight bend in the RNA would readily allow contacts to be made with essential amino acids located around the α-helix (*Figure 7*). The observation that the RNA changes conformation upon binding suggests that the bound RNA structure is stabilized in the complex and the finding that removing any single essential amino acid almost entirely eliminates RNA-binding specificity (78) suggests that all contacts may be needed simultaneously to stabilize the bound RNA structure. The RNA structure may be further stabilized by additional 'non-specific' electrostatic contacts, as described for the tat–TAR complex.

5. Other peptide–RNA model systems: what might be learned?

It seems clear from the two examples described above that considerable RNA-binding specificity can be achieved by relatively small peptides and that peptide–RNA model systems can provide substantial information about protein–RNA recognition. Even with simple peptides that appear to be related at the sequence level (the 'arginine-rich' motif), it is apparent that considerable structural diversity will exist, both in peptide structure and in RNA structure. The similarities seen in these systems may reflect features common to other systems or general principles of recognition. For example, recognition of groups in the major groove and distortion of the backbone near bulges, conformational changes in the RNA, and contributions of electrostatics to specificity are likely to be general features and some already have been observed in larger protein–RNA complexes (7, 8).

Peptide models are likely to be derived from other members of the arginine-rich class of RNA-binding proteins, in part because the binding domains appear to be highly localized and because the charged nature of the domain suggests that RNA-binding affinity will be high. Several candidates are obvious. tat and rev proteins from related lentiviruses, including HIV-2 and simian immunodeficiency virus (SIV), already are being studied and electrostatic contributions to binding specificity have been found (42, 97). A 17 amino acid arginine-rich peptide from the BIV tat protein binds to its RNA site as an unstructured peptide, but unlike the HIV tat peptide, the BIV peptide uses several amino acids for specific recognition and binds with high specificity to an unusual RNA structure (L. Chen and A. Frankel, in press). The rex protein from HTLV-I is functionally related to rev and contains an arginine-rich domain that clearly has been implicated in RNA binding (98, 99). Viral coat proteins are attractive candidates for peptide studies. A 25 amino acid arginine-rich peptide from the coat protein of cowpea chlorotic mottle virus (CCMV) adopts an α-helical conformation in the presence of oligophosphates (100) and is likely to bind to a specific packaging sequence in the viral RNA, although its binding site has not yet been defined. A basic (although not arginine-rich) peptide from alfalfa mosaic virus (AlMV) coat protein has been shown to bind to a specific site on the viral RNA and induce a change in RNA conformation (L. Gehrke, personal communication). Other proteins originally identified by Lazinski et al. (11) as members of the arginine-rich class of RNA-binding proteins, including phage antiterminators, retroviral gag proteins, and ribosomal proteins, are good candidates for peptide–RNA model studies.

It will be interesting to compare the tat–TAR and rev–RRE complexes with other peptide–RNA systems, both within the arginine-rich class and from other protein families. How common is major groove binding? Can β-strands or other defined structures bind specifically to RNA? Do unstructured peptides adopt defined conformations upon binding? How important are backbone contacts with phosphate and hydroxyl groups? What are the sequence-specific amino acid–base and amino acid–backbone interactions? How general is recognition of alternative RNA conformations? How dynamic are peptide–RNA complexes? What types of tertiary interactions are observed in RNA structures? These and many more questions will be answered through detailed structural analyses of many protein- and peptide–RNA complexes, just as for protein–DNA recognition. Reductionist approaches may be expected to contribute further to this effort.

References

1. Frankel, A. D. and Kim, P. S. (1991) Modular structure of transcription factors: implications for gene regulation. *Cell*, **65**, 717.
2. Steitz, T. A. (1990) Structural studies of protein–nucleic acid interaction: the sources of sequence-specific binding. *Q. Rev. Biophys.*, **23**, 205.
3. Harrison, S. C. (1991) A structural taxonomy of DNA-binding domains. *Nature*, **353**, 715.
4. Pabo, C. O. and Sauer, R. T. (1992) Transcription factors: structural families and principles of DNA recognition. *Ann. Rev. Biochem.*, **61**, 1053.
5. Pabo, C. O., Aggarwal, A. K., Jordan, S. R., Beamer, L. J., Obeysekare, U. R., and Harrison, S. C. (1990) Conserved residues make similar contacts in two repressor–operator complexes. *Science*, **247**, 1210.
6. Mattaj, I. W. (1993) RNA recognition: a family matter? *Cell*, **73**, 837.
7. Rould, M. A., Perona, J. J., Söll, D., and Steitz, T. A. (1989) Structure of *E. coli* glutaminyl-tRNA synthetase complexed with tRNA(Gln) and ATP at 2.8 Å resolution. *Science*, **246**, 1135.
8. Cavarelli, J., Rees, B., Ruff, M., Thierry, J. C., and Moras, D. (1993) Yeast tRNA(Asp) recognition by its cognate class II aminoacyl-tRNA synthetase. *Nature*, **362**, 181.
9. Puglisi, J. D., Tan, R., Calnan, B. J., Frankel, A. D., and Williamson, J. R. (1992) Conformation of the TAR RNA–arginine complex by NMR spectroscopy. *Science*, **257**, 76.
10. Steitz, T. A. (1993) Similarities and differences between RNA and DNA recognition by proteins. In *The RNA World*. Gesteland, R. F. and Atkins, J. F. (eds). Cold Spring Harbor Press, Cold Spring Harbor, NY, p. 219.
11. Lazinski, D., Grzadzielska, E., and Das, A. (1989) Sequence-specific recognition of RNA hairpins by bacteriophage antiterminators requires a conserved arginine-rich motif. *Cell*, **59**, 207.
12. Frankel, A. D. (1992) Activation of HIV transcription by Tat. *Curr. Opin. Genet. Devel.*, **2**, 293.
13. Sodroski, J., Patarca, R., Rosen, C., Wong-Staal, F., and Haseltine, W. (1985) Location of the *trans*-activating region on the genome of human T-cell lymphotropic virus type III. *Science*, **229**, 74.
14. Arya, S. K., Guo, C., Josephs, S. F., and Wong-Staal, F. (1985) *Trans*-activator gene of human T-lymphotropic virus type III (HTLV-III). *Science*, **229**, 69.
15. Rosen, C. A., Sodroski, J. G., and Haseltine, W. A. (1985) The location of *cis*-acting regulatory sequences in the human T cell lymphotropic virus type III (HTLV-III/LAV) long terminal repeat. *Cell*, **41**, 813.
16. Jakobovits, A., Smith, D. H., Jakobovits, E. B., and Capon, D. J. (1988) A discrete element 3′ of human immunodeficiency virus 1 (HIV-1) and HIV-2 mRNA initiation sites mediates transcriptional activation by an HIV *trans* activator. *Mol. Cell Biol.*, **8**, 2555.
17. Selby, M. J., Bain, E. S., Luciw, P. A., and Peterlin, B. M. (1989) Structure, sequence, and position of the stem–loop in TAR determine transcriptional elongation by tat through the HIV-1 long terminal repeat. *Genes Devel.*, **3**, 547.
18. Muesing, M. A., Smith, D. H., and Capon, D. J. (1987) Regulation of mRNA accumulation by a human immunodeficiency virus *trans*-activator protein. *Cell*, **48**, 691.
19. Garcia, J. A., Harrich, D., Soultanakis, E., Wu, F., Mitsuyasu, R., and Gaynor, R. B.

(1989) Human immunodeficiency virus type 1 LTR TATA and TAR region sequences required for transcriptional regulation. *EMBO J.*, **8**, 765.
20. Roy, S., Parkin, N. T., Rosen, C., Itovitch, J., and Sonenberg, N. (1990) Structural requirements for *trans* activation of human immunodeficiency virus type 1 long terminal repeat-directed gene expression by tat: importance of base pairing, loop sequence, and bulges in the tat-responsive sequence. *J. Virol.*, **64**, 1402.
21. Berkhout, B., Silverman, R. H., and Jeang, K. T. (1989) Tat *trans*-activates the human immunodeficiency virus through a nascent RNA target. *Cell*, **59**, 273.
22. Frankel, A. D., Bredt, D. S., and Pabo, C. O. (1988) Tat protein from human immunodeficiency virus forms a metal-linked dimer. *Science*, **240**, 70.
23. Dingwall, C., Ernberg, I., Gait, M. J., Green, S. M., Heaphy, S., Karn, J., Lowe, A. D., Singh, M., Skinner, M. A., and Valerio, R. (1989) Human immunodeficiency virus 1 tat protein binds *trans*-activation-responsive region (TAR) RNA *in vitro*. *Proc. Natl Acad. Sci. USA*, **86**, 6925.
24. Weeks, K. M., Ampe, C., Schultz, S. C., Steitz, T. A., and Crothers, D. M. (1990) Fragments of the HIV-1 Tat protein specifically bind TAR RNA. *Science*, **249**, 1281.
25. Cordingley, M. G., LaFemina, R. L., Callahan, P. L., Condra, J. H., Sardana, V. V., Graham, D. J., Nguyen, T. M., LeGrow, K., Gotlib, L., and Schlabach, A. J. (1990) Sequence-specific interaction of Tat protein and Tat peptides with the *trans*activation-responsive sequence element of human immunodeficiency virus type 1 *in vitro*. *Proc. Natl Acad. Sci. USA*, **87**, 8985.
26. Roy, S., Delling, U., Chen, C. H., Rosen, C. A., and Sonenberg, N. (1990) A bulge structure in HIV-1 TAR RNA is required for Tat binding and Tat-mediated *trans*-activation. *Genes Devel.*, **4**, 1365.
27. Calnan, B. J., Biancalana, S., Hudson, D., and Frankel, A. D. (1991) Analysis of arginine-rich peptides from the HIV Tat protein reveals unusual features of RNA–protein recognition. *Genes Devel.*, **5**, 201.
28. Calnan, B. J., Tidor, B., Biancalana, S., Hudson, D., and Frankel, A. D. (1991) Arginine-mediated RNA recognition: the arginine fork. *Science*, **252**, 1167.
29. Delling, U., Roy, S., Sumner-Smith, M., Barnett, R., Reid, L., Rosen, C. A., and Sonenberg, N. (1991) The number of positively charged amino acids in the basic domain of Tat is critical for *trans*-activation and complex formation with TAR RNA. *Proc. Natl Acad. Sci. USA*, **88**, 6234.
30. Feng, S. and Holland, E. C. (1988) HIV-1 tat *trans*-activation requires the loop sequence within tar. *Nature*, **334**, 165.
31. Berkhout, B. and Jeang, K. T. (1989) *Trans* activation of human immunodeficiency virus type 1 is sequence specific for both the single-stranded bulge and loop of the *trans*-acting-responsive hairpin: a quantitative analysis. *J. Virol.*, **63**, 5501.
32. Weeks, K. M. and Crothers, D. M. (1991) RNA recognition by Tat-derived peptides: interaction in the major groove? *Cell*, **66**, 577.
33. Sumner-Smith, M., Roy, S., Barnett, R., Reid, L. S., Kuperman, R., Delling, U., and Sonenberg, N. (1991) Critical chemical features in *trans*-acting-responsive RNA are required for interaction with human immunodeficiency virus type 1 Tat protein. *J. Virol.*, **65**, 5196.
34. Hamy, F., Asseline, U., Grasby, J., Iwai, S., Pritchard, C., Slim, G., Butler, P. J., Karn, J., and Gait, M. J. (1993) Hydrogen-bonding contacts in the major groove are required for human immunodeficiency virus type-1 tat protein recognition of TAR RNA. *J. Mol. Biol.*, **230**, 111.

35. Delling, U., Reid, L. S., Barnett, R. W., Ma, M. Y., Climie, S., Sumner-Smith, M., and Sonenberg, N. (1992) Conserved nucleotides in the TAR RNA stem of human immunodeficiency virus type 1 are critical for Tat binding and *trans* activation: model for TAR RNA tertiary structure. *J. Virol.*, **66**, 3018.
36. Churcher, M. J., Lamont, C., Hamy, F., Dingwall, C., Green, S. M., Lowe, A. D., Butler, J. G., Gait, M. J., and Karn, J. (1993) High affinity binding of TAR RNA by the human immunodeficiency virus type-1 tat protein requires base-pairs in the RNA stem and amino acid residues flanking the basic region. *J. Mol. Biol.*, **230**, 90.
37. Subramanian, T., Govindarajan, R., and Chinnadurai, G. (1991) Heterologous basic domain substitutions in the HIV-1 Tat protein reveal an arginine-rich motif required for *trans*activation. *EMBO J.*, **10**, 2311.
38. Yarus, M. (1988) A specific amino acid binding site composed of RNA. *Science*, **240**, 1751.
39. Tao, J. and Frankel, A. D. (1992) Specific binding of arginine to TAR RNA. *Proc. Natl Acad. Sci. USA*, **89**, 2723.
40. Tan, R. and Frankel, A. D. (1992) Circular dichroism studies suggest that TAR RNA changes conformation upon specific binding of arginine or guanidine. *Biochemistry*, **31**, 10 288.
41. Dingwall, C., Ernberg, I., Gait, M. J., Green, S. M., Heaphy, S., Karn, J., Lowe, A. D., Singh, M., and Skinner, M. A. (1990) HIV-1 tat protein stimulates transcription by binding to a U-rich bulge in the stem of the TAR RNA structure. *EMBO J.*, **9**, 4145.
42. Tao, J. and Frankel, A. D. (1993) Electrostatic interactions modulate the RNA-binding and *trans*activation specificities of the human immunodeficiency virus and simian immunodeficiency virus Tat proteins. *Proc. Natl Acad. Sci. USA*, **90**, 1571.
43. Weeks, K. M. and Crothers, D. M. (1992) RNA binding assays for Tat-derived peptides: implications for specificity. *Biochemistry*, **31**, 10 281.
44. Clarke, N. D., Beamer, L. J., Goldberg, H. R., Berkower, C., and Pabo, C. O. (1991) The DNA binding arm of lambda repressor: critical contacts from a flexible region. *Science*, **254**, 267.
45. Talanian, R. V., McKnight, C. J., and Kim, P. S. (1990) Sequence-specific DNA binding by a short peptide dimer. *Science*, **249**, 769.
46. Luo, Y. and Peterlin, B. M. (1993) Juxtaposition between activation and basic domains of human immunodeficiency virus type 1 Tat is required for optimal interactions between Tat and TAR. *J. Virol.*, **67**, 3441.
47. Selby, M. J. and Peterlin, B. M. (1990) *Trans*-activation by HIV-1 Tat via a heterologous RNA binding protein. *Cell*, **62**, 769.
48. Berkhout, B. and Jeang, K. T. (1991) Detailed mutational analysis of TAR RNA: critical spacing between the bulge and loop recognition domains. *Nucleic Acids Res.*, **19**, 6169.
49. Luo, Y., Madore, S. J., Parslow, T. G., Cullen, B. R., and Peterlin, B. M. (1993) Functional analysis of interactions between Tat and the *trans*-activation response element of human immunodeficiency virus type 1 in cells. *J. Virol.*, **67**, 5617.
50. Marciniak, R. A., Garcia-Blanco, M. A., and Sharp, P. A. (1990) Identification and characterization of a HeLa nuclear protein that specifically binds to the *trans*-activation-response (TAR) element of human immunodeficiency virus. *Proc. Natl Acad. Sci. USA*, **87**, 3624.
51. Sheline, C. T., Milocco, L. H., and Jones, K. A. (1991) Two distinct nuclear transcription factors recognize loop and bulge residues of the HIV-1 TAR RNA hairpin. *Genes Devel.*, **5**, 2508.

52. Wu, F., Garcia, J., Sigman, D., and Gaynor, R. (1991) Tat regulates binding of the human immunodeficiency virus *trans*-activating region RNA loop-binding protein TRP-185. *Genes Devel.*, **5**, 2128.
53. Alonso, A., Derse, D., and Peterlin, B. M. (1992) Human chromosome 12 is required for optimal interactions between Tat and TAR of human immunodeficiency virus type 1 in rodent cells. *J. Virol.*, **66**, 4617.
54. Madore, S. J. and Cullen, B. R. (1993) Genetic analysis of the cofactor requirement for human immunodeficiency virus type 1 Tat function. *J. Virol.*, **67**, 3703.
55. Hart, C. E., Galphin, J. C., Westhafer, M. A., and Schochetman, G. (1993) TAR loop-dependent human immunodeficiency virus *trans* activation requires factors encoded on human chromosome 12. *J. Virol.*, **67**, 5020.
56. Puglisi, J. D., Chen, L., Frankel, A. D., and Williamson, J. R. (1993) Role of RNA structure in arginine recognition of TAR RNA. *Proc. Natl Acad. Sci. USA*, **90**, 3680.
57. Riordan, F. A., Bhattacharyya, A., McAteer, S., and Lilley, D. M. (1992) Kinking of RNA helices by bulged bases, and the structure of the human immunodeficiency virus *trans*activator response elément. *J. Mol. Biol.*, **226**, 305.
58. Colvin, R. A., White, S. W., Garcia-Blanco, M. A., and Hoffman, D. W. (1993) Structural features of an RNA containing the CUGGGA loop of the human immunodeficiency virus type 1 *trans*-activation response element. *Biochemistry*, **32**, 1105.
59. Michnicka, M. J., Harper, J. W., and King, G. C. (1993) Selective isotopic enrichment of synthetic RNA: application to the HIV-1 TAR element. *Biochemistry*, **32**, 395.
60. Seeman, N. C., Rosenberg, J. M., and Rich, A. (1976) Sequence-specific recognition of double helical nucleic acids by proteins. *Proc. Natl Acad. Sci. USA*, **73**, 804.
61. Pavletich, N. P. and Pabo, C. O. (1991) Zinc finger-DNA recognition: crystal structure of a Zif268–DNA complex at 2.1 Å. *Science*, **252**, 809.
62. Weeks, K. M. and Crothers, D. M. (1993) Major groove accessibility of RNA. *Science*, **261**, 1574.
63. Wu, H. N. and Uhlenbeck, O. C. (1987) Role of a bulged A residue in a specific RNA–protein interaction. *Biochemistry*, **26**, 8221.
64. Turner, D. H. (1992) Bulges in nucleic acids. *Curr. Opin. Struct. Biol.*, **2**, 334.
65. Cullen, B. R. and Malim, M. H. (1991) The HIV-1 Rev protein: prototype of a novel class of eukaryotic post-transcriptional regulators. *Trends Biochem. Sci.*, **16**, 346.
66. Feinberg, M. B., Jarrett, R. F., Aldovini, A., Gallo, R. C., and Wong-Staal, F. (1986) HTLV-III expression and production involve complex regulation at the levels of splicing and translation of viral RNA. *Cell*, **46**, 807.
67. Malim, M. H., Hauber, J., Le, S. Y., Maizel, J. V., and Cullen, B. R. (1989) The HIV-1 rev *trans*-activator acts through a structured target sequence to activate nuclear export of unspliced viral mRNA. *Nature*, **338**, 254.
68. Daly, T. J., Cook, K. S., Gray, G. S., Maione, T. E., and Rusche, J. R. (1989) Specific binding of HIV-1 recombinant Rev protein to the Rev-responsive element *in vitro*. *Nature*, **342**, 816.
69. Zapp, M. L. and Green, M. R. (1989) Sequence-specific RNA binding by the HIV-1 Rev protein. *Nature*, **342**, 714.
70. Malim, M. H., Tiley, L. S., McCarn, D. F., Rusche, J. R., Hauber, J., and Cullen, B. R. (1990) HIV-1 structural gene expression requires binding of the Rev *trans*-activator to its RNA target sequence. *Cell*, **60**, 675.
71. Daefler, S., Klotman, M. E., and Wong-Staal, F. (1990) *Trans*-activating Rev protein of

the human immunodeficiency virus 1 interacts directly and specifically with its target RNA. *Proc. Natl Acad. Sci. USA*, **87**, 4571.

72. Cochrane, A. W., Chen, C. H., and Rosen, C. A. (1990) Specific interaction of the human immunodeficiency virus Rev protein with a structured region in the env mRNA. *Proc. Natl Acad. Sci. USA*, **87**, 1198.

73. Heaphy, S., Dingwall, C., Ernberg, I., Gait, M. J., Green, S. M., Karn, J., Lowe, A. D., Singh, M. and Skinner, M. A. (1990) HIV-1 regulator of virion expression (Rev) protein binds to an RNA stem–loop structure located within the Rev response element region. *Cell*, **60**, 685.

74. Hope, T. J., McDonald, D., Huang, X. J., Low, J., and Parslow, T. G. (1990) Mutational analysis of the human immunodeficiency virus type 1 Rev *trans*activator: essential residues near the amino terminus. *J. Virol.*, **64**, 5360.

75. Olsen, H. S., Cochrane, A. W., Dillon, P. J., Nalin, C. M., and Rosen, C. A. (1990) Interaction of the human immunodeficiency virus type 1 Rev protein with a structured region in env mRNA is dependent on multimer formation mediated through a basic stretch of amino acids. *Genes Devel.*, **4**, 1357.

76. Zapp, M. L., Hope, T. J., Parslow, T. G., and Green, M. R. (1991) Oligomerization and RNA binding domains of the type 1 human immunodeficiency virus Rev protein: a dual function for an arginine-rich binding motif. *Proc. Natl Acad. Sci. USA*, **88**, 7734.

77. Bohnlein, E., Berger, J., and Hauber, J. (1991) Functional mapping of the human immunodeficiency virus type 1 Rev RNA binding domain: new insights into the domain structure of Rev and Rex. *J. Virol.*, **65**, 7051.

78. Tan, R., Chen, L., Buettner, J. A., Hudson, D., and Frankel, A. D. (1993) RNA recognition by an isolated α helix. *Cell*, **73**, 1031.

79. Kjems, J., Calnan, B. J., Frankel, A. D., and Sharp, P. A. (1992) Specific binding of a basic peptide from HIV-1 Rev. *EMBO J.*, **11**, 1119.

80. Kjems, J., Frankel, A. D., and Sharp, P. A. (1991) Specific regulation of mRNA splicing *in vitro* by a peptide from HIV-1 Rev. *Cell*, **67**, 169.

81. Kjems, J. and Sharp, P. A. (1993) The basic domain of Rev from human immunodeficiency virus type 1 specifically blocks the entry of U4/U6.U5 small nuclear ribonucleoprotein in spliceosome assembly. *J. Virol.*, **67**, 4769.

82. Kjems, J., Brown, M., Chang, D. D., and Sharp, P. A. (1991) Structural analysis of the interaction between the human immunodeficiency virus Rev protein and the Rev response element. *Proc. Natl Acad. Sci. USA*, **88**, 683.

83. Huang, X., Hope, T. J., Bond, B. L., McDonald, D., Grahl, K., and Parslow, T. G. (1991) Minimal Rev-response element for Type 1 human immunodeficiency virus. *J. Virol.*, **65**, 2131.

84. Tiley, L. S., Malim, M. H., Tewary, H. K., Stockley, P. G., and Cullen, B. R. (1992) Identification of a high-affinity RNA-binding site for the human immunodeficiency virus type 1 Rev protein. *Proc. Natl Acad. Sci. USA*, **89**, 758.

85. Heaphy, S., Finch, J. T., Gait, M. J., Karn, J., and Singh, M. (1991) Human immunodeficiency virus type 1 regulator of virion expression, rev, forms nucleoprotein filaments after binding to a purine-rich 'bubble' located within the rev-responsive region of viral mRNAs. *Proc. Natl Acad. Sci. USA*, **88**, 7366.

86. Cook, K. S., Fisk, G. J., Hauber, J., Usman, N., Daly, T. J., and Rusche, J. R. (1991) Characterization of HIV-1 REV protein: binding stoichiometry and minimal RNA substrate. *Nucleic Acids Res.*, **19**, 1577.

87. Bartel, D. P., Zapp, M. L., Green, M. R., and Szostak, J. W. (1991) HIV-1 Rev

regulation involves recognition of non-Watson–Crick base pairs in viral RNA. *Cell*, **67**, 529.

88. Holland, S. M., Chavez, M., Gerstberger, S., and Venkatesan, S. (1992) A specific sequence with a bulged guanosine residue(s) in a stem–bulge–stem structure of Rev-responsive element RNA is required for *trans* activation by human immunodeficiency virus type 1 Rev. *J. Virol.*, **66**, 3699.

89. Iwai, S., Pritchard, C., Mann, D. A., Karn, J., and Gait, M. J. (1992) Recognition of the high affinity binding site in rev-response element RNA by the human immunodeficiency virus type-1 rev protein. *Nucleic Acids Res.*, **20**, 6465.

90. Daly, T. J., Rennert, P., Lynch, P., Barry, J. K., Dundas, M., Rusche, J. R., Doten, R. C., Auer, M., and Farrington, G. K. (1993) Perturbation of the carboxy terminus of HIV-1 Rev affects multimerization on the Rev responsive element. *Biochemistry*, **32**, 8945.

91. Malim, M. H. and Cullen, B. R. (1991) HIV-1 structural gene expression requires the binding of multiple Rev monomers to the viral RRE: implications for HIV-1 latency. *Cell*, **65**, 241.

92. McDonald, D., Hope, T. J., and Parslow, T. G. (1992) Posttranscriptional regulation by the human immunodeficiency virus type 1 Rev and human T-cell leukemia virus type I Rex proteins through a heterologous RNA binding site. *J. Virol.*, **66**, 7232.

93. Venkatesan, S., Gerstberger, S. M., Park, H., Holland, S. M., and Nam, Y. (1992) Human immunodeficiency virus type 1 Rev activation can be achieved without Rev-responsive element RNA if Rev is directed to the target as a Rev/MS2 fusion protein which tethers the MS2 operator RNA. *J. Virol.*, **66**, 7469.

94. Daly, T. J., Rusche, J. R., Maione, T. E., and Frankel, A. D. (1990) Circular dichroism studies of the HIV-1 Rev protein and its specific RNA binding site. *Biochemistry*, **29**, 9791.

95. Wimberly, B., Varani, G., and Tinoco, I., Jr (1993) The conformation of loop E of eukaryotic 5S ribosomal RNA. *Biochemistry*, **32**, 1078.

96. Wyatt, J. R. and Tinoco, I., Jr (1993) RNA structural elements and RNA function. In *The RNA World*. Gesteland, R. F. and Atkins, J. F. (eds). Cold Spring Harbor Laboratory Press, Cold Spring Harbor, NY, p. 465.

97. Elangovan, B., Subramanian, T., and Chinnadurai, G. (1992) Functional comparison of the basic domains of the Tat proteins of human immunodeficiency virus types 1 and 2 in *trans* activation. *J. Virol.*, **66**, 2031.

98. Hofer, L., Weichselbraun, I., Quick, S., Farrington, G. K., Bohnlein, E., and Hauber, J. (1991) Mutational analysis of the human T-cell leukemia virus type I *trans*-acting rex gene product. *J. Virol.*, **65**, 3379.

99. Hammes, S. R. and Greene, W. C. (1993) Multiple arginine residues within the basic domain of HTLV-I Rex are required for specific RNA binding and function. *Virology*, **193**, 41.

100. van der Graaf, M., Scheek, R. M., van der Linden, C. C., and Hemminga, M. A. (1992) Conformation of a pentacosapeptide representing the RNA-binding N-terminus of cowpea chlorotic mottle virus coat protein in the presence of oligophosphates: a two-dimensional proton nuclear magnetic resonance and distance geometry study. *Biochemistry*, **31**, 9177.

11 | Study of RNA–protein recognition by *in vitro* selection

DAVID P. BARTEL and JACK W. SZOSTAK

A major focus of biochemistry and genetics has been to deduce the determinants of molecular recognition and catalysis. *In vitro* selection of nucleic acids is a method that represents a synthesis of these two disciplines—a method that retains both the control of *in vitro* biochemical manipulation and the power of genetic selection. This new tool has been used successfully in many laboratories for applications ranging from defining DNA-binding sites of transcription factors to isolating new ribozymes. The basic features and applications of *in vitro* selection of nucleic acids have been reviewed extensively (1–6). After a brief overview of these features and applications, this chapter focuses on practical considerations for designing, implementing, and interpreting *in vitro* selection schemes for the characterization of RNA–protein interactions.

1. Overview
1.1 Fundamentals of *in vitro* selection

The typical goal of an *in vitro* selection experiment is to generate a set of RNA or DNA sequences able to perform an interesting function, such as binding to a ligand or catalysing a reaction. Comparative analysis of the set of functional sequences then allows determination of the primary and sometimes secondary or tertiary structural features of the DNA or RNA that are important for function. There are three fundamental steps in any *in vitro* selection protocol. First a large pool of randomized sequences is generated, usually by chemical synthesis of DNA followed by enzymatic synthesis of the complementary strand. (When RNA molecules are to be selected this DNA is transcribed *in vitro* into RNA.) Second, the pool of molecules is fractionated based on the ability to perform the specified biochemical function. Third, the small amount of material that co-purifies with the desired activity is amplified *in vitro*. The amplified material is then subjected to additional rounds of selection and amplification until the pool is enriched in active sequences to the

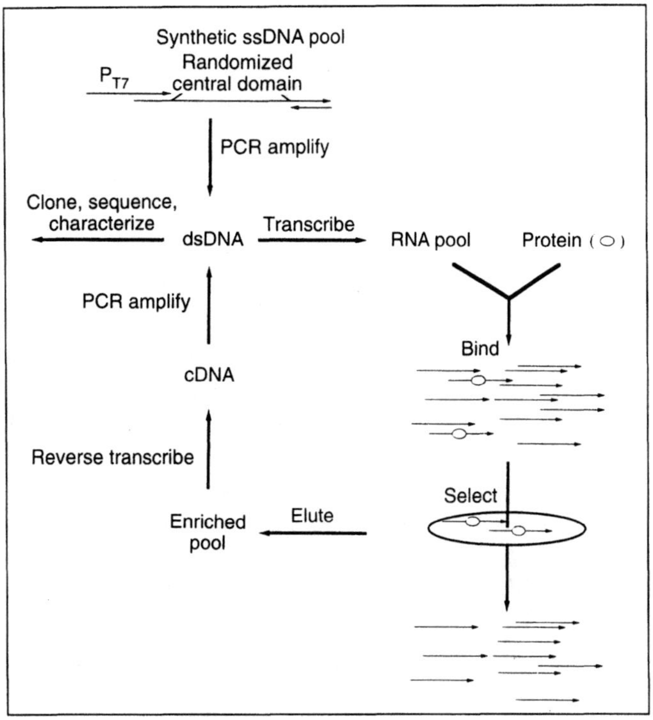

Fig. 1 An *in vitro* scheme utilizing filter binding to isolate RNAs that bind to a given purified protein. Typically three to ten cycles of selection and amplification are performed prior to cloning and characterizing the binding variants.

point that most of the pool molecules bind the ligand or catalyse the reaction. At this point the pool can be cloned, sequenced, and characterized. A typical selection scheme for RNAs that bind a protein is shown in *Figure 1*.

The idea of *in vitro* selection is based upon the fact that the nucleic acid molecules that function *in vitro* also contain the coding information for their replication *in vitro*. Put in genetic terms, nucleic acid molecules have both an *in vitro* phenotype and an *in vitro* genotype. Therefore, the three fundamentals of *in vivo* genetic selections are accomplished by *in vitro* biochemistry: a pool of heritable variation is generated, variants within the pool with a specified phenotype are selected, and survivors are preferentially propagated. The fact that the basic principles of genetic selection are employed entirely in the test-tube to isolate molecules with very rare phenotypic properties has led to the process being referred to as '*in vitro* genetics'.

1.2 Development of *in vitro* selection methods

The invasion of biochemistry by this kind of genetics was incremental and depended upon a series of innovations developed in many laboratories. The creation of a pool of biochemical diversity and the *in vitro* separation of functional from non-functional molecules dates back to the methylation interference assay. More recently, random or degenerate synthetic DNA molecules have been used to generate *heritable* diversity *in vitro* that was then used for *in vivo* selections (7–11). Creating mutant populations with randomized DNA oligonucleotides was critical for achieving controlled, targeted mutagenesis. Struhl and colleagues (12) were the first to combine this *in vitro* randomization with *in vitro* selection. With four serial bindings to a GCN4 affinity column they were able to select sequences from a random dsDNA pool that bound to the GCN4 transcription factor. The selected sequences were then cloned and amplified *in vivo* using bacteria, but Struhl and colleagues (12) proposed that more stringent selections would be feasible with repeated cycles of *in vitro* selection and PCR amplification. Kinzler and Vogelstein (13) were the first to combine *in vitro* selection and *in vitro* amplification. Human DNA fragments were selected by co-immunoprecipitation with transcription factor IIIA (TFIIIA), then amplified by the polymerase chain reaction (PCR). After two rounds of selection and amplification 10% of the pool bound to TFIIIA. Kinzler and Vogelstein (14) used a similar approach to identify the consensus binding site of a putative zinc finger domain of the GLI protein. The complete fruition of *in vitro* genetics came as at least six laboratories independently developed and published successful schemes in which all three steps, randomization, selection, and amplification, were accomplished *in vitro* (15–22).

1.3 Comparison of *in vitro* and *in vivo* genetic selections

It was only when the need for cells was completely circumvented that the enormous potential of *in vitro* selection became apparent. Much of this potential stems from the fact that much larger pools can be utilized *in vitro* than is possible *in vivo* — the feasibility of finding an active variant is no longer limited by the efficiency of bacterial transformation or the difficulties of handling large quantities of cells. Nucleic acid pools with greater than 10^{13} initial variants can be manipulated in microfuge tubes and pools of up to 10^{16} variants can be handled without undue difficulty in laboratory-scale experiments. The ability to sample such large numbers of initial variants has not only facilitated the identification and characterization of binding elements recognized by DNA- and RNA-binding proteins, but has also allowed the isolation of heretofore unknown activities from completely random-sequence pools. RNA or DNA sequences have been isolated that specifically bind a variety of smaller molecules, including organic dyes (17, 23), L-tryptophan (24), arginine (25), and adenosine (26). Sequences have also been found that bind antibodies that had been raised to a peptide antigen (27) and other selections have produced molecules that bind and specifically inhibit thrombin (*Figure 2*; 28–30) and HIV reverse transcriptase (*Figure 3*; 31). Nucleic acids that have been selected from pools of random sequences on the basis of binding to specific ligands are sometimes called aptamers, from the Latin *aptus*, to fit. The ability to isolate new binding activities from large pools of random sequences led to the supposition that new catalysts might also be isolated from random sequences (16, 17). We have recently isolated new ribozymes that catalyse an RNA ligation reaction analogous to one cycle of nucleotide addition during RNA polymerization (32).

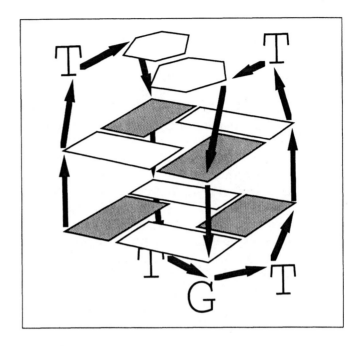

Fig. 2 Schematic model for the structure of a thrombin-binding DNA aptamer (5'-GGTTGGTGTGGTTGG) as determined by NMR (redrawn from Macaya et al. (29) and Wang et al. (30)). Residues involved in hydrogen-bond interactions with other residues are shown as shapes rather than letters (hexagon, T; rectangle, G; shaded rectangle, *syn* G). Thrombin-binding aptamers have K_d values less than 200 nM and inhibit clot formation in human plasma (28).

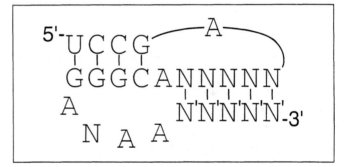

Fig. 3 Minimal consensus secondary structure of the RNA molecules selected for binding to HIV-1 reverse transcriptase (redrawn from Tuerk et al. (31) and Gold et al. (39)). Some of the selected RNA ligands have K_d values of 5 nM and specifically inhibit HIV reverse transcriptase activity *in vitro*.

The ability to perform genetic studies outside of the confines of the cell enables much more flexibility in setting selection criteria and conditions. Beutel and Gold (33) were able to select for sequences that kink DNA using an *in vitro* selection that differentiated sequences based on mobility in native gels. It is difficult to imagine how a selection for this phenotype could have been achieved within cells. In a similar vein, Rittner et al. (34) were able to select HIV antisense RNA fragments *in vitro* on the basis of their speed of hybridization to HIV transcripts and Harada and Orgel (35) were able to select for optimal DNA substrates for T4 RNA ligase. The flexibility of *in vitro* selection conditions has been particularly important for ribozyme selections. Group I ribozymes have a specific requirement for Mg^{2+} or Mn^{2+} for activity. Lehman and Joyce (36) used an *in vitro* selection to isolate group I ribozyme variants that can use Ca^{2+} to circumvent this Mg^{2+}/Mn^{2+} requirement. Such a selection, which demanded activity in the absence of Mg^{2+} or Mn^{2+}, would have been impossible *in vivo* because the cellular levels of divalent cations cannot be substantially varied. Similarly, Pan and Uhlenbeck (37) isolated ribozymes from a pool of tRNA variants that use Pb^{2+} to cleave themselves. The levels of Pb^{2+} used in this study would be toxic to cells. Now that genetics is in the realm of *in vitro* biochemistry the choice of selectable phenotypes is more limited by the interest and imagination of the researcher than by the restrictive conditions that support life.

Most investigators that use *in vitro* selections to probe protein–RNA interactions are concerned that their findings be relevant *in vivo* and as such they are not as interested in using buffer conditions that vary significantly from the intracellular milieu. Still, *in vitro* selection offers advantages beyond the previously mentioned capability of sampling large numbers of variants. The use of purified components *in vitro* helps eliminate effects that may confound *in vivo* analysis, such as differential stabilities to nucleases or differential binding to other competing cellular factors. Selecting directly for binding also bypasses any artefacts associated with the indirect read-out of *in vivo* reporter schemes. We outline below some of the options and considerations for designing schemes to harness the power, simplicity, and control of this new genetic system for the study of RNA–protein interactions.

2. Choosing the length and degree of randomization

The first consideration when designing an *in vitro* selection experiment is the extent of pool randomization. This is an important decision since the randomization strategy can have an effect on the results and interpretation of the selection. *Figure 4* shows the results of selections for HIV rev-binding sequences from four different types of pools (38–40).

2.1 Completely random pools

Pools of completely random sequences have been used when the protein has no natural RNA target sequence (27, 28, 39) or when the natural RNA target is unknown (41). Selections from completely random pools have also identified critical recognition determinants of known RNA elements that bind the U1 snRNP U1A protein (42), the bacteriophage R17 coat protein (43), and the transcriptional termination ρ factor of *Escherichia coli* (44). Although in some cases the wild-type target may emerge as a result of selection from a pool of random sequences, in other cases the aptamers may not be recognizable in relation to the wild-type sequence and, hence, may contribute little to the understanding of the natural binding mode. When designing a completely random-sequence pool, the main consideration is the length of the random region. Many dsDNA–protein selections utilize pools with only ~ 20 random positions. Pools as short as those used for DNA may also be sufficient for some RNA selections, but there are several compelling reasons for using longer pools.

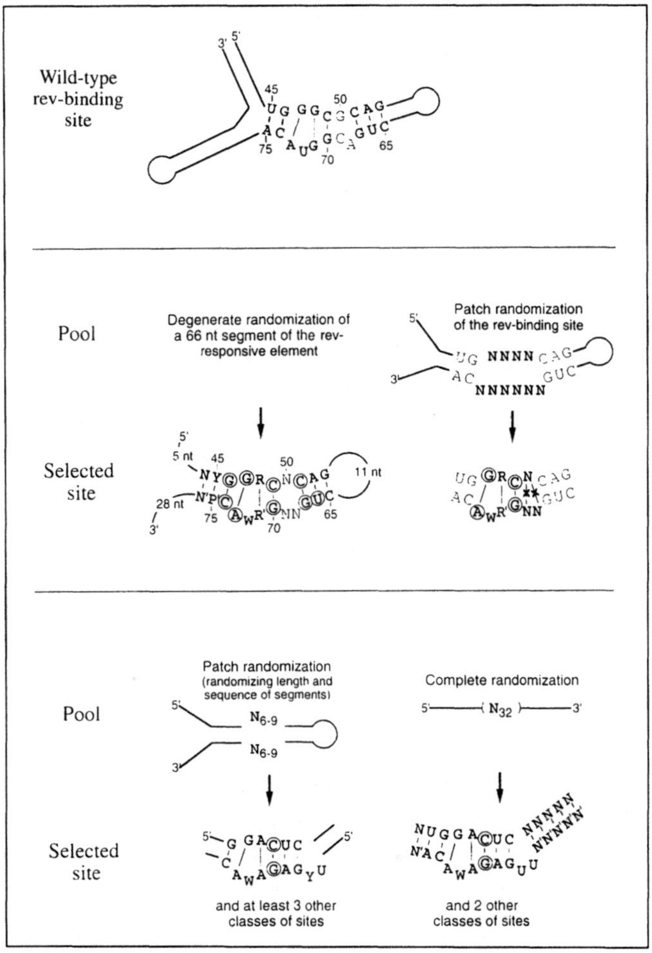

Fig. 4 Abridged summary of HIV rev-binding sequences selected from four different types of randomized pools (redrawn from Bartel et al. (38), Gold et al. (39), and Giver et al. (40)). Short, thin lines connecting residues indicate a highly conserved potential for Watson–Crick pairing, longer lines indicate a highly conserved potential for non-Watson–Crick pairing, and an X through a line indicates a strong bias against base combinations that can form Watson–Crick pairs. Circled residues are invariant in all sequenced aptamers from the particular selection. Y, pyrimidine; R, purine; W, A or U.

2.1.1 Accommodation of the RNA-binding element structure

Unlike DNA promoter elements, most RNA structures do not fit comfortably within 20 nucleotides (nt). For example, the HIV-1 rev protein binds a 20 nt element within the viral RNA, but the 9 and 11 nt segments of this core element must be present within a stable stem–internal loop–stem secondary structure that is difficult to construct with < 30 nt (*Figure 4*). T4 DNA polymerase binds a 36 nt site within its mRNA that contains an 18 nt stem–loop and another 18 bases of flanking sequence (45). Thus, to have the possibility of selecting structures more complex than a simple RNA stem–loop, pools used for RNA selections should be larger than those typically used for dsDNA selections.

2.1.2 Issues of sequence space coverage

Many people are uncomfortable starting with a random RNA pool that does not contain at least one representative of every possible sequence. For example, a pool with one copy of all possible 30 nt sequences ($4^{30} = 1.2 \times 10^{18}$) would have a mass of ~ 10 mg, assuming the pool has ~ 30 constant nucleotides for primer binding. (In practice, the sequence distribution is not even, but follows a binomial distribution, so that 10 mg of RNA would miss ~ 37% of the sequences and it would take ~ 40 mg of pool RNA to represent more than 98% of the sequences at least once.) It is impractical to make and use 10 mg pools. For instance, transcribing such a pool *in vitro* using the standard template concentration (0.5 µM) would require a 1 litre reaction. All *in vitro* selections have started with <10 mg of RNA; often as little as 10 µg of RNA is used in the initial round, so that only a small fraction of the possible sequences are represented.

As it turns out, the inability to represent all possible sequences does not compromise the effectiveness of RNA–protein selections. This is because RNA structures that bind proteins, particularly the more complex structures requiring 30 nt or more, do not require strict sequence identity at all positions. For instance, RNA-binding elements often contain base-paired regions in which any of the four standard base pairs will suffice or there are single-stranded regions that merely serve to link two parts of the element (for example, loop sequences that are not bound by the protein) in which any of the bases will serve equally well. In many cases there are so many possibilities for neutral changes within the binding element that it is impractical to sequence every permitted combination of these changes. Therefore, it is not critical that all the combinations of neutral changes exist within the initial pool.

2.1.3 Increasing effective sequence complexity with pool length

When the active binding structures may be complex and therefore rare it is best to use pools with considerably more than 30 random positions. Increasing the length of the random region provides a linear advantage in finding a specific linear motif. For instance, a hypothetical motif requiring 20 contiguous bases is ~ 7.4 times more likely to be present in a 100 nt random sequence than a 30 nt sequence. (There are 11 registers in which to find the 20 nt motif in the stretch of 30 random bases and 81 registers in 100 random bases.) More importantly, the increase in effective sequence complexity provided by this register consideration is much more significant when RNA structure is taken in account. A specific 8 bp stem is 50 times more likely to be present in the pool with 100 random positions, if one assumes that the loop region that joins the two stem segments can be of any sequence or length. (Assuming that the loop region can be of any sequence or length is clearly an oversimplification; some sequences would interfere with the formation of the 8 bp stem, but many, if not most, sequences would be compatible.) Because longer random regions provide the possibility of selecting complex motifs while increasing the probability of selecting shorter motifs, the main justification for using a pool with only 30 random positions would be in instances when the binding sequence was known to be simple or relatively non-specific. In these instances the shorter random region would simplify the sequence comparisons and structural analysis of the binding molecules. In addition, when the protein may bind cooperatively to the RNA, a shorter pool may be used out of concern that molecules with multiple low-affinity sites may interfere with the identification of molecules with high-affinity elements.

A final reason for using larger random regions is that it is easier to handle larger nucleic acid molecules. This consideration is particularly important during PCR amplification. Many PCR primer pairs yield primer artefacts that co-migrate on agarose gels with PCR products less than 100 bp in length. Pool molecules that have longer random regions have total lengths larger than 100 bp and can be easily separated from the primer artefacts. Longer PCR DNA is also less apt to denature at temperatures used to melt gel slices or inactivate enzymes. Denatured strands of a pool with high sequence complexity do not find their complementary strands, instead they remain single stranded or form mismatched hybrids with related but not fully complementary strands. If such heteroduplex DNA is repaired by the bacteria after transformation the resulting clones will contain hybrid sequences that may not be active.

2.2 Partially random pools

Since it has been possible to isolate protein-binding RNA sequences from completely random-sequence pools, it may be argued that the most elegant experiments will always start from completely random pools so that there will not be any reliance on assumptions (perhaps flawed) concerning the sequence requirements for protein binding. However, selection from completely random pools may provide several classes of structures that bind the protein, some of which may have little resemblance to the wild-type binding element. Therefore, biasing the pool towards the wild-type sequence may be the most efficient route to gaining insight into the wild-type binding interaction. Comparative sequence analysis of the functional sequences generated by *in vitro* selection is also easiest and most powerful when examining a set of related sequences of known 'ancestry'.

2.2.1 Artificial phylogeny

The set of functional, related molecules selected from partially randomized pools is sometimes referred to as an 'artificial phylogeny' to highlight the common ancestry of the molecules and the utility of the sequences for gaining structural information by sequence comparison (18, 38). Comparative sequence analysis of true phylogenies has been an indispensable tool for predicting the features important for the function of structured RNAs (46–48). Conservation of sequence motifs identifies segments most important for function; covarying residues within and flanking these motifs indicate required secondary and tertiary interactions. Analysis of an artificial phylogeny is as powerful as analysis of true phylogeny and, in some respects, more straightforward. All the variants of an artificial phylogeny are selected for binding to the same protein, whereas RNA variants in natural phylogeny may coincide with changes in the protein. Natural phylogeny may contain cryptic sequence variants that require accessory factors for protein binding or have even lost binding activity altogether. In addition, the ancestry of all the variants of an artificial phylogeny is known—all variants have arisen independently from the same 'ancestral' sequence. In contrast, when comparing natural sequences, the significance of the repeated occurrence of the same covariation hinges on establishing that it represents multiple independent examples of simultaneous changes, as opposed to a set of changes all inherited from one common ancestor.

2.2.2 Patch randomization

The most common partial-randomization strategy is to randomize only a subset of the binding element positions (16, 31, 40, 42). We refer to this type of randomization as 'patch randomization'. For instance, in the first RNA-protein *in vitro* selection, Tuerk and Gold (16) randomized only the eight loop bases of the T4 DNA polymerase binding element—the other 28 bases were not randomized. This allowed for the 4^8 (65 536) loop possibilities to be exhaustively sampled within the wild type context. The patch randomization approach is best suited for study of binding elements that have already been identified and mapped. RNA primary sequence and higher-order structural features important for protein binding can be identified or confirmed very cleanly by this method, provided that the randomization is based on accurate assumptions about the binding element structure. Some knowledge of the structure allows all the bases of a substructure of the binding element to be randomized simultaneously, permitting the detection of more potentially covarying positions.

2.2.3 Degenerate randomization

If the precise location of the binding element is not known or the secondary structure of the element is still controversial, then a partial randomization approach is more useful than patch randomization. In this second approach, essentially a high frequency mutagenesis, the initial pool consists of degenerate versions of a parental sequence known to contain the binding element. For example, we used partial randomization and *in vitro* selection to map and characterize the RNA element within a 66 nt segment of HIV-I mRNA that was known to bind to the viral regulatory protein rev (38). We generated a pool of variants of this 66 nt segment, such that a mean of 65% of the positions in a given variant were wild-type, 30% were a non-wild-type base, and 5% were deleted. This degree of degeneracy allows interacting residues to change in a concerted manner at a reasonable frequency, but ensures that most of the binding species bear sufficient resemblance to the wild-type element to allow straightforward sequence alignment. At this degree of randomization an essential Watson–Crick pair that can be replaced with any of the other three Watson–Crick alternatives will be replaced in ~ 7% of the functional binding variants—the fraction of the initial pool with a wild-type pair is 0.65×0.65; the fraction of the initial pool with any of the suppressor pairs is $3(0.10 \times 0.10)$. If the mutagenesis was only half as severe, such that 82.5% of positions were wild type, then this Watson–Crick pair would only be replaced in ~ 1% of the active variants.

It is worth noting that biasing the pool sequences toward the wild-type sequence or structure does not completely eliminate the possibility that the active sequences will incorporate non-wild-type structural features. Pan and Uhlenbeck (37) used patch randomization to randomize all the tRNA^{Phe} residues directly involved in tertiary interactions. Their selection yielded many interesting non-tRNA-like structures (37,49) that, like tRNA^{Phe}, were able to self-cleave in the presence of Pb^{2+}.

2.2.4 Whole-genome PCR

When a protein is thought to have a natural RNA target it may be possible to use an adaptation of Kinzler and Vogelstein's (13) 'whole-genome PCR' method of pool construction to identify its natural targets. For example, if the protein is thought to bind a mature mRNA it may be desirable to construct a pool from cDNA fragments that could be transcribed into an RNA pool. Such an approach may facilitate the identification or cloning of the natural RNA target, but use of a pool in which the members vary in length limits the ability to purify the pool on the basis of size. Size purification is sometimes required to eliminate PCR artefacts.

3. Creating randomized pools

In vitro selection techniques have progressed to the point that a single active sequence can be selected from an initial pool of variants. For some applications, the probability of success is therefore limited only by the sequence complexity of the initial pool. If the initial pool complexity is too small to contain variants with the desired phenotype, the selection will fail. In other applications, particularly applications involving patch randomization, a small amount of pool RNA (ng–μg) contains many copies of all possible variants. The only concern in constructing these pools is to ensure the proper distribution of bases at each randomized position.

3.1 Constructing very large, complex pools

3.1.1 Limits of standard DNA synthesis

Large complex pools are most useful when selecting from completely random pools or when selecting for increased or novel activities from degenerate pools. If the desired pool length is much greater than 100 residues, the yield and quality of the synthetic DNA oligonucleotide can become limiting. For instance, a 1 μmol synthesis of a 150 nt oligomer has a theoretical yield of 2.5 mg (with 98% stepwise coupling). However, the yield of gel-purified material is typically much lower (∼ 100 μg). This lower yield can be ascribed to at least two factors: depurination and branching. The extended time of exposure to the acidic deblocking reagent leads to depurination, which in turn promotes strand cleavage during the ammonia deprotection procedure. A shorter deblocking step at each synthetic cycle improves yields. A shorter deblocking step also may lead to point deletions in a fraction of the molecules, particularly at the position immediately 5′ of pyrimidine residues. However, low levels of point deletions are not problematic in completely random pools and may be considered another useful mutagenesis technique when constructing degenerate pools. In addition to the shorter products resulting from strand cleavage and abortive synthesis, a large amount of product migrates anomalously slowly on the polyacrylamide gel. This class of side products is thought to be the result of stable coupling to sites in the interior of the molecule and subsequent extension of this aberrant branch during the following synthesis cycles (R. Green, D. P. Bartel, and J. W. Szostak, unpublished data).

The 100 μg obtained from gel purification of a 150-mer would still represent a very large pool of molecules. However, primer extension experiments comparing synthetic DNA to enzymatically generated DNA have shown that long synthetic molecules are much poorer templates for second-strand synthesis (1). Typically over 60% of gel-purified synthetic 110-mer and 98% of gel-purified 150-mer contain lesions that block second-strand synthesis. The lesions do not appear to be the result of incomplete deprotection and the results are similar with DNA phosphoramidites from several suppliers. Therefore, synthesis of a 150 nt pool can yield less than 2 μg of DNA that can be effectively amplified and transcribed.

3.1.2 Combinatorial pool construction

Pools that start with synthetic DNA 100–110 nt in length (with 70–80 randomized bases) represent a good compromise between the limits of DNA synthesis and the effective complexity advantage of longer random regions. There may be some situations, though, when a longer pool is required. For example, some RNA-binding elements have only been localized to RNA segments much larger than 80 nt, so use of a degenerate pool to map the element would require a correspondingly longer pool, or, when isolating molecules with a new activity from completely random sequence, the complexity advantage gained from a much longer random region may be required because the functional sequences may be exceedingly rare.

Beaudry and Joyce (50) have used a combinatorial method to make a pool of 394 nt group I ribozyme variants. Degenerate oligonucleotides corresponding to the template strand for RNA transcription were annealed to full-length ssDNA that represented the complement of the template strand. The template oligonucleotides were then joined with DNA ligase and transcribed directly. We have recently used a different combinatorial construction method to generate a 274 nt RNA pool with 220 completely random positions (32). Double-stranded DNA fragments from smaller pools were linked to generate the template DNA for this large pool. Both of these combinatorial methods permit the construction of very long, complex pools, but they require short (4–6 nt) constant segments punctuating the randomized region to ensure efficient ligation of the template DNA.

3.2 Large-scale PCR

When pools are constructed from shorter DNA oligomers the factors limiting pool sequence complexity change from the yield and quality of the DNA synthesis to the volume of the PCR and *in vitro* transcription reactions. For example, if a 110-mer synthesis yields 250 µg of gel-purified DNA, 40% of which is lesion-free, then a 20 ml PCR reaction would be required just for second-strand synthesis. (This calculation is based on the observation that the yield of a PCR amplification generally plateaus when the product concentration reaches $1\,\mu g\,100\,\mu^{-1}$.) A 160 ml reaction would be required to amplify this pool another three cycles to obtain eight copies of each molecule so that the pool could be utilized for more than one *in vitro* transcription. Most selections do not require such a large initial sequence complexity and, therefore, it is only necessary to amplify a fraction of the synthetic material. When high sequence complexity is thought to be required, the PCR volume can be scaled-up using multiple 15 ml polypropylene tubes and water baths, with manual transfer of racks of tubes between water baths at different temperatures.

3.3 Base distribution
3.3.1 Differential coupling rates
We find that random sequences made from an equimolar mixture of phosphoramidites have an A:C:G:T ratio of approximately 2:2:3:3 in the synthetic strand. Making compensatory changes in the initial phosphoramidite ratio yields a more even distribution of bases. This sequence bias has occurred with several pools and with phosphoramidites from two different suppliers, so it presumably reflects faster coupling rates of G and T. Regardless of the mechanism of this sequence bias, the possibility of such a bias highlights the importance of sequencing the starting pool to determine the initial sequence bias so it can be taken into account during analysis of the selected sequences.

3.3.2 Point deletions
Introducing point deletions provides a sometimes overlooked dimension of partial randomization. As noted above, decreasing the length of the deblocking step during pool synthesis can introduce deletions, particularly at the positions immediately 5' of pyrimidine residues. Other changes in the DNA synthesis protocol can be employed to introduce a more even distribution of point deletions. The capping step can be omitted so that any growing chains that fail to couple to the phosphoramidite in a particular synthesis cycle will be able to be extended in later cycles (38). To further increase the number of point deletions the coupling efficiency can be decreased by shortening the coupling time.

4. Enrichment and amplification of functional sequences
4.1 Enrichment techniques
4.1.1 Filter binding
Any biochemical technique that physically separates active from inactive sequences can be used in an *in vitro* selection scheme. Co-immunoprecipitation, gel-mobility

shift, and column/bead immobilization have all been successfully utilized for the enrichment step of protein-binding selections, but the most popular procedure for RNA–protein selections has been filter binding (*Figure 1*; 16, 31, 38–40, 43, 44). The RNA pool is first allowed to bind to the protein in solution under conditions of RNA excess. The reaction mixture is then passed through a filter that retains the protein but not the free RNA. RNA co-retained with the protein on the filter is eluted and then amplified. Filter-binding selections are simple, fast, and require no special reagents other than the purified active protein.

4.1.2 Ligand immobilization on columns or beads

Columns with immobilized ligand can retain binding species with low affinity to the ligand (K_d up to 10 mM), so they provide the enrichment method of choice for particularly challenging binding selections. Affinity columns have been successfully employed to select sequences that bind to small molecules, such as L-tryptophan, arginine, and ATP (24–26). Column-binding selections are ideal for selecting sequences with moderate-binding affinity (K_d of 0.1–100 μM), but the fact that columns retain sequences with low or moderate affinity to the ligand compromises the utility of affinity columns to discriminate between tight- and moderate-binding sequences. After the sequence pool is bound to the column, the column must be washed extensively to remove sequences with moderate affinity. For example, a column with 100 μM accessible immobilized ligand must be washed for more than 100 column volumes to adequately discriminate between a molecule with a 0.1 μM K_d for the ligand and a molecule with a 1 μM K_d.

Immobilized ligand can be utilized in a non-column format in order to distinguish tight- from moderate-binding sequences more efficiently. For instance, the pool can be incubated with a small amount of beads such that binding sequences are in large molar excess over the ligand, which forces pool molecules to compete for binding to the ligand. High-stringency washes are also easier to perform in a non-column format — a small volume of beads is simply mixed with a large volume of wash buffer. Keene and colleagues (27, 51) have selected RNA antibody epitopes from an RNA pool by using high-stringency washes of antibodies immobilized on protein A-Sepharose beads. They also used a co-immunoprecipitation procedure to select for RNAs that bind to purified U1 snRNP-A protein (42) and purified Hel-N1 protein (41).

When an antibody is used to select the desired protein–nucleic acid complex, the protein need not be purified prior to the immunoprecipitation procedure (22). While use of unpurified cellular extracts may not be practical for protein–RNA selections, due to the likelihood that RNases in the extract would degrade the RNA pool, use of cellular extracts expands the utility of DNA *in vitro* selections to applications in which some of the components of the protein–DNA complex are unknown. Chittenden *et al.* (52) exploited cellular extracts to provide factor(s) that mediate the interaction between an immobilized retinoblastoma protein and DNA. By determining the DNA-binding specificity of the complex they were able to propose that the complex contains the E2F transcription factor or a protein with similar DNA-binding properties.

4.1.3 Gel-mobility shift

The gel-mobility shift assay is a very popular tool for detecting RNA–protein interactions. Though enrichment by gel shift has been used extensively in DNA–protein selections, this method has yet to be used for RNA–protein selections. The reason for the underutilization of gel-shift enrichment for RNA–protein selections may be that single-stranded RNA sequences, unlike double-stranded DNA sequences, are prone to aggregation by intermolecular hybridization and through the formation of G-quartet structures. Thus, the gel-shift procedure may inadvertently enrich for molecules that migrate aberrantly due to RNA–RNA interactions rather than RNA–protein interaction.

4.2 Selection stringency

Early rounds of selection are typically performed under relatively permissive binding conditions, to ensure that all active sequences survive the initial round of selection. A very stringent initial selection would impose a bottleneck that could eliminate some active sequences that are initially present at very low frequencies. As functional sequences replace non-functional sequences the selection stringency can be safely increased without fear of losing interesting sequences. Once the pool has at least marginal activity, high selection stringency is required to efficiently discriminate between the best and suboptimal sequences. Irvine *et al.* (53) provide a detailed mathematical description of the optimal ligand and pool concentrations that provide maximum enrichment without significant loss of the initially rare molecules that bind well.

Any of a number of familiar biochemical techniques can be used to increase binding stringency. Most enrichment strategies allow for the use of limiting protein so that sequences can compete with each other for binding. Non-amplifiable, competitor molecules, such as tRNA or whole-cell RNA are also often added to saturate non-specific binding sites or to compete off less active pool sequences. In addition, if the wild-type binding element is known, a non-amplifiable form of this RNA (that is, a sequence without primer-binding sites for reverse transcription and PCR) can be added as a chase following the initial binding incubation (38, 40). The chase RNA binds most vacated protein sites so that the pool molecules that remain bound for the duration of the chase are preferentially selected. Use of a chase increases the importance of slower off-rates (as opposed to faster on-rates) in the discrimination of pool molecules. Similar selection for slower off-rates can be achieved by extensive washing of complexes that have been immobilized by immunoprecipitation or other methods.

4.3 Amplification
4.3.1 Amplification methods

Nearly all *in vitro* selections employ PCR and transcription to amplify the selected sequences. Transcription-based amplification system (TAS; (54)) has also been employed by Joyce and colleagues (15, 36, 50). This alternative amplification scheme involves incubating the small amount of selected RNA with reverse transcriptase and RNA polymerase and the appropriate primers and nucleotides. The cDNA copy of a given sequence is transcribed many times by RNA polymerase and reverse transcription of the newly transcribed RNA provides additional templates for RNA transcription, accelerating the amplification process. In a single manipulation up to a million-fold amplification can be achieved in 1 h at 37 °C.

The most important criterion for choosing an amplification method is not simplicity or speed but rather the ability to amplify all sequences at equivalent rates. This is particularly important when starting from completely randomized pools which may contain sequences with disparate replication efficiencies. Unfortunately, there has not been a systematic study comparing sequence distributions before and after amplification in either the TAS or the PCR-based system. Each system is likely to amplify some sequences significantly more slowly than the median rate. One might imagine that TAS is more likely than PCR to amplify some sequences significantly more efficiently than the median rate, leading to a more skewed sequence distribution in the selected pool. Amplification at 37 °C rather than at 72 °C may also increase the role of secondary structure in causing replication-rate differences between sequences, though the recent commercial availability of thermal-stable RNA polymerase would permit TAS to be performed at higher temperatures. On the other hand, it is possible that PCR leads to more skewed sequence distributions than does TAS, because many more amplification cycles are required in PCR. For example, after 65 PCR cycles (five cycles for pool construction, plus 15 cycles in each of four rounds of selection and amplification) two equivalent-binding sequences will be present in a ratio of nearly 1:1000 if one of the sequences replicates only 90% as well as the other in each PCR cycle.

4.3.2 *In vitro* evolution

The first *in vitro* evolution experiments were performed over 25 years ago. Using a purified RNA-dependent RNA polymerase from Qβ phage, Spiegelman and colleagues were able to evolve RNA template sequences with enhanced replication rates (55) and templates that could be replicated in the presence of ethidium bromide (56). Since the Qβ polymerase replicates RNA with a high error rate, mutation, selection, and amplification could be accomplished *in vitro* in a single continuous process. With the advent of *in vitro* selection much more general evolution schemes can be developed simply by introducing an *in vitro* mutagenesis step prior to each selection cycle (57). The fittest molecular variants are selected and propagated and then serve as the starting material for further mutagenesis. The polymerases used during *in vitro* selections for reverse transcription, PCR, and transcription lack proofreading activities and are known to lead to mutations under standard conditions, so it could be argued that all *in vitro* selections are in fact *in vitro* evolution. However, Beaudry and Joyce (50) were the first to use conditions that lower polymerase fidelity with the purpose of increasing the contribution of this added genetic diversity. Using *in vitro* evolution it is possible to select for sequences that were not present in the initial pool but are instead more active variants of functional ancestral molecules (32).

The value of *in vitro* evolution is tied to the ability of the mutagenesis to generate sequences more active than any sequences already present in the initial pool of variants. As such, the utility *in vitro* evolution is often hampered by the low mutation rate of error-prone PCR. Amplification conditions that provide an even representation of mutation types yield a very low rate of mutagenesis (0.0066 per residue for a reaction involving ten doublings of input DNA; 58). Variations on this protocol that give more useful mutation rates skew the distribution of mutation types (32, 58).

4.4 Selection and amplification artefacts

The power of *in vitro* selection comes from its iterative nature. In many binding interactions the specific target sequence has only a 100–1000-fold *in vitro* binding advantage over non-specific targets. Thus, it is only with repeated selection cycles that rare target sequences can be isolated from a random pool. Unfortunately, the power of iterative selection can also be harnessed by unwanted sequences that are enriched in unanticipated ways.

4.3.1 Enrichment-specific artefacts

Each of the enrichment strategies listed above may lead to the isolation of artefact sequences instead of the desired species. As already mentioned, gel-shift selections are prone to enrichment of sequences that aggregate. Filter-binding selections have yielded sequences that bind nitrocellulose filters (31). Similarly, column selections have yielded a variety of sequences that bind the column matrix with or without ligand (23). The successful isolation of RNA epitopes that are cross-reactive with a peptide immunogen (27) suggests that co-immunoprecipitation enrichment could also lead to selection of analogous artefact sequences. The class of artefact sequences that is selected in the absence of ligand can sometimes be held at bay by alternating negative selection with positive selection. For example, affinity-column selections often utilize a pre-column made of the column matrix without ligand (23); only RNA that flows quickly through the pre-column is applied to the column with ligand. Affinity elution of the column with soluble ligand also provides an important enrichment for molecules that utilize contacts to the desired ligand for at least a large part of their binding to the column (25, 26). Finally, it may be desirable to change enrichment techniques, for example by switching from filter binding to co-immunoprecipitation when nitrocellulose binders become problematic.

4.3.2 Wild-type contamination

Other types of artefacts are more general to all enrichment strategies. When using *in vitro* selection to characterize a wild-type binding sequence there is a danger that wild-type molecules will contaminate the pool. A very small number of contaminating wild-type molecules, from used gel electrophoresis boxes or pipetting devices, for instance, can take over the pool during the course of the selection, particularly if the contamination occurs during pool construction or in the initial round when functional sequences are very rare. The first line of defence against contamination is to ensure that the pool primer-binding sites have little similarity to sequences present in the laboratory that contain functional binding elements, so that contaminating sequences are less likely to be amplified. Even with this precaution, wild-type sequences may recombine into pool sequences during reverse transcription and PCR, particularly if the pool was constructed by partial randomization of a wild-type sequence. When selecting from patch-randomized pools it may be difficult to distinguish between such recombinants and sequences that were present in the initial pool. To identify these recombinants it is useful to incorporate defined base changes within the pool molecules in a region of the sequence known not to be important for binding. The presence of these signature residues in a selected molecule indicates that the molecule, regardless of its resemblance to the wild-type sequence in the randomized patch, indeed arose from the randomized pool. Signature residues that create or change a restriction site facilitate the monitoring of possible pool contamination (38, 59).

4.3.3 Aggregation

Undesirable RNA–RNA interactions can be problematic even in selections that do not employ gel-mobility shift for enrichment. When attempting to select for a very rare activity it may be desirable to incubate the pool in buffers that are known to stabilize marginally stable RNA structures, for example buffers with high levels of Mg^{2+} or spermidine. However, incubation of random-sequence RNA in conditions that favour RNA intramolecular structure also stabilizes intermolecular contacts and extensive aggregation can result (32). We minimized the problem of aggregation by immobilizing the RNA on a column prior to adding Mg^{2+} and selecting for catalysis. However, binding selections require that the RNA be in solution, so temperature, salt concentrations, and incubation duration should be adjusted to minimize aggregation. A sizing column (for example, Sephacryl-S400) can be used to monitor aggregation following various trial incubation conditions.

5. Analysis of selected sequences

Once the pool is enriched in binding species to the point that it binds with the desired affinity, members of the pool are cloned and sequenced. In artificial phylogeny studies comparative sequence analysis can then indicate the residues important for binding as well as critical secondary and tertiary interactions. When patch mutagenesis is employed sequence conservation can be quickly monitored by batch sequencing the selected pool (16, 18). When binding molecules are selected from completely random sequences the comparative analysis is more difficult, but often the sequences can be grouped into classes and conservation and covariation within each class can be used to deduce critical structural features (23, 31, 39). In other instances all the molecules appear to be in different classes and conventional mapping or a second *in vitro* selection experiment can be used to determine the important residues (17). Often conclusions made from comparative sequence analysis have been tested *in vivo* or used to engineer minimal binding elements.

Because molecules can be inadvertently selected on criteria other than binding the ligand of interest, it is critical to assay binding of at least some of the more interesting selected sequences. In particular, conclusions concerning binding affinities cannot be made on sequence abundance (or absence) alone; the most abundant selected sequence may not be the tightest binder but instead may be replicated slightly more efficiently than other binding sequences. If a wild-type sequence is in hand, it is important to show that selected sequences can compete with the wild type for binding to the protein.

Although the abundance of each selected sequence must be interpreted carefully it is sometimes interesting to follow the changes in population structure with successive rounds of selection. When the initial pool was created by degenerate mutagenesis a massive sequencing effort is required to follow these changes (36). However, when the activity is selected from long stretches of random sequence, analysis utilizing restriction fragment length polymorphisms can conveniently monitor the effect of each round of selection on the relative abundances of sequences (32).

6. Future prospects

In vitro genetics is now a firmly established tool for characterizing RNA sequences that catalyse reactions and bind ligands, including proteins. As such it promises to reduce the time and expense devoted to site-directed mutagenesis experiments since *in vitro* selection can quickly limit the number of candidates for the most interesting residues involved in RNA–protein and RNA–RNA interactions. *In vitro* genetics is also emerging as a method for generating completely novel activities. In this capacity the potential of *in vitro* genetics is still unknown. Can feasible therapeutic agents be generated by this method? Can RNA or DNA sequences be isolated that efficiently catalyse reactions that differ markedly from the reactions catalysed by known natural ribozymes? Judging from the excitement that *in vitro* selection has generated in the 'biotech' community (60), we will have the answers to these questions soon.

References

1. Green, R., Ellington, A. D., Bartel, D. P., and Szostak, J. W. (1991) *In vitro* genetic analysis: selection and amplification of rare functional nucleic acids. *Methods Comp. Methods Enzymol.*, **2**, 75.

2. Szostak, J. W. (1991) *In vitro* genetics. *Trends Biochem. Sci.*, **17**, 89.
3. Joyce, G. F. (1992) Directed molecular evolution. *Sci. Am.*, **267**, 90.
4. Famulok, M. and Szostak, J. W. (1992) *In vitro* selection of specific ligand-binding nucleic acids. *Angew. Chem.*, **31**, 979.
5. Burke, J. M. and Berzal-Herranz, A. (1993) *In vitro* selection and evolution of RNA: applications for catalytic RNA, molecular recognition, and drug discovery. *FASEB J.*, **7**, 106.
6. Wright, W. E. and Funk, W. D. (1993) CASTing for multicomponent DNA-binding complexes. *Trends Biochem. Sci.*, **18**, 77.
7. Horwitz, M. S. Z. and Loeb, L. A. (1986) Promoters selected from random DNA sequences. *Proc. Natl Acad. Sci. USA*, **83**, 7405.
8. Gronostajski, R. M. (1986) Site-specific DNA binding of nuclear factor I: effect of the spacer region. *Nucleic Acids Res.*, **15**, 5545.
9. Kaiser, C. A., Preuss, D., Grisafi, P., and Botstein, D. (1987) Many random sequences functionally replace the secretion signal sequence of yeast invertase. *Science*, **235**, 312.
10. Ma, J. and Ptashne, M. (1987) A new class of yeast transcriptional activators. *Cell*, **51**, 113.
11. Oliphant, A. R. and Struhl, K. (1987) Use of random-sequence oligonucleotides for determining consensus sequences. *Methods Enzymol.*, **155**, 568.
12. Oliphant, A. R., Brandl, C. J., and Struhl, K. (1989) Defining the sequence specificity of DNA-binding proteins by selecting binding sites from random-sequence oligonucleotides: analysis of yeast GCN4 protein. *Mol. Cell. Biol.*, **9**, 2944.
13. Kinzler, K. W. and Vogelstein, B. (1989) Whole genome PCR: application to the identification of sequences bound by gene regulatory proteins. *Nucleic Acids Res.*, **17**, 3645.
14. Kinzler, K. W. and Vogelstein, B. (1990) The GL1 gene encodes a nuclear protein which binds specific sequences in the human genome. *Mol. Cell. Biol.*, **10**, 634.
15. Robertson, D. L. and Joyce, G. F. (1990) Selection *in vitro* of an RNA enzyme that specifically cleaves single-stranded DNA. *Nature*, **344**, 467.
16. Tuerk, C. and Gold, L. (1990) Systematic evolution of ligands by exponential enrichment: RNA ligands to bacteriophage T4 DNA polymerase. *Science*, **249**, 505.
17. Ellington, A. D. and Szostak, J. W. (1990) *In vitro* selection of RNA molecules that bind specific ligands. *Nature*, **346**, 818.
18. Green, R., Ellington, A. D., and Szostak, J. W. (1990) *In vitro* genetic analysis of the *Tetrahymena* self-splicing intron. *Nature*, **347**, 406.
19. Thiesen, H.-J. and Bach, C. (1990) Target detection assay (TDA): a versatile procedure to determine DNA binding sites as demonstrated on SP1 protein. *Nucleic Acids Res.*, **18**, 3203.
20. Blackwell, T. K. and Weintraub, H. (1990) Differences and similarities in DNA-binding preferences of MyoD and E2A protein complexes revealed by binding site selection. *Science*, **250**, 1104.
21. Blackwell, T. K., Kretzner, L., Blackwood, E. M., Eisenman, R. N., and Weintraub, H. (1990) Sequence-specific DNA binding by the c-Myc protein. *Science*, **250**, 1149.
22. Pollock, R. and Treisman, R. (1990). A sensitive method for the determination of protein–DNA binding specificities. *Nucleic Acids Res.*, **18**, 6197.
23. Ellington, A. D. and Szostak, J. W. (1992) Selection *in vitro* of single-stranded DNA molecules that fold into specific ligand-binding structures. *Nature*, **355**, 850.
24. Famulok, M. and Szostak, J. W. (1992) Stereospecific recognition of tryptophan agarose by *in vitro* selected RNA. *J. Am. Chem. Soc.*, **114**, 3990.

25. Connell, G. J., Illangesekare, M., and Yarus, M. (1993) Three small ribonucleotides with specific arginine sites. *Biochemistry*, **32**, 5497.
26. Sassanfar, M. and Szostak, J. W. (1993) Receptor–ligand interactions with oligonucleotides: an RNA motif that binds ATP. *Nature*, **364**, 550.
27. Tsai, D. E., Kenan, D. J., and Keene, J. D. (1992) *In vitro* selection of an RNA epitope immunologically cross-reactive with a peptide. *Proc. Natl Acad. Sci. USA*, **89**, 8864.
28. Bock, L. C., Griffin, L. C., Latham, J. A., Vermass, E. H., and Toole, J. J. (1992) Selection of single-stranded DNA molecules that bind and inhibit human thrombin. *Nature*, **355**, 564.
29. Macaya, R. F., Schultze, P., Smith, F. W., Roe, J. A., and Feigon, J. (1993) Thrombin-binding DNA aptamer forms a unimolecular quadruplex structure in solution. *Proc. Natl Acad. Sci. USA*, **90**, 3745.
30. Wang, K. Y., McCurdy, S., Shea, R. G., Swaminathan, S., and Bolton, P. H. (1993) A DNA aptamer which binds to and inhibits thrombin exhibits a new structural motif for DNA. *Biochemistry*, **32**, 1899.
31. Tuerk, C., MacDougal, S., and Gold, L. (1992) RNA pseudoknots that inhibit human immunodeficiency virus type 1 reverse transcriptase. *Proc. Natl Acad. Sci. USA*, **89**, 6988.
32. Bartel, D. P. and Szostak, J. W. (1993) Isolation of new ribozymes from a large pool of random sequences. *Science*, **261**, 1411.
33. Beutel, B. A. and Gold, L. (1992) *In vitro* evolution of intrinsically bent DNA. *J. Mol. Biol.*, **228**, 803.
34. Rittner, K., Burmester, C., and Sczakiel, G. (1993) *In vitro* selection of fast-hybridizing and effective antisense RNAs directed against the human immunodeficiency virus type 1. *Nucleic Acids Res.*, **21**, 1381.
35. Harada, K. and Orgel, L. E. (1993) *In vitro* selection of optimal DNA substrates for T4 RNA ligase. *Proc. Natl Acad. Sci. USA*, **90**, 1576.
36. Lehman, N. and Joyce. G. F. (1993) Evolution *in vitro* of an RNA enzyme with altered metal dependence. *Nature*, **361**, 182.
37. Pan, T. and Uhlenbeck, O. C. (1992) *In vitro* selection of RNAs that undergo autolytic cleavage with Pb^{2+}. *Biochemistry*, **31**, 3887.
38. Bartel, D. P., Zapp, M. L., Green, M. R., and Szostak, J. W. (1991) HIV-1 Rev regulation involves recognition of non-Watson–Crick base pairs in viral RNA. *Cell*, **67**, 529.
39. Gold, L., Allen, P., Binkley, J., Brown, D., Schneider, D., Eddy, S. R., Tuerk, C., Green, L., MacDougal, S., and Tasset, D. (1993) RNA: the shape of things to come. In *The RNA World*. Gesteland, R. F. and Atkins, J. F. (eds). Cold Spring Harbor Laboratory Press, Cold Spring Harbor, NY, p. 497.
40. Giver, L., Bartel, D. P., Zapp, M. L., Pawul, A., Green, M. R., and Ellington, A. D. (1993) Selective optimization of the Rev-binding element of HIV-1. *Nucleic Acids Res.*, **21**, 5509.
41. Levine, T. D., Gao, F., King, P. H., Andrews, L. G., and Keene, J. D. (1993) Hel-N1: an autoimmune RNA-binding protein with specificity for 3' uridylate-rich untranslated regions of growth factor mRNAs. *Mol. Cell. Biol.*, **13**, 3494.
42. Tsai, D. E., Harper, D. S., and Keene, J. D. (1991) U1-snRNP-A protein selects a ten nucleotide consensus sequence from a degenerate RNA pool presented in various structural contexts. *Nucleic Acids Res.*, **19**, 4931.
43. Schneider, D., Gold, L., and Platt, T. (1993) Selective enrichment of RNA species for tight binding to *Escherichia coli* rho factor. *FASEB J.*, **7**, 201.

44. Schneider, D., Tuerk, C., and Gold, L. (1992) Selection of high affinity RNA ligands to the bacteriophage R17 coat protein. *J. Mol. Biol.*, **228**, 862.
45. Tuerk, C., Eddy, S., Parma, D., and Gold, L. (1990) Autogenous translational operator recognized by bacteriophage T4 DNA polymerase. *J. Mol. Biol.*, **213**, 749.
46. Levitt, M. (1969) Detailed molecular model for transfer ribonucleic acid. *Nature*, **224**, 759.
47. Woese, C. R., Gutell, R., Gupta, R., and Noller, H. F. (1983) Detailed analysis of the higher-order structure of 16S-like ribosomal ribonucleic acids. *Microbiol. Rev.*, **47**, 621.
48. Michel, F. and Westhof, E. (1990) Modelling of the three-dimensional architecture of group I catalytic introns based on comparative sequence analysis. *J. Mol. Biol.*, **216**, 585.
49. Pan, T. and Uhlenbeck, O. C. (1993) A small metalloribozyme with a two-step mechanism. *Nature*, **358**, 560.
50. Beaudry, A. A. and Joyce, G. F. (1992) Directed evolution of an RNA enzyme. *Science*, **257**, 635.
51. Tsai, D. E. and Keene, J. D. (1993) *In vitro* selection of RNA epitopes using autoimmune patient serum. *J. Immunol.*, **150**, 1137.
52. Chittenden, T., Livingston, M., William, J., and Kaelin, G. (1991) The T/E1A-binding domain of the retinoblastoma product can interact selectively with a sequence-specific DNA-binding protein. *Cell*, **65**, 1073.
53. Irvine, D., Tuerk, C., and Gold, L. (1991) SELEXION systematic evolution of ligands by exponential enrichment with integrated optimization by non-linear analysis. *J. Mol. Biol.*, **222**, 739.
54. Kwoh, D. Y., Davis, G. R., Whitefield, K. M., Chappelle, H. L., DiMichele. L. J., and Gingeras, T. R. (1989) Transcription-based amplification system and detection of amplified human immunodeficiency virus type 1 with a bead-based sandwich hybridization format. *Proc. Natl Acad. Sci. USA*, **86**, 1173.
55. Mills, D. R., Peterson, R. L., and Spiegelman, S. (1967) An extracellular Darwinian experiment with a self-duplicating nucleic acid molecule. *Proc. Natl Acad. Sci. USA*, **58**, 217.
56. Kramer, F. R., Mills, D. R., Cole, P. E., Nishihara, R., and Spiegelman, S. (1974) Evolution *in vitro*: sequence and phenotype of a mutant RNA resistant to ethidium bromide. *J. Mol. Biol.*, **89**, 719.
57. Joyce, G. F. (1989) Amplification, mutation and selection of catalytic RNA. *Gene*, **82**, 83.
58. Cadwell, R. C. and Joyce, G. F. (1992) Randomization of genes by PCR mutagenesis. *PCR Methods Appl.*, **2**, 28.
59. Green, R. and Szostak, J. W. (1993) *In vitro* genetic analysis of the hinge region between helical elements P5-P4-P6 and P7-P3-P8 in the sun Y group I self-splicing intron. *J. Mol. Biol.*, **235**, 140.
60. Edgington, S. M. (1993) Shape space: is biopharmaceutical discovery entering a new evolutionary stage? *Bio/Technology*, **11**, 285.

Index

A-form helix 9, 10, 30, 89, 180
α-helix 169, 179, 236–7, 238, 240
α-sarcin 13
acceptor stem 54, 67
ADDFRAG 44
adenine-bulge 11
adenosine binding RNA 250
alanyl-tRNA synthetase 58, 59
amber suppressor 57, 65
amino acid sequence of C5 protein 118
aminoacyl-adenylate 60
aminoacyl-tRNA synthetase 52, 58–76
amplification of RNA for *in vitro* selection 261
amplification artefacts 263
anti 32, 207
anti–anti G–A base pair 13
anticodon 54, 57, 67
anticodon arm 54, 57
anticodon loop 26, 33, 62, 71
antiparallel β-sheet 58, 71, 76, 75, 179, 169
antisense oligonucleotide 115
aptamer 250
archaebacteria, 88, 90
arginine-binding RNA 250
arginine-rich domain 197, 221–41
artificial phylogeny 255
asparaginyl-tRNA synthetase 74
aspartyl-tRNA synthetase 59, 67–74
autoimmune disorders 156
AUU triple 17
auxiliary domain 137

B-form helix 9, 10
β-barrel 61, 62
β-sheet 133
βαββαβ 132
base stacking 53
base triple 17, 40, 181
bind and chew experiment 85, 88
biotinylated RNA 162, 163
branch point 151

branch point binding site 154
bubble *see* internal loop
bulge 2, 11, 89, 180, 194–5, 224, 229, 233, 238

C–A base pair 38, 39
carbon-13 labelling of protein and RNA 3
C5 protein 103–26
CCA end 62, 67, 76, 107
chemical modification of RNA 182
chemical mutagenesis of RNA 113
chemical synthesis of RNA 3
class I aminoacyl-tRNA synthetases 58–60, 61, 76
  group a 59, 60
  group b 59, 60
class II aminoacyl-tRNA synthetases 58, 60, 67–74, 76
classification of aminoacyl-tRNA synthetases 59
cleavage site 108
cloverleaf structure 54, 63
co-axial stacking 19
codon–anticodon interaction 26
combinatorial pool construction 258
common protein 156
comparative sequence analysis 37–42
compensatory base change 43, 92
conformational change of tRNA 74, 75
cooperative folding 45–7
core hnRNP proteins 127
core proteins (G, F, E, D1, D2, D3, 69K) 153, 157, 165
correlation spectroscopy (COSY, TOCSY) 5
cross-peaks 6
crosslinking 43
crystal structure
  of ribosomal proteins 95, 96
crystallography 131, 169
cysteinyl tRNA synthetase 58, 59

D arm 54, 57
D loop 57, 66, 67, 71
deletion mutagenesis 183
ΔG of C5 protein–M1 RNA interaction 110, 112
ΔH of C5 protein–M1 RNA interaction 110, 112
diethylpyrocarbonate (DEPC) 86, 196, 205, 225, 230
dimethylsulphate (DMS) 86, 113, 164, 205, 225
divalent ions 84, 106, 110, 111
ΔS of C5 protein–M1 RNA interaction 110, 112

E loop of 5S RNA 13, 14, 89
elbow of tRNA 54
electron microscopy of snRNPs 164–6
elongation factor 82
energy minimization 37
enrichment of binding RNA 259
env 202–3, 234
ethylation 174, 194
ethylnitrosourea 89, 174, 194
eubacteria 90
eukaryote 88
evolution of RNase P 119
Ewing's sarcomas 136

FAR (finger associated repeat) 188
FAX (finger associated box) 188
filter binding assay 83, 162, 259
footprint 85, 89, 91, 113, 183, 235
fragile X syndrome (FMR-1) 136
FRAGMENT 44
FRODO 45

G–A base pair 205, 239, 240
G–G base pair 205, 239, 240
G-quartets 18, 19, 261

G–U base pair 38, 39
gag 202–3
GCAA tetraloop 16
GCC⁺ triple 17
gel-mobility shift 83, 115, 162, 195, 259, 261
gel-retardation analysis *see* gel-mobility shift
gel-shift assay *see* gel mobility shift
glutamine-rich region 138
glutaminyl-adenylate 61
glutaminyl-tRNA synthetase 60, 61–5
glycosidic angle (χ) 8
glycyl-tRNA synthetase 59
GNRA tetraloop 15, 26, 31, 33, 39, 41, 181
group I introns 17, 27–9
  self-splicing 38, 39, 41
group I ribozyme 251, 258
group II self-splicing introns 38

H1 RNA 109
hairpin 2, 14, 179, 167–72, 201
hammerhead ribozyme 3, 19
helix 9
hetero-nuclear multidimensional NMR 6
HIGH motif 58–60, 65
histidyl-tRNA synthetase 59
HIV 192–247
hnRNP protein A1 129, 131, 134, 137, 139, 140
hnRNP protein A2/B1 129, 131–3
hnRNP protein C 134, 138
hmRNP protein C1/C2 129, 131–3, 138–9, 140
hnRNP protein I 129, 131
hnRNP protein K 129, 136, 137
hnRNP protein L 129, 131
hnRNP protein M 129, 131
hnRNP protein U 129, 134, 138, 140
hnRNP proteins 127–49, 209
hnRNP proteins, abundance of 139
Hoogsteen base pair 19, 31, 32, 53
hrp40.1 130
human immunodeficiency virus *see* HIV
hydroxyl radical 86

identity determinants of tRNA 56
immuno-affinity chromatography 152

immunoprecipitation 259
in vitro evolution 262
in vitro selection 248–65
in vitro transcription 259
induced-fit 112
instability sequence (INS) 203
interference experiment 86
internal loop 2, 12, 179, 172–3, 204, 205, 207, 208, 236, 253
isoacceptor tRNA 52
isotope labelling (^{13}C, ^{15}N)
  ^{13}C-labelling of RNA and proteins 3
  ^{15}N-labelling of RNA and proteins 3

junction 91
junction helical 94

k_{cat} of RNase P 105
k_d of RNase P 110
K_i of RNase P 106
K_m of RNase P 105, 106
Karplus equation 7
kethoxal 86, 164
KH domain 130, 136, 137
kinetic parameters of RNase P 105, 110
KMSKS motif 58, 60, 65

L3 protein 95
L6 protein 95
L7/L12 protein 95, 96
L11 protein 88, 90, 91, 109
L24 protein 93
L25 protein 89, 92
L30 protein 95, 96
λN protein 198, 222
large subunit (ribosome) 82
lariat intron 151
ligand immobilization 260
long terminal repeat (LTR) *see* LTR
loop E 89
loop–loop interaction 36
LTR 192, 194, 196, 198–201, 203, 209–11
lysyl-tRNA synthetase 59, 74

M1 RNA 103–26
major groove 9, 10, 67, 196, 205, 206, 240
major groove recognition 76
methylphosphonate 197, 207
Mg^{2+} ion 92, 110, 111
minor groove 9, 10
minor groove recognition 76
mischarging 65
missing nucleoside experiments 184
mithionyl-tRNA synthetase 60, 65, 66
modelling of tertiary structure 44–5
modified nucleotides 152
  2'-O-alkyl 163
  2'-O-methyl 163
  4-thiouridine 86
  7-deaza-dA 194
  7-deaza-dG 194, 205, 206
  2,2,7-trimethyl-guanosine (m3G) 152
  N^6-methyl-dA 195, 198, 205, 206
  N^7-methyl guanosine 152
  O^4-methyl-dT 194
  O^6-methyl-dG 205, 206
modified nucleotide in tRNA 53
molecular recognition 248
MOLSCRIPT 54
monovalent ions 111
motif 1 67
motif 2 67, 71
myxoid liposarcomas 136

NAB2 130
NAHELIX 44
NF-κB 213
NMR 95, 132, 229–30, 239
  2-D NMR 180
NMR-solution-structure determination 4–9
NOE 8
NOESY spectrum 7
non-specific binding 84
non-Watson–Crick base pair 12, 206–7, 239, 253
nuclear localization signal (NLS) 137
nuclear magnetic resonance *see* NMR
nuclear overhauser effect (NOE) 4
nucleolin 131, 134
nucleoplasmin 137
NUCLIN-NUCLSQ 45

oligomerization 208
organic dye 250

P protein 109
P RNA 109
p43 178, 185, 188
packaging of HIV RNA 19, 209
parallel β-sheet 58, 76
patch randomization 256
Pb²⁺ cleavage 252
PCR 249, 258
peptidyl transferase 82
phenylalanyl-tRNA synthetase 58, 59, 74
phosphodiester linkage (2'-3') 150, 151
phosphorus-31 (^{31}P) resonance 8
phylogenetic tree 42
poly-(A) binding protein (PABP) 134
polyadenylation 172
pre-mRNA 128
precursor 4.5S RNA 103, 104
precursor tRNA (ptRNA) 103, 104
primary-binding ribosomal protein 83
primer extension 164
prolyl-tRNA synthetase 59
protein binding site in RNA 2
protein binding site in snRNP 160
protein components of snRNP 152–8
protein composition of snRNPs 157
proton–proton distance restraint 8
PSEUDOKNOT 44
pseudoknot 18, 27, 34, 35
Pub1P 130

Qβ phage 262
Qβ polymerase 262

R17 coat protein 109, 117, 238, 252
radomized pool 257–9
randomization 249–50, 252–7
recognition element of tRNA 55
regulator of virion expression (rev) see rev
resonance assignments 4
rev 192, 202–13, 234–41, 253, 256
reverse Hoogsteen A–U base pair 13

reverse Hoogsteen base pair 53
reverse transcriptase 250, 251
reverse Watson–Crick 53
RGG box 130, 134–6
ribonucleoprotein particle (7S RNP) 178, 182–5, 187
ribosomal proteins 82–102
ribosomal RNA 82–102
 5S rRNA 13, 38, 83, 89, 178–88
 16S rRNA 83, 85, 88, 93, 109
 23S rRNA 13, 38, 83, 90, 109
ribosome
 30S ribosome 93
 50S ribosome 93
ricin 13
RNA
 10Sa RNA 103
RNA-annealing activity 140
RNA-binding specificity 138
RNA helix 26
RNA ligase (T4) 251
RNA polymerase (SP6, T7, T3) 3
RNA polymerase II 127, 152, 199
RNA polymerase III 152
RNA recognition motif (RRM) 96
RNA synthesis 3
RNase H 163
RNase P M1 RNA 38, 46
RNase P 103–26
RNase P holoenzyme 116
RNase T$_1$ 93, 94, 113, 114
RNP consensus sequence 130, 131, 141
RNP motif 130, 131
RNP-CS RBD see RNP motif
RNP1 (octamer) 130–2, 167–70, 173
RNP2 (hexamer) 130–2, 167–70, 173
RNP80 see RNP consensus
rnpA gene 113
rnpB gene 113
Rossmann fold 75
RRE 193, 202–13, 234–41
RS (SR) motif 138

S4 protein 93, 94, 109
S4 protein binding site 85, 89
S5 protein 95
S8 protein 88, 89, 117
S15 protein 116
S17 protein 95
S20 protein 89
secondary structure of ribosomal RNA 86

secondary structure of RNA 9–17, 26
secondary structure prediction 37
selection stringency 261
selection/amplification experiment 138, 248–65
sequence-specific recognition 2
seryl-tRNA synthetase 60, 73, 74
SF2/ASF 140
Sm proteins 153, see also core protein
Sm site 153–6, 158–60
small subunit (ribosome) 82
snRNPs 150–77
splice site 151, 202
splicing of HIV RNA 234
splicing reaction 150
stem-loop 253, see also hairpin
structure motifs 129
sugar conformation
 2'-endo sugar 9
 3'-endo sugar 9
suppressor tRNA 57
syn 32, 207
systemic lupus erythematosus 156

T arm 54, 71
T cell 212
T-loop 57
T7 promoter 249
T7 RNA polymerase 162
TAR 11, 109, 193–203, 205, 211, 213, 222–36
TAR RNA see TAR
tat 109, 117, 192–204, 205, 210–12, 222–42
telomeric DNA 19
tertiary motif 29–31
tertiary structure of RNA 17–19
tertiary motif of RNA 27
tetraloops 15
TFIIIB 186
TFIIIC 186
thermodynamic parameter of M1 RNA-C5 protei interaction 110
thermodynamic parameters of RNA folding 30
three-dimensional (3-D) NMR 5
threonyl-tRNA synthetase 59
thrombin binding RNA 251
thymine loop 33
TOCSY 5
torsion angles (α, β, γ, δ, ε, ζ) 8

trans Watson–Crick base pair 31
trans-activator protein (tat) *see* tat
transcription factor IID (TFIID) 199
transcription factor IIIA (TFIIIA)
    178–88
transcription termination ρ factor 252
transport of mature mRNA 129
transport, pre-mRNA 140, 166
trimethyl-guanine cap (2,2,7-
    trimethyl-guanosine) (m3G) 152
triple helix 32
tRNA 26, 38, 52, 53–8, 252
tRNA U33 sharp turn 31
tRNA^{Ala} 57
tRNA^{Arg} 57
tRNA^{Asp} 29, 53, 67–70
tRNA^{Cys} 57
tRNA^{Gln} 57, 61–5
tRNA^{Glu} 57
tRNA^{Ile} 57
tRNA^{Met} 53, 57, 66
tRNA^{Phe} 29, 53, 57, 257
tRNA^{Tyr} 56, 66
tRNA^{Val} 57
tryptophan (L-tryptophan) 250
tryptophanyl-tRNA synthetase 58, 59

two-dimensional gel 128
two-dimensional (2-D) NMR 5
TψC-arm 57
tyrosyl-tRNA synthetase 60, 66

U1 70K protein 156, 157, 160, 161, 165, 166
U1 snRNA 150, 158–60
U1 snRNP 153, 156, 157, 164–5
U1A protein 14, 118, 156, 157, 160–2, 165, 166, 167–73, 252, 260
    structure of 132
U1A small nuclear ribonucleoprotein *see* U1A protein
U1C 157, 160–2, 165, 166
U2 snRNA 150, 158–60
U2 snRNP 154, 156, 157, 162–4, 165–7
U2A' protein 156, 157, 162–3, 167, 169
U2AF 139
U2B" protein 14, 118, 156, 157, 162, 167–70
U2B" small nuclear ribonucleoprotein 162, 167–70, *see* U2B" protein

U4 snRNA 150, 158–60
U4/U6 snRNP 155, 158, 164, 165, 166
U4/U6-U5 tri-snRNP 157
U5 snRNA 150, 156, 158–60
U5 snRNP 164, 165, 166
U6 snRNA 150, 158–60
UNCG tetraloop 15, 16, 31
untranslated region (3') 172
UUCG tetraloop 16
UUNG tetraloop 33
UV cross-linking 13, 86, 116, 128

variable loop 54

water-mediated hydrogen bond 12
Watson–Crick base pair 38
whole-genome PCR 257
wobble base pair 57

zinc finger 178, 185–8
zone-interference electrophoresis 84

Printed in the United States
204490BV00004B/15/A